Exact and Heuristic Scheduling Algorithms

Exact and Heuristic Scheduling Algorithms

Special Issue Editors

Frank Werner
Larysa Burtseva
Yuri Sotskov

MDPI • Basel • Beijing • Wuhan • Barcelona • Belgrade • Manchester • Tokyo • Cluj • Tianjin

Special Issue Editors

Frank Werner
Otto-von-Guericke-University
Magdeburg
Germany

Larysa Burtseva
Universidad Autonoma de Baja
California Mexicali
Mexico

Yuri Sotskov
National Academy of Sciences of
Belarus
Belarus

Editorial Office
MDPI
St. Alban-Anlage 66
4052 Basel, Switzerland

This is a reprint of articles from the Special Issue published online in the open access journal *Algorithms* (ISSN 1999-4893) (available at: https://www.mdpi.com/journal/algorithms/special_issues/Scheduling_Algorithms).

For citation purposes, cite each article independently as indicated on the article page online and as indicated below:

LastName, A.A.; LastName, B.B.; LastName, C.C. Article Title. *Journal Name* **Year**, *Article Number*, Page Range.

ISBN 978-3-03928-468-9 (Pbk)
ISBN 978-3-03928-469-6 (PDF)

Contents

About the Special Issue Editors

Frank Werner studied Mathematics from 1975–1980 and graduated from the Technical University Magdeburg (Germany) with honors. He defended his Ph.D. thesis on the solution of special scheduling problems in 1984 with 'summa cum laude' and his habilitation thesis in 1989. In 1992, he received a grant from the Alexander-von-Humboldt Foundation. Currently, he works as an extraordinary professor at the Faculty of Mathematics of the Otto-von-Guericke University Magdeburg (Germany). He is the author or editor of five books and has published more than 280 papers in international journals. He is on the Editorial Board of 17 journals, in particular, he is the Editor-in-Chief of *Algorithms* and an Associate Editor of the *International Journal of Production Research* and *Journal of Scheduling*. He was a member of the Program Committee of more than 75 international conferences. His research interests are operations research, combinatorial optimization, and scheduling.

Larysa Burtseva graduated in Economic Cybernetics (1975) at the Rostov State University (Russia) and defended her Ph.D. thesis on Control in Technical Systems (Technical Cybernetics), 1989, at the Radioelectronics University of Kharkov (Ukraine). Since 2000, she works as a professor researcher at the Engineering Institute of Universidad Autónoma de Baja California, Mexicali, Mexico. She has led the Laboratory of Scientific Computation since 2005. She has more than 70 publications in international journals and proceedings of congresses and 6 book chapters. Her research interests are discrete optimization, particularly combinatorial optimization, scheduling, and also packing problems for different applications. She is a member of the National System of Researchers of the National Council for Science and Technology of Mexico (SNI CONACYT) and a regular member of The Mexican Academy of Computation (AMEXCOMP).

Yuri N. Sotskov, Prof. D.Sc., finished secondary school with a gold medal in 1966 and graduated from the Faculty of Applied Mathematics of the Belarusian State University in Minsk in 1971. In 1980, he defended his Ph.D. thesis at the Institute of Mathematics of the National Academy of Sciences of Belarus in Minsk. In 1991, he defended his D.Sc. thesis at the Institute of Cybernetics of the National Academy of Sciences of Ukraine in Kyiv. He has the title of Professor in Application of Mathematical Models and Methods in Scientific Research (Russian Academy of Sciences, 1994). He currently works as a principal researcher at the United Institute of Informatics Problems of the National Academy of Sciences of Belarus in Minsk. He has more than 350 publications: five scientific monographs, two textbooks, more than 200 papers in international journals, books, and conference proceedings in English on applied mathematics, operations research, graph theory, and scheduling. He is on the Editorial Board of five journals. He has supervised eight Ph.D. scholars (defended). In 1998, he received the National Prize of Belarus in Science and Engineering.

Preface to "Exact and Heuristic Scheduling Algorithms"

Optimal scheduling is an important area of operations research and in particular of decision making. This research field covers a large variety of different solution approaches designed for particular problems. It is clear that efficient algorithms are highly desirable for large-size real-world scheduling problems. Due to the NP-hardness of the majority of scheduling problems, practitioners often prefer to use rather simple scheduling algorithms since large-size problems usually cannot be solved exactly in an acceptable time. However, fast heuristics may produce schedules the function values of which might be far away from the optimal ones.

This book is based on a Special Issue entitled "Exact and Heuristic Scheduling Algorithms". It is a follow-up book of the booklet, "Algorithms for Scheduling Problems" published in 2018 (ISBN 978-3-03897-119.1) and contains both theoretical and practical aspects in the area of designing efficient exact and heuristic scheduling algorithms. We hope that the articles contained in this book will stimulate researchers to find new practical directions for implementing their scheduling algorithms. In response to the Call for Papers, we selected eight submissions, all of which are of high quality, reflecting the interest in the area of developing exact and heuristic algorithms to solve real-world production planning and scheduling problems. As a rule, all submissions have been reviewed by three experts in the Discrete Optimization and Scheduling area.

The first article deals with a real-world distribution problem arising in the vehicle production industry in Mexico. This problem includes both the loading and optimal routing of each auto-carrier, where both the capacity constraints and the time windows have to be taken into account. The authors suggest a two-stage heuristic algorithm. First, a route is constructed by an optimal insertion heuristic. Then a feasible loading is determined. For the computational experiments, two scenarios were generated with 11 different instances of the demand. Here one instance describes a real problem of a logistics company in Mexico. The algorithm allowed to obtain the routes and the loading of the vehicles for problems with 4400 vehicles and 44 dealerships considered. The results showed that the developed algorithm was successful in minimizing total traveling distance, loading/unloading operations and transportation costs.

The second article is devoted to the heuristic solution of job shop scheduling problems with blocking constraints and minimization of total tardiness. This problem has many applications in manufacturing, but also in railway scheduling. Due to the NP-hardness of the problem, the development of heuristics is of high relevance. In particular, a permutation-based heuristic is derived. The authors use three interchange- and shift-based transition schemes. However, typical neighborhoods applied to job shop problems often generate blocking infeasible solutions. Therefore, the core of this paper is to present two repair mechanisms that always generate neighbors being feasible with respect to the blocking constraints. Characteristics for the complex neighborhoods such as the average distance of a neighbor are analyzed. The suggested neighborhoods are embedded into a simulated annealing algorithm. Detailed computational results have been given for the modified Lawrance instances with up to 30 jobs and 15 machines as well as for hard train inspired instances with up to 20 jobs and 11 machines. A computational comparison was made with the MIP formulation given in an earlier paper by Lange and Werner in *Journal of Scheduling* (2018). It turned out that for small instances, the heuristic often obtained an optimal or near-optimal solution, while for larger instances several of the best known solutions by the MIP solver have been improved.

The third article deals with scheduling jobs on a set of uniform parallel machines subject to given unavailability intervals with the objective to minimize the makespan. For this NP-hard problem, a new quadratic model is developed. In addition, the authors present a two-stage heuristic procedure. In the first phase, a modified Longest-Processing-Time rule is used to construct a schedule, which is then improved by pairwise interchanges of jobs between the machines. Computational results have been presented for instances with up to 2001 jobs and 1000 machines. The experiments showed that the quadratic model can solve small- and medium-size instances with roughly up to 140 jobs and 70 machines within a time limit of one hour. The heuristic algorithm presented in this paper was very fast and outperformed also an earlier heuristic by Kaabi and Herrath (2019).

In the fourth article, some special cases of flow and job shop problems with so-called sprecedence constraints are addressed. This means that a task of a job cannot start before the task which precedes it has started. Polynomial algorithms are given for three special cases, namely a two-machine job shop problem with two jobs and allowed recirculation, the two-machine flow shop problem and an m-machine flow shop problem with two jobs, each with the objective to minimize the makespan. Finally, some special cases with open complexity status are mentioned.

The fifth article analyzes the connections between usual scheduling criteria typically applied to flow shop problems like the makespan or idle time and customary shop floor performance measures such as work-in-progress or throughput. The authors setup a deep experimental analysis consisting in finding optimal or near-optimal schedules under several scheduling criteria and then investigating how these schedules behave in terms of different shop floor performance measures for several instances with different structures of the processing times. In particular, detailed computational results have been presented for instances with up to 200 jobs and 50 machines. It turned out that some of the scheduling criteria are poorly related to shop floor performance measures. In particular, the makespan performed only well with respect to throughput. On the other hand, the minimization of total completion time appeared to be better balanced in terms of shop floor performance. The article finishes with suggesting some aspects for future work.

The sixth article addresses a two-machine job shop scheduling problem, where the job duration may take any real value from a given segment. A stability approach is applied to this uncertain scheduling problem. The scheduling decisions in the stability approach may consist of two successive phases: the first off-line phase, which is finished before starting the realization of a schedule, and the second on-line phase of scheduling, which is started with the beginning of the schedule realization. Using the information on the lower and upper bounds for each job duration available at the off-line phase, a scheduler can determine a minimal dominant set (DS) of schedules based on sufficient conditions for schedule dominance. The DS optimally covers all possible scenarios of the job durations in the sense that for each possible scenario, there exists at least one schedule in the DS which is optimal. Based on the DS, a scheduler can choose a schedule, which is optimal for the majority of possible scenarios. Polynomial algorithms have been developed for testing a set of conditions for schedule dominance. The conducted computational experiments on the randomly generated instances have shown the effectiveness of the developed algorithms. Most instances from the nine tested classes were optimally solved. If the maximum error was not greater than 20%, then more than 80% of the tested instances were optimally solved. If the maximum error was equal to 50%, then 45% of the tested instances from the nine classes were optimally solved.

The seventh article deals with the Euclidean version of the traveling salesman problem (TSP), where the locations of the cities are points in the two-dimensional Euclidean space and the distances are Euclidean ones. For this problem, the authors suggest fast and easily implementable heuristics. They consist of three phases: construction, insertion and improvement. The first two phases run in O(n) time with n being the number of points, and the number of improvement repetitions in the third phase is bounded by a small constant. The practical behavior of the suggested heuristics has been tested on 218 benchmark instances from several well-known libraries for TSP instances. In particular, the authors grouped the instances into small ones (up to 199 points), medium-size ones (between 200 and 9,999 points) and large instances (between 10,000 and 250,000 points), and results for two very large instances with 498,378 and 744,410 points were also given. Although for most of the tested benchmark instances, the best known values have not been improved, nevertheless the computational times were smaller than the best known values earlier reported, and the heuristic is also efficient with respect to the required memory.

In the last article, a new mathematical model for a resource leveling problem with variable job durations is proposed, where the problem includes both scheduling and resource management decisions within a fixed planning horizon. The objective is to minimize total overload costs necessary for executing all jobs by the given deadline. In particular, the authors consider three different approaches for representing the scheduling constraints and decision variables, and they choose after some experiments a step (start/end) formulation of the scheduling constraints. Both the theoretical difference and relationships between the generalized modeling presented in this paper and the aggregated fraction model are discussed. The new formulation was compared to other models of the resource leveling problem from the literature on benchmark instances with up to 10 resource types and 30 jobs. Although the generalized modeling uses more variables and constraints, it nevertheless provided much better final solutions.

The editors would like to thank the authors for submitting their interesting works to this collection of articles about new scheduling algorithms, the reviewers for their timely and insightful comments on the submitted articles, and the editorial staff of the MDPI Journal *Algorithms* for their assistance in managing the review process in a prompt manner.

Frank Werner, Larysa Burtseva, Yuri Sotskov
Special Issue Editors

Article

A Heuristic Algorithm for the Routing and Scheduling Problem with Time Windows: A Case Study of the Automotive Industry in Mexico

Marco Antonio Juárez Pérez [1,*], Rodolfo Eleazar Pérez Loaiza [1],
Perfecto Malaquias Quintero Flores [1], Oscar Atriano Ponce [2] and Carolina Flores Peralta [1]

[1] Tecnológico Nacional de México, Instituto Tecnológico de Apizaco, Apizaco 90300, Mexico;
rodolfo.pl@apizaco.tecnm.mx (R.E.P.L.); perfecto.qf@apizaco.tecnm.mx (P.M.Q.F.);
carolinaflxp@gmail.com (C.F.P.)
[2] Smartsoft America BA, Chiautempan 90802, Mexico; oatriano@smartsoftamerica.com.mx
* Correspondence: marcoantoniojuarezperez@gmail.com; Tel.: +52-241-149-6529

Received: 3 May 2019; Accepted: 22 May 2019; Published: 25 May 2019

Abstract: This paper investigates a real-world distribution problem arising in the vehicle production industry, particularly in a logistics company, in which cars and vans must be loaded on auto-carriers and then delivered to dealerships. A solution to the problem involves the loading and optimal routing, without violating the capacity and time window constraints for each auto-carrier. A two-phase heuristic algorithm was implemented to solve the problem. In the first phase the heuristic builds a route with an optimal insertion procedure, and in the second phase the determination of a feasible loading. The experimental results show that the purposed algorithm can be used to tackle the transportation problem in terms of minimizing total traveling distance, loading/unloading operations and transportation costs, facilitating a decision-making process for the logistics company.

Keywords: heuristic; time windows; feasible loading; auto-carrier transportation problem (ACTP)

1. Introduction

Vehicle production in Mexico has been increasing in recent years [1,2]. As well as the number of imported vehicles, generating one of the main tasks to be solved by logistics companies: the transport of vehicles to dealerships. Currently, there are several commercial offers that provide a solution to route planning and fleet management. However, the cost of these applications is significantly high because they depend on the number of auto-carriers to route, making the acquisition of application difficult to afford.

This paper presents a vehicle-routing problem with time windows (VRPTW) in the real-world proposed by a logistics company in Mexico. In the problem, a heterogeneous fleet of auto-carriers departs from the new car storage yard (NCSY), delivers and unloads vehicles in the dealerships within predefined time windows, and finishes at the NCSY as shown in Figure 1. The objective of this research is to design and develop a logistic software to solve it. Considering the restrictions on the transport of vehicles imposed by Mexican traffic regulations [3], capacity, and allocation constraints, optimal performance with delivery time windows and proper planning of transportation routes. The software has a two-phase heuristic algorithm: in the first phase, the heuristic [4] is implemented to design the auto-carrier routes and an algorithm is proposed for the allocation of the vehicles in the auto-carriers. For experimentation, we used a real database of approximately 4000 vehicles, more than 600 auto-carriers, and 44 different dealerships as a destination, obtaining results with the proposed algorithm in a reasonable time.

The structure of this paper is as follows. A review the relevant literature is provided in Section 2. An overview of the importance of research and the problem are shown in Section 3. The VRPTW is defined and mathematical formulations are presented in Section 4. The proposed solution and the developed methodology are described in Section 5. The experimental results of the algorithm are presented and analyzed with real-world instances in Section 6. Finally, in Section 7 conclusions and future research work are given.

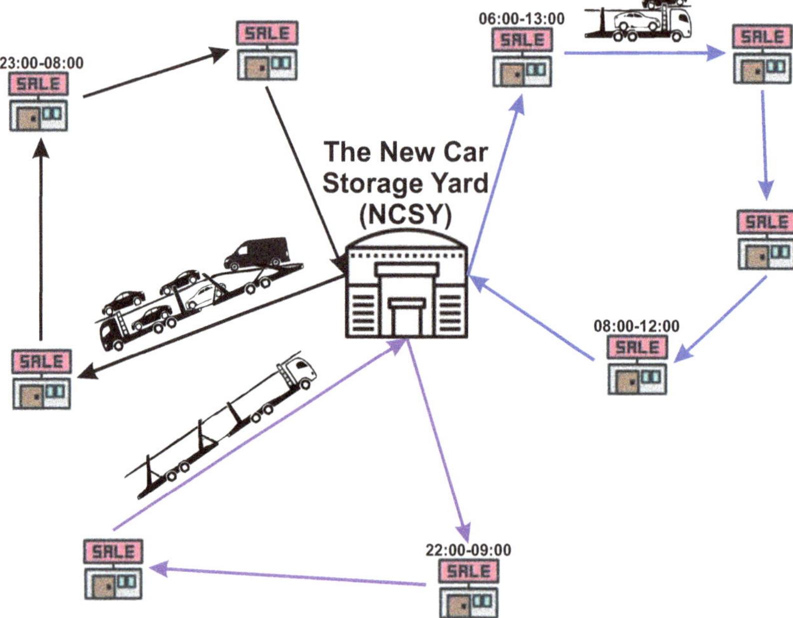

Figure 1. An illustration for the proposed problem by a logistics company.

2. Literature Review

The definition of the vehicle-routing problem (VRP) has its origins in the formulation of the traveling salesman problem (TSP) [5]. This section first reviews proposed algorithms and methods for the VRP and its variants. Then, focusing on the revision of the VRPTW and finally conclude with the review of the auto-carrier transportation problem (ACTP).

An important part of optimization systems are heuristics, which have multiple applications. From the extraction of features for a voice evaluation mechanism [6] to the generation of feasible VRP solutions. Arnau et al. [7] studied VRP with dynamic travel times, considering inputs of a dynamic nature and re-evaluating travel times dynamically as the solution was being developed. They proposed a learnt heuristic-based approach that integrates statistical learning techniques within a metaheuristic framework. Cassettari et al. [8] investigated the capacitated vehicle-routing problem (CVRP) applied to natural gas distribution networks. The authors introduced an algorithm based on the saving algorithm heuristic approach to solve it. Zhao and Lu [9] presented an electric vehicle-routing problem (EVRP) raised by a logistics company. They developed a heuristic approach based on the adaptive large neighborhood search

(ALNS) and integer programming, specifically designed a charging station heuristic adjustment and other one for the departure time decreasing the total operational cost.

Some well-known heuristic algorithms have been inspired by natural physical phenomena. Stodola [10] addressed the modified multi-depot vehicle-routing problem (MDVRP). He developed a metaheuristic algorithm based on the ant colony optimization (ACO) improved by a deterministic optimization process that is executed repeatedly within the ACO algorithm iterations. Połap and Woźniak [11] proposed a polar bear optimization algorithm (PBO) which imitates the survival and hunting behaviors of polar bears for local and global search. The authors presented a novel birth and death mechanism to control the population. Chen et al. [12] proposed a monarch butterfly optimization (MBO) algorithm to solve the dynamic vehicle-routing problem (DVRP) using a greedy strategy. Ahmed and Sun [13] designed a bilayer local search-based particle swarm optimization (BLS-PSO) algorithm to solve CVRP.

Currently, one of the most studied variants of the VRP is with time windows, in the research by Desrochers et al. [14] introduced an optimization algorithm to solve a VRPTW, using dynamic programming. Tan et al. [15] explored simulated annealing (SA), tabu search (TS) and a genetic algorithm (GA) heuristics to solve it. In another study, Yu et al. [16] proposed a hybrid approach, consisting of the use of the ACO and TS algorithms, for the VRPTW. To improve the performance of the ACO algorithm, they introduced a neighborhood search and a TS algorithm to maintain the diversity of the ACO algorithm and explore new solutions. Taner et al. [17] developed two metaheuristic algorithms to solve the VRPTW, the SA algorithm and an iterated local search (ILS). Sripriya et al. [18] designed a hybrid genetic search with diversity control using a GA to solve the VRPTW, using the Pareto approach and two mutation operators to find the optimal solution set.

Tadei et al. [19] investigated and defined a variant of the VRP, called the ACTP, proposed a three-step heuristic procedure that considers the loading, vehicle selection, and routing aspects for a solution to the problem. In other research, B. M. Miller [20] addressed the ACTP for collection and delivery with limitations in the delivery times and the capacity of the auto-carrier, for new and used vehicles. The author proposed a constructive heuristic to solve the problem. Dell'Amico, et al. [21] defined the ACTP as a combinatorial problem of the CVRP. The authors presented a study of a real case and implemented an ILS algorithm for the routing and mathematical techniques for the loading of vehicles. On the other hand, Tran et al. [22] implemented a heuristic algorithm for location of alternative-fuel stations. Hosseinabadi et al. [23] developed a method called TIME_GELS that uses the gravitational emulation local search algorithm (GELS) for solving the multiobjective flexible dynamic job-shop scheduling problem.

The VRP is widely studied in the areas of operations research and computer sciences, due to its computational complexity and its multiple applications. The variants of the VRP allow the use of time window restrictions and vehicle capacity, among others, these restrictions allow solving problems with solutions closer to the optimum of real-world cases, the ACTP is a result from this. As it has been described in the literature, several authors have proposed algorithms and methods to solve this problem. Nevertheless, the characteristics of our problem, motivate us to implement a heuristic approach that contemplates the restrictions imposed by the logistics company.

3. Importance of the Problem

The automotive industry has been one of the most important engines for the development and economic growth of Mexico [2,24]. Hence, the importance of promoting the insertion of technology in the sectors that are parts of it. For this reason, it is necessary to implement technology in the process of transporting new vehicles within the country. In addition, with it to diminish one of the most expensive processes for the companies of transport, the routing and scheduling of auto-carriers.

In addition to having a positive impact on the operating expenses for the transportation companies, decreasing the amount kilometers of the traveled route from each of the auto-carrier also represents a positive environmental impact because downward the harmful emissions to the environment produced by diesel motors. According to [25] most of the auto-carriers use this fuel and the main characteristic of diesel emissions is that particles are produced in a proportion 20 times higher than gasoline engines.

Nitrogen oxides (NO_x) are considered an important source of air pollution and contribute greatly to photochemical smog, acid rain, depletion of the ozone layer and the greenhouse effect. Diesel exhaust gases are generally composed of more than 90% NO_x [26]. Therefore, one of the main contributors to emissions of NO_x and sulfur oxides are diesel engines [27], so it is important to reduce these emissions. Optimizing the routes of the auto-carriers that generate these emissions are a good way to do it.

According to the National Institute of Statistics and Geography [24], the Mexican automotive industry is important because:

- It is ranked as the second most important activity in manufacturing after the food industry
- Because its exports were ranked fourth in the world in 2014
- When demanding inputs to carry out its production, it generates impacts on 157 economic activities out of a total of 259, according to the input-output matrix

The production of the automotive industry has increased its relative importance in the economy. Before the North American Free Trade Agreement (NAFTA) came into force, this industry represented 1.9% of gross domestic product (GDP) in Mexico, while in 2014 it was 3.0% [24]. This increase was due to the implementation of new technologies in the last decade. Both in the automotive sector and in the rest of economic activities that are suppliers of this, one of them the transportation of vehicles.

Finally, the transportation of vehicles is an important field in operations research (OR), which has attracted increasing interest in recent years, due to the expected benefits of substantial cost reduction and efficient consumption of resources. The VRPTW has multiple applications such as supermarkets, cement plants, hospitals, etc., though its main applications are in the industry.

4. Problem Definition and Mathematical Model

A logistics company distributes new vehicles in Mexico, manufactured in another country. It carries out the delivery of thousands of vehicles, according to the demand of each of the dealerships responsible for the sale of vehicles. Currently, the logistics company uses an empirical allocation and routing method for the auto-carriers.

The empirical method consists of the design of the route according to the experience of the operator of the auto-carrier, based on the vehicles that will be transported without the use of a heuristic or similar method for the optimization of the route. Similarly, the allocation corresponds to a simulation with fictitious vehicles of the load of the auto-carrier, positioning the vehicles in different levels of the auto-carrier, considering the dimensional restrictions.

The process of the routing and loading of the auto-carriers, begins at the moment that the operators receive a list of the vehicles to be delivered to the different dealerships. In the NCSY, the operators confirm the vehicles to be transported with the manager of the NCSY, who to complete the loading process verifies that the vehicles in the auto-carrier correspond to the request of the dealerships.

Empirical routing is inefficient because it does not consider restrictions as the time windows. The time windows are the hours in which an auto-carrier can perform the unloading of vehicles at a dealership, who defines an initial time and an end time to carry it out. Time windows are defined to not violate local traffic laws, thus avoiding monetary penalties.

Avoiding various penalties and monetary losses for companies, are some reasons of importance for the VRPTW and its applications. An example of its application is in the cement industry, if the concrete mixer trucks do not arrive within the stipulated time window. It may be that part of the concrete dries, becoming unusable, and the work stops. In the case of the logistics company, if an auto-carrier arrives at the dealership at a time outside the time window, it causes a time penalty that is, the operator must wait at the dealership to unload vehicles.

In addition to the time window restrictions, this case study includes a total of 44 dealerships, a demand with approximately 4000 vehicles of different dimensions (which add three restrictions to the allocation) and a variable number of auto-carriers of different capacity load (3, 6, 7, 10 and 11 vehicles). This paper describes the algorithm developed for a real-world problem of a logistics company; the problem can be summarized as follows:

> given a heterogeneous fleet of auto-carriers based at a NCSY and a set of dealerships each requiring a set of vehicles, the loading of the vehicles into the auto-carriers and route the auto-carriers through the road network to deliver all dealerships with minimum cost (total number of kilometers traveled) that start and ends in the NCSY, considering the restrictions of time windows, a LIFO policy for the loading/unloading of vehicles and maximizing the total use of the capacity of each auto-carrier

The characteristics of the dealerships (time windows), the NCSY and the auto-carriers (capacity), as well as different operational restrictions on the routes, bring forth the VRPTW, several authors [14–18] have worked on this variant of the VRP. In this case, study the term vehicle denotes a transported item (e.g., a car, a van), the term auto-carrier denotes a truck transporting vehicles, and the term dealership denotes a delivery point (i.e., a customer requiring one or more vehicles). With the previously mentioned elements, the model can be described as follows [21]:

- Network: Given a complete graph $G = (V, E)$, where $V = 0, 1, \ldots, n$ is the set of vertices and E the set of edges connecting each vertex pair. Vertex 0 corresponds to the NCSY, whereas vertices $1, \ldots, n$ correspond to the n dealerships to be served. The edge is connecting vertices i and j is denoted by (i, j) and has an associated routing cost $c_{ij}(i, j \in V)$ shown in Figure 2. The distance and times matrices are symmetric.

Figure 2. Example for routing cost between dealership i and j.

- Fleet: Given a heterogeneous fleet of auto-carriers, composed by a set T of auto-carrier types. Each auto-carrier type $t(t \in T)$ has a maximum vehicles capacity W_t and is formed $A_k^{1,2}$ by loading platforms (levels, shown in Figure 3). There are K_t auto-carriers available for each type t.

Figure 3. Auto-carrier levels.

- Demand: The demand of dealership i consists of a set M of vehicles ($i \in V\backslash\{0\}$). Each vehicle $m_i \in M$ demanded by dealership i belongs to a vehicle type (or vehicle model) shown in Figure 4, which is defined by a height h_m and a vehicle identification number (VIN).

Figure 4. Vehicle types (vehicle, vehicle, van).

In this VRPTW, each dealership $i \in V\backslash\{0\}$ has an associated time window $[e_i, l_i]$, with a time allowed service for arriving auto-carriers to it and service time or delay d_i. If (i, j) is an arc of the solution and a_i and a_j are the arrival times to the dealerships i and j, time window imply that necessarily must be fulfilled $a_i \leq l_i$ and $a_j \leq l_j$. On the other hand, if $a_i \leq e_i$, then the auto-carrier must wait until the dealership "opens" so necessarily $a_j = e_i + d_i + c_{ij}$.

Using the nodes 0 and $n+1$ to represent the NCSY and the set K to represent the auto-carriers, the problem is formulated for a heterogeneous fleet of auto-carriers, according to [28]:

$$\min \sum_{k \in K} \sum_{(i,j) \in E} c_{ij}^k x_{ij}^k \tag{1}$$

subject to

$$\sum_{k \in K} \sum_{j \in \Delta-(i)} x_{ij}^k = 1 \qquad \forall i \in V\backslash\{0, n+1\} \tag{2}$$

$$\sum_{j \in \Delta+(0)} x_{0j}^k = 1 \qquad \forall k \in K \tag{3}$$

$$\sum_{j \in \Delta+(i)} x_{ij}^k - \sum_{j \in \Delta-(i)} x_{ji}^k = 0 \qquad \forall k \in K, i \in V\backslash\{0, n+1\} \tag{4}$$

$$\sum_{i \in V\backslash\{0, n+1\}} x_{ij}^k - \sum_{j \in \Delta+(i)} x_{ji}^k \leq W^k \qquad \forall k \in K \tag{5}$$

$$y_j^k - y_i^k \geq d_i + a_{ij}^k - H(1 - x_{ij}^k) \qquad \forall i, j \in V\backslash\{0, n+1\}, k \in K \tag{6}$$

$$e_i \leq y_i^k \leq l_i \qquad \forall i \in V\backslash\{0, n+1\}, k \in K \tag{7}$$

$$x_{ij}^k \in 0, 1 \qquad \forall (i, j) \in E, k \in K$$

$$y_i^k \geq 0 \qquad \forall i \in V\backslash\{0, n+1\}, k \in K$$

The x_{ij}^k variables indicate if the arc (i,j) is used by the auto-carrier k. The y_i^k variables indicate the arrival time at the dealership i when it is visited by the k auto-carrier (if the dealership is not visited by the auto-carrier, the variable has no meaning). The objective function (1) minimizes the total routing cost.

Constraint (2) state that each dealership is visited exactly once, while constraints (3) and (4) determine that each auto-carrier $k \in K$ goes through a path of 0 to $n+1$. The capacity of each auto-carrier is imposed in (5). Since H is a sufficiently large constant, restriction (6) ensures that if an k auto-carrier travels from i to j, it cannot reach j before $y_i + d_i + a_{ij}^k$. These constraints also eliminate subtours and constraints (7) enforce time windows restriction.

The use of a heterogeneous fleet and the nature of the demand (vehicles and vans) impose the allocation constraints:

$$h_m > 2.5 = (A_k^{1-2}, W_k^3) \qquad \forall m_i \in M, k \in K \tag{8}$$

$$1.8 < h_m < 2.5 = (A_k^1, W_k^1) \qquad \forall m_i \in M, k \in K \tag{9}$$

$$h_m < 1.8 = (A_k^{1,2}, W_k^1) \qquad \forall m_i \in M, k \in K \tag{10}$$

The constraint (8) considers the assignment of a vehicle m with a height h_m greater than 2.5 m (meters) that corresponds to a van, which occupies an allocated space in level A_k^1 and two spaces on level A_k^2, using three spaces of the capacity W of the auto-carrier k. A vehicle m with height h_m greater than 1.8 m and less than 2.5 m, its assignment corresponds to a space W_k^1 and can only be accommodated at level A_k^1, i.e., the constraint (9). The last assignment constraint (10) defines that a vehicle m with a height h_m less than 1.8 m corresponds to an allocation space W_k^1 and can be accommodated in either of the two levels A_k^1 or A_k^2.

5. Methodology

The use of a heuristic methodology allows obtaining the solution to the routing problem of the auto-carriers at a reasonable time, meaning a representative change versus the empirical methodology previously used by the logistics company. A graphic illustration of the comparison of the insertion heuristic I1 is made in [29,30]. Regarding its comparison with other methods is presented in [31] and its computational complexity of the proposed algorithm is $O(n^2 \log n^2)$. Hereafter, the approach and development of the heuristic algorithm are described.

5.1. Heuristic Approach

The development of the solution is divided into two phases, the first one is to generate the route of the auto-carrier and the second one the vehicle allocation in the auto-carrier, both phases are part of a main algorithm. In the first phase, the routes are obtained with the implementation of the Solomon I1 insertion heuristic, due to the logistics company is needed to obtain a solution to the VRPTW in fairly necessary time, given that the VRPTW is an NP-complete problem [28]. This routing process applies a methodology of cluster first, route second, i.e., first group by the dealership, to then build the route, which starts with the dealership that has the shortest and earliest time window, considering the allocation and capacity constraints. The following describes the application of the Solomon I1 insertion heuristic [4] and the vehicle loading process for allocation phase on this VRPTW.

The routing algorithm builds a feasible solution by constructing one route at a time. At each iteration the algorithm decides which new dealership $u^* \in U$ has to be inserted in the current solution, and between which adjacent dealerships $i(u^*)$ and $j(u^*)$ the new dealership u^* has to be inserted on the current route. When choosing u^*, the algorithm takes into account both the cost increase associated with the insertion of u^*, and the delay in service time at dealerships following u^* on the route. The three steps of the algorithm are:

Step 0. *(Initialization)*. The first route is initially $R_1 = \{0, i, 0\}$, where i is the dealership with the shortest and earliest time window. In the allocation phase, if vehicle m of dealership i has a feasible assignment, then set $k = 1$, otherwise get the next vehicle from dealership i or next dealership with the shortest and earliest time window, until the allocation phase of the vehicle m is feasible.

Step 1. Let $R_k = \{i_0, i_1, \ldots, i_m\}$ be the current route, where $i_0 = i_m = 0$, i.e., the NCSY. Set

$$f1(i_{p-1}, u, i_p) = \alpha(r_{i_{p-1}u} + r_{ui_p} - \mu r_{i_{p-1}i_p}) + (1 - \alpha)(b_{i_p}^u - b_{i_p}) \tag{11}$$

where $0 \leq \alpha \leq 1$, $\mu \geq 0$ and $b_{i_p}^u$ is the time when service begins at dealership i_p provided that dealership u is inserted between i_{p-1} and i_p. For each unrouted dealership u, compute its best feasible insertion position in route R_k as:

$$f1(i(u), u, j(u)) = \min_{p=1,\ldots,m} f1(i_{p-1}, u, i_p)$$

where $i(u)$ and $j(u)$ are the two adjacent vertices of the current route between which u should be inserted. Determine the best unrouted customer u^* to be inserted yielding.

$$f2(i(u^*), u^*, j(u^*)) = \max_u \{f2(i(u), u, j(u))\}$$

where

$$f2(i(u), u, j(u)) = \lambda r_{0_u} - f1(i(u), u, j(u)) \tag{12}$$

with $\lambda \geq 0$.

Step 2. Insert dealership u^* in route R_k between $i(u^*)$ and $j(u^*)$, in the allocation phase, if vehicle m of dealership u^* has a feasible assignment, then go back to **Step 1**, otherwise get the next vehicle from dealership u^* until the allocation phase of the vehicle m is feasible. If u^* does not exist, but there are still unrouted dealerships, set $k = k + 1$, initialize a new route R_k (as in **Step 0**) and go back to **Step 1**. Otherwise, **STOP**, a feasible solution has been found.

The insertion heuristic tries to maximize the benefit obtained when servicing a dealership on the current route rather than on an individual route. For example, when $\mu = \alpha = \lambda = 1$, Equation (12) corresponds to the saving in distance from servicing dealership u on the same route as dealerships i and j rather than using an individual route. The best feasible insertion place of an unrouted dealership is determined by minimizing a measure, defined by the Equation (11), of the extra distance and the extra time required to visit it. Different values of the parameters μ, α and λ lead to different possible criteria for selecting the dealership to be inserted and its best position in the current route.

After starting a new route R_k or inserting a dealership u^* in the current route, the vehicle allocation phase is responsible for obtaining the feasible load of the auto-carrier, considering the constraints imposed by the logistics company, which are listed below by rank:

- Vehicle $h_m > 2.5$ m: It uses three spaces of the capacity of the auto-carrier k, i.e., a space in level A_k^1 and two spaces on level A_k^2, this is shown in Figure 5. To maximize the use of the capacity of the auto-carrier, another allocation is to occupy one space above (A_k^2) and two below (A_k^1).

Figure 5. Constraint of vehicles with $h > 2.5$ m.

- Vehicle 1.8 m $< h_m < 2.5$ m: It uses a space of the capacity of the auto-carrier k and can only be assigned in level A_k^1, as shown in Figure 6.

Figure 6. Constraint of vehicles with 1.8m $< h < 2.5$ m.

- Vehicle $h_m < 1.8$ m: It uses a space of the capacity of the auto-carrier k and can be assigned in any available space to it, as shown in Figure 7.

Figure 7. Constraint of vehicles with $h < 1.8$ m.

- Policy Last In First Out (LIFO): Last vehicle loaded, first vehicle unloaded. For example, if the first dealership to visit is d_2 on the current route, the vehicles of d_2 should be the last to be loaded on the auto-carrier.

If the assignment of a vehicle m is not feasible and there are still vehicles on demand, then go back to the routing phase, while the vehicle m will be assigned to the next auto-carrier route. The development of the heuristic algorithm is described in the following subsection.

5.2. Development of the Two-Phase Heuristic

To implement the heuristic algorithm, it was necessary to create a distance matrix, with the distance information (in kilometers) among the 44 dealerships, as shown in Table 1, the NCSY is represented by d_0, e.g., a trip from the NCSY (d_0) to dealership 2 (d_2) has a cost of 1373 km, while the trip from dealership 2 (d_2) to dealership 44 (d_{44}) would represent a route of 1245 km.

Table 1. Distance (in kilometers) between dealerships.

Dealership	d_0	d_1	d_2	$d_{...}$	d_{44}
d_0	0	885	1373	...	114
d_1	885	0	498	...	752
d_2	1373	498	0	...	1245
$d_{\vdots}$	$\vdots$	$\vdots$	$\vdots$	0	$\vdots$
d_{44}	114	752	1245	...	0

A time matrix is also required for the implementation of the heuristic algorithm. Table 2 shows the duration in minutes of the travel times between the dealerships, for example, the duration of the trip from NCSY (d_0) to dealership 1 (d_1) is 568 min, in other words, 9 h and 28 min. A route from the dealership 44 (d_{44}) to dealership 2 (d_2) is 13 h and 49 min of travel.

Table 2. Travel times (in minutes) between dealerships.

Dealership	d_0	d_1	d_2	$d_{...}$	d_{44}
d_0	0	568	880	...	92
d_1	568	0	319	...	490
d_2	880	319	0	...	829
$d_{\vdots}$	$\vdots$	$\vdots$	$\vdots$	0	$\vdots$
d_{44}	92	490	829	...	0

The distances (Table 1) and times (Table 2) matrices are symmetric, but in the time windows, a matrix was created with the earliest (e_i) and the latest (l_i) time window, using a 24-h time format, as shown in Table 3. The NCSY (d_0) does not have a time window established, therefore, e_i = 00:00 and l_i = 23:59. Figure 1 shows an example of the dealerships who have established a time window, otherwise they do not have a time window established as the NCSY.

Table 3. Time windows for the dealerships in 24-h time format.

	TimeWindow	
Dealership	e_i	l_i
d_0	00:00	23:59
d_1	06:00	13:00
d_2	08:00	12:00
$d_{\vdots}$	$\vdots$	$\vdots$
d_{44}	22:00	09:00

The heuristic is described in Algorithm 1, first phase is responsible for generating the route for the auto-carriers and the second of the feasible load of vehicles in the auto-carrier. This algorithm is codified in JAVA language, it has as input a matrix M with the demand of vehicles to be transported and a matrix with auto-carriers K available for the delivery of vehicles. The first phase clusters the demand M according to the dealerships to visit (U), then perform the sorting of the dealerships with the shortest and earliest time window, considering that the execution time (*current time, CT*) of the algorithm influences this ordination, i.e., the execution of the algorithm at different times of the day with the same data produces different routing outputs and vehicle accommodation.

Algorithm 1: Two-phase heuristic

   Data: M (*demand*), K (auto-carriers).

   Output: The K auto-carriers with A_k arrangement, R_k route and delivery schedules. A vector with
         the remaining M, if any.

 1 **begin**

 2    $U \longleftarrow$ get *dealerships* from demand

 3    $CT \longleftarrow$ get *current time*

 4   **while** $M > 0$ **and** $K > 0$ **do**

 5   Initialization a new route R_k

 6       $W_t \longleftarrow$ get *capacity* from k

 7       $A_k \longleftarrow$ generate a new arrangement with capacity W_t

 8       $u \longleftarrow$ get *first dealership* to visit from U

 9   **while** $W_t > 0$ **and** $u^* \in U$ **do**

10   **while** $m_u \in M$ **do**

11   **if** $Allocation(m_u, A_k)$ **then**

12   Update demand M, capacity W_t and route R_k

13   **end**

14   **end**

15          $u^* \longleftarrow$ get *next dealership*(R_k, CT) to visit and update U

16   **end**

17   **if** $K \leq 0$ **and** $M > 0$ **then**

18   Get *remaining* from M

19   **end**

20   Add to $Solution(R_k, A_k)$, $k = k + 1$

21   **end**

22 **end**

The first loop is the demand M and the auto-carriers K, while there are vehicles to load and auto-carriers, a new route R_k is initialized with time and distance counters, the auto-carrier $k \in K$ is obtained, assigns W_t according to k, which is the vehicle load capacity of k for its type t, then a new accommodation A_k based on capacity W_t is generated. After determining the *first dealership* to visit u, i.e., Step 0 of the heuristic, this is shown in line 8 of Algorithm 1.

With the first dealership u to be selected, start the loop of the auto-carrier k with capacity W_t and loop of the existing demand M of the dealership u, carrying out the loading of the vehicle m_u in the auto-carrier k in the second phase of the algorithm, *the allocation*, this is shown in line 11 of Algorithm 1. The allocation algorithm receives as parameters the vector A_k of the current arrangement and the vehicle m_u to be loaded (see Algorithm 2). If the load is successful, then update the demand M (eliminating m_u), the number of available spaces of the capacity W_t and the route R_k. In the case that the auto-carrier k is not filled or the vehicle m_u does not comply with the assignment restrictions, obtain the next vehicle $m_u + 1$ to load, until the auto-carrier k is full.

Algorithm 2: Allocation

Data: m_u (vehicle), A_k (arrangement).

Output: The A_k arrangement with m_u assigned if it is feasible.

1 **begin**
2      $W_t \longleftarrow$ size of A_k
3      $level \longleftarrow \frac{W_t}{2}$
4   **for** $A_{k_i} \in A_k$ **do**
5   **if** $h_{m_u} > 2.5$ **and** $i < level$ **and** *parity* **then**
6   **if** $A_{k_i} \in A_k$ **and** $A_{k_{i+1}} \in A_k$ **and** $A_{k_{i+level}} \in A_k$ **then**
7   Allocate m_u vehicle to spaces A_{k_i}, $A_{k_{i+1}}$ and $A_{k_{i+level}}$
8   **else**
9   **if** $A_{k_i} \in A_k$ **and** $A_{k_{i+(level-1)}} \in A_k$ **and** $A_{k_{i+level}} \in A_k$ **then**
10   Allocate m_u vehicle to spaces A_{k_i}, $A_{k_{i+(level-1)}}$ and $A_{k_{i+level}}$
11   **end**
12   **end**
13   **else**
14   **if** $A_{k_i} \in A_k$ **and** $A_{k_{i+1}} \in A_k$ **and** $A_{k_{i+(level+1)}} \in A_k$ **then**
15   Allocate m_u vehicle to spaces A_{k_i}, $A_{k_{i+1}}$ and $A_{k_{i+(level+1)}}$
16   **else**
17   **if** $A_{k_i} \in A_k$ **and** $A_{k_{i+level}} \in A_k$ **and** $A_{k_{i+(level+1)}} \in A_k$ **then**
18   Allocate m_u vehicle to spaces A_{k_i}, $A_{k_{i+level}}$ and $A_{k_{i+(level+1)}}$
19   **end**
20   **end**
21   **end**
22   **if** $h_{m_u} > 1.8$ **and** $h_{m_u} < 2.5$ **and** $i > level$ **and** $A_{k_i} \in A_k$ **then**
23   Allocate m_u vehicle in space A_{k_i}
24   **end**
25   **if** $h_{m_u} < 1.8$ **and** $A_{k_i} \in A_k$ **then**
26   Allocate m_u vehicle in space A_{k_i}
27   **end**
28   **end**
29   **return** A_k
30 **end**

If the auto-carrier k still has available spaces W_t, but the demand of the dealership u does not comply with the assignment restrictions, update accumulators of time and distance to obtain the next dealership u^* to visit, this is Step 1 of the heuristic and is observed on line 15 of Algorithm 1. To obtain u^* the route R_k built so far and the current time CT are received as parameters, finally enter the demand loop M of the dealership u^*.

Once selected u^*, in the capacity loop W_t the vehicle m_{u^*} allocation phase of the auto-carrier k starts, this is Step 2 of the heuristic algorithm. If the accommodation of m_{u^*} is feasible and there are still spaces of W_t, return to Step 1 of the heuristic algorithm, to obtain the next vehicle $m_{u^*} + 1$, the capacity loop ends when W_t is equal to 0 or does not exist u^*. Then, the route R_k and the arrangement A_k of the auto-carrier k add to the *Solution*, and finally a new route R_{k+1} is started.

A requirement of the logistics company is to add to the solution the remaining demand M, in the case that the number of auto-carriers K were not enough to transport the demand M, line 17. Algorithm 1 ends when there is no demand M or auto-carriers K available for routing.

In the allocation phase (Algorithm 2), to accommodate the vehicles in the auto-carrier, A_k is abstracted as a vector of size W_t (capacity of the t-type auto-carrier), to simulate and delimit the levels of the auto-carrier a variable called *level* is created. If t is pair, the index i of the upper level initializes at $i = 0$ and ends at $i = level - 1$, while the lower level initializes at $i = level$ and ends at $i = W_t$, as shown in Figure 8a. If t is odd, the index i of the upper level initializes at $i = 0$ and ends at $i = level$, while the lower level initializes at $i = level + 1$ and ends at $i = W_t$, as shown in Figure 8b.

Once A_k is defined, the load and assignment of the vehicle m_u is defined by its height h_{m_u}, Algorithm 2 starts by obtaining $level = \frac{W_t}{2}$. If $h_{m_u} > 2.5$ m, the vehicle m_u will occupy spaces in the two levels of the auto-carrier, then first determine if there is available space in the lower level $i < level$, otherwise m_u is assigned in the next auto-carrier with available space, the case that is met $i < level$, *parity* verifies if t is even or odd, depending on this result obtain the spaces that comply with Equation (8) and if are available, perform the allocation of m_u (e.g., see Figure 5) to these spaces.

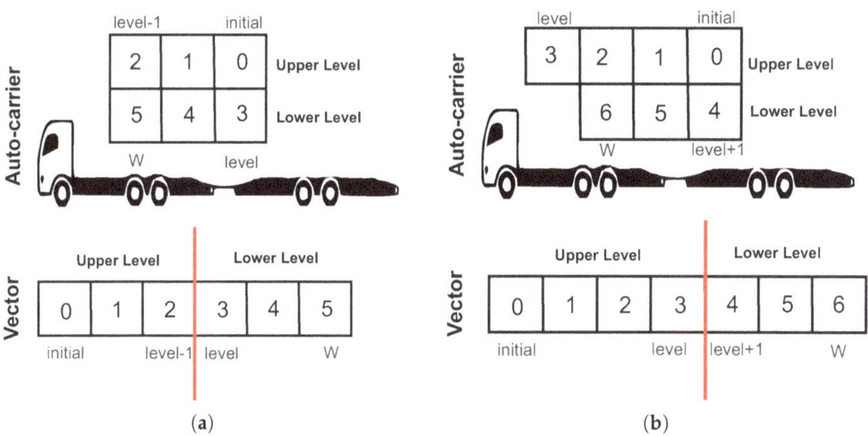

Figure 8. Vector of auto-carrier capacity: (**a**) Pair (**b**) Odd.

If $h_{m_u} > 1.8$ m and $h_{m_u} < 2.5$ m, to assign the vehicle m_u verify if there is available space in the lower level $i > level$ as shown in line 22 of Algorithm 2, otherwise m_u is assigned to the next auto-carrier with available space in the lower level. Finally, for $h_{m_u} < 1.8$ m it is only determined if there is space available in the auto-carrier and m_u is allocated, with the assignment of m_u in any of the cases and the return of A_k, the allocation phase ends.

6. Results and Analysis

6.1. Experimental Results

To evaluate the performance of the auto-carriers routing algorithm with time windows, two scenarios were designed with 11 different instances of the demand, the last instance corresponds to the real problem of the logistics company. The scenarios are the following:

- *Random Dealerships with Time Windows (RDTW)*—context in which most of the dealerships (34 of 44) were set different time windows for vehicle unloading.
- *Main Dealerships with Time Windows (MDTW)*—context that corresponds to the case of the logistics company, only the dealerships (14 of 44) that are located in the main cities of the country establish a time window for the unloading of vehicles.

The configuration of the most important parameters for the implementation of the proposed heuristic algorithm is shown in Table 4. Next, the content of the instances is described in Table 5, the first column corresponds to the instance number, the second is the demand size, and in the following columns the content of this in terms of vehicles (cars, partners) and vans (managers). It is necessary to emphasize the number of vans because they use more spaces in the auto-carrier compared to the vehicles. For both scenarios, the same instances demand was used to perform tests and compare the results of the total distance of the generated routes.

Table 4. Parameters of the proposed heuristic algorithm.

Parameter	Value
Algorithm 1	
mu	1
alpha	0.9
lamda	1
time_unloading	15
k	1
Algorithm 2	
initial	0
W_t	It depends on the auto-carrier K
level	$\frac{W_t}{2}$
parity	It depends on the auto-carrier K

Table 5. Test Instances.

Instance	Demand Size	Cars	Partners	Managers
1	20	6	14	0
2	50	46	1	3
3	100	32	66	2
4	200	108	87	5
5	500	206	270	24
6	1000	456	502	42
7	1500	654	767	79
8	2000	941	974	85
9	2500	1164	1224	112
10	3000	1433	1431	136
11	3884	1810	1906	168

With the instances of Table 5, a total of 132 tests were made in the two scenarios to the Algorithm 1, as a result of each of the tests the routes were obtained (auto-carriers, each route corresponds to one scenario previously mentioned) and the accommodation of the auto-carrier considering the allocation restrictions, in order to present all the results, these are grouped according to the capacity of the auto-carriers (3, 6, 7, 10, 11, and a heterogeneous fleet with these). Each table shows a comparison of the two scenarios for the eleventh instance, each table contains the column **Routes (K=Auto-carriers)**, this shows the number of routes generated for the eleventh instance, the **Distance (KM)** column contains the cost in terms of kilometers of the routes and the **Time (Min)** column shows the cost in minutes of the same.

Table 6 concentrates the results of the routing of the instances using auto-carriers with a capacity 10 and 11 vehicles in the MDTW scenario. The use of capacity auto-carriers 11 obtains a 13% decrease in total distance and total time compared to capacity 10. In addition, 60 less auto-carriers were used to route the demand of the eleventh instance.

Table 6. Results of the main dealerships with different auto-carriers.

	Auto-Carrier with $W = 10$			Auto-Carrier with $W = 11$		
Instance	Routes (K)	Distance (Km)	Time (Min)	Routes (K)	Distance (Km)	Time (Min)
1	4	19,263	12,660	4	19,263	12,660
2	7	18,329	12,481	7	18,338	12,493
3	18	71,366	48,256	14	62,571	41,809
4	24	81,429	55,950	21	71,231	48,997
5	69	148,588	102,617	59	130,604	87,928
6	137	323,696	213,737	118	270,872	179,242
7	195	417,909	277,747	170	364,385	241,743
8	254	550,555	367,032	220	471,456	312,449
9	318	681,336	458,476	278	585,245	389,968
10	366	794,553	525,284	323	696,878	459,669
11	478	1,037,633	686,562	418	900,562	597,259

Algorithm 2 was designed to work with a heterogeneous fleet of auto-carriers, the results of the tests in the two scenarios are shown in Table 7. In this table the results obtained from the tests are compared with the 11 instances in the two scenarios. Sometimes obtaining a smaller number of routes does not guarantee that it is the lowest total distance of the routes, e.g., in the row of instance 5 for the RDTW scenario, 95 routes are generated. In comparison with the 102 routes obtained in the MDTW scenario, but the total travel distance is 41,218 km smaller in this scenario (MDTW).

Table 7. Results of heterogeneous fleet (3,6,7,10 & 11).

	Random Dealerships			Main Dealerships		
Instance	Routes (K)	Distance (Km)	Time (Min)	Routes (K)	Distance (Km)	Time (Min)
1	4	17,458	11,416	5	25,649	16,875
2	10	27,626	18,854	9	22,964	15,622
3	26	94,510	62,905	22	91,525	61,156
4	32	111,317	75,206	37	124,226	84,035
5	95	240,376	160,016	102	199,158	134,659
6	180	436,777	284,208	190	446,823	294,392
7	278	563,385	367,985	278	572,442	381,464
8	350	724,204	476,686	350	764,601	511,309
9	444	932,767	611,969	450	933,942	623,494
10	490	1,091,345	710,645	495	1,049,780	696,274
11	655	1,389,356	907,976	660	1,416,358	943,464

Regarding the assignment of vehicles, Algorithm 2 returned feasible loads as illustrated in Figure 9. Table 8, concentrates the data of 18 vehicles (VIN and height (h_k)) before entering the allocation phase.

Table 8. Vehicle data.

VIN	Vehicle h_k	VIN	Vehicle h_k
1	2.52 m	10	1.47 m
2	1.47 m	11	1.47 m
3	1.47 m	12	1.47 m
4	1.47 m	13	1.47 m
5	1.47 m	14	1.47 m
6	1.47 m	15	1.47 m
7	2.52 m	16	1.47 m
8	1.87 m	17	1.47 m
9	1.47 m	18	1.47 m

From Table 8, Figure 9a shows the output of Algorithm 2 using the auto-carriers of capacity 11 (odd), in which we can observe the allocation of two vans (VIN1, VIN7) using three spaces in both auto-carriers in their platforms, the rest of spaces are occupied by other vehicles. On the other hand, Figure 9b shows the output of the allocation of the vans in auto-carriers of capacity 6.

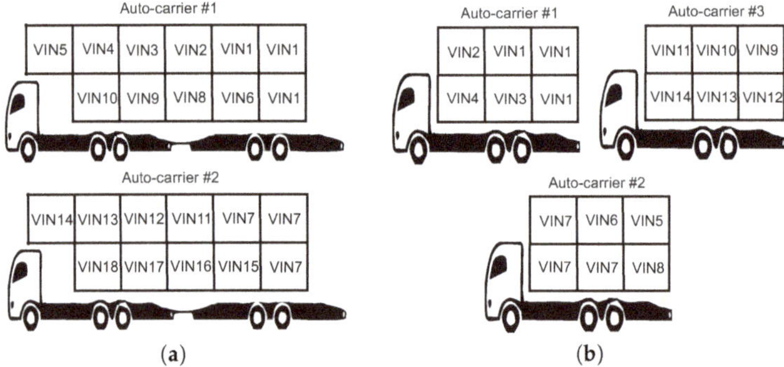

Figure 9. Output of vehicles allocation in auto-carriers: (**a**) Odd capacity (**b**) Pair capacity.

In Table 9 concentrates the results obtained to perform the routing of the demand in the MDTW scenario, using different capacities of auto-carriers and heterogeneous fleet, as it is highlighted in the capacity row of 11 vehicles, this shows the best results as regards distance (km) and time (min), as well as a lower number of auto-carriers (418) employed to carry out the routing of the real demand of the logistics company. A computer with Intel Core i5 7600K@3.8 GHz processor and 16 GB of RAM were used to perform the tests.

Table 9. Results of routing instance 11 with different auto-carriers.

Auto-Carrier (Capacity)	Main Dealerships		
	Routes (K)	Distance (Km)	Time (Min)
3	2043	4,146,291.8	2,720,157
6	936	2,018,547.9	1,332,574
7	688	1,488,546.3	980,416
10	478	1,037,633.2	686,562
11	**418**	**900,562.8**	**597,259**
3, 6, 7, 10 & 11	660	1,416,358.8	943,464

6.2. Analysis of the Results

The logistics company before the implementation of Algorithm 1 routed the auto-carriers empirically, i.e., the personnel in charge of this process did it without the assistance of some planning or optimization software, meaning the construction of inefficient routes [17]. Similarly, the loading of the vehicles in the auto-carrier was the responsibility of the operators, a process prone to damage during the unloading of the vehicles upon arrival at the dealership. Due to the absence of a LIFO policy that considers the route of the auto-carrier in the process of vehicle allocation.

The planning and routing of the auto-carriers of the logistics company were favorably impacted by the implementation of Algorithm 1, obtaining as output the generated routes (e.g., $d_0-> d_{44}-> d_{35}-> d_0$) for the auto-carriers, the unloading vehicles in each dealership and auto-carrier schedules, as shown in Table 10. These results were possible to obtain thanks to the heuristic routing algorithm that considers the time (CT variable and time matrix). In addition, the proposed Algorithm 2 allowed to automate the process of allocation of vehicles in the auto-carrier, making it easier for operators to load the vehicles, examples of the output of this Algorithm 2 are shown in Figure 9. The allocation constraints can be adjusted to the requirements of different vehicles to be transported, but retaining the allocation logic.

Table 10. Example of schedule of the auto-carrier route.

Dealership	Arrival	Departure	Unloaded Vehicles
d0	–:–	18:19	0
d44	22:59	23:14	6
d35	08:21	08:36	5
d0	21:55	–:–	0

The efficiency of the proposed algorithm was demonstrated by being able to generate planning (routes, schedules, vehicle accommodations) in a reasonable time for more than 2000 routes, using auto-carriers of 3 vehicles of capacity. With the same performance, the results were obtained in a real case of the logistics company, using a heterogeneous fleet were generated 660 routes as shown in Table 9. As a consequence of the size of the demand, the routes constructed by Algorithm 1 contain from one dealership to four dealerships in their planning.

7. Conclusions and Future Research Works

The results of the experimental work with the proposed heuristic algorithm were satisfactory. These show the ability to route and obtain feasible loads for the auto-carriers with the demand of the logistics company. The allocation of vehicles using restrictions reduced the likelihood that the vehicles suffer some damage during the loading/unloading in the dealership, in addition to complying with the traffic guidelines that govern the auto-carriers in Mexico.

The implementation of the algorithm allowed obtaining the planning of the routes and the feasible loading of vehicles at a reasonable time, considering a demand of 4000 vehicles and 44 dealerships as a destination, which translates into thousands of kilometers diminished, i.e., a saving of fuel, money, and time for the logistics company, while polluting emissions are reduced. Impacting favorably in the decision-making regarding the planning and programming of the routing of the auto-carriers that has its service.

Future work is to develop tests with other auto-carrier capabilities, in addition to developing a metaheuristic algorithm, with the combination of the heuristic Algorithm 1 to obtain an initial solution and a PSO to improve the current solution. In addition to implementing a dynamic routing according to a series of variables that can be presented in the current route of the auto-carrier, such as blocked routes or the transport of vehicles from one dealership to another.

Author Contributions: Conceptualization, O.A.P., P.M.Q.F. and R.E.P.L.; methodology and supervision, R.E.P.L. and P.M.Q.F.; data curation, software and writing–original draft preparation, M.A.J.P. and C.F.P.; investigation and writing–review and editing, M.A.J.P. and R.E.P.L.; validation and project administration, P.M.Q.F. and O.A.P.; all authors have read and approved the final manuscript.

Funding: This research received no external funding.

Acknowledgments: The first two authors acknowledge support from CONACyT through a scholarship to complete the Master in Computational Systems program at Tecnológico Nacional de México (TecNM). The authors appreciate the company for providing real data and the facilities granted for this research. In addition, the authors would like to thank the referees for their useful suggestions that helped to improve the quality of this paper.

Conflicts of Interest: The authors declare no conflict of interest.

References

1. Bonassa, A.C.; Cunha, C.B.; Isler, C.A. An exact formulation for the multi-period auto-carrier loading and transportation problem in Brazil. *Comput. Ind. Eng.* **2019**, *129*, 144–155. [CrossRef]
2. Asociación Mexicana de la Industria Automotriz. *AMIA-BOLETIN DE PRENSA/HISTORICO/BOLETIN HISTORICO/2005 2018*; AMIA: Ciudad de México, Mexico, 2018. Available online: http://www.amia.com.mx/descargarb.html (accessed on 22 July 2018).
3. Secretaría de Comunicaciones y Transportes. *PROYECTO de Norma Oficial Mexicana PROY-NOM-012-SCT-2-2017*; DOF—Diario Oficial de la Federación/SCT: México city, Mexico, 2017. Available online: https://www.dof.gob.mx/nota_detalle.php?codigo=5508944&fecha=26/12/2017 (accessed on 15 March 2018).
4. Ghiani, G.; Laporte, G.; Musmanno, R. *Introduction to Logistics Systems Planning and Control*; Online Edition; John Wiley & Sons Ltd: Chichester, UK, 2004; pp. 273–278, ISBN 0-470-84916-9.
5. Dantzig, G.; Fulkerson, R.; Johnson, S. Solution of a large-scale traveling-salesman problem. *J. Op. Res. Soc. Am.* **1954**, *2*, 393–410. [CrossRef]
6. Połap, D.; Woźniak, M.; Damaševičius, R.; Maskeliūnas, R. Bio-inspired voice evaluation mechanism. *Appl. Soft Comput.* **2019**, *80*, 342–357. [CrossRef]
7. Arnau, Q.; Juan, A.A.; Serra, I. On the Use of Learnheuristics in Vehicle Routing Optimization Problems with Dynamic Inputs. *Algorithms* **2018**, *11*, 208. [CrossRef]
8. Cassettari, L.; Demartini, M.; Mosca, R.; Revetria, R.; Tonelli, F. A Multi-Stage Algorithm for a Capacitated Vehicle Routing Problem with Time Constraints. *Algorithms* **2018**, *11*, 69. [CrossRef]
9. Zhao, M.; Lu, Y. A Heuristic Approach for a Real-World Electric Vehicle Routing Problem. *Algorithms* **2019**, *12*, 45. [CrossRef]
10. Stodola, P. Using Metaheuristics on the Multi-Depot Vehicle Routing Problem with Modified Optimization Criterion. *Algorithms* **2018**, *11*, 74. [CrossRef]
11. Połap, D.; Woźniak, M. Polar Bear Optimization Algorithm: Meta-Heuristic with Fast Population Movement and Dynamic Birth and Death Mechanism. *Symmetry* **2017**, *9*, 203. [CrossRef]
12. Chen, S.; Chen, R.; Gao, J. A Monarch Butterfly Optimization for the Dynamic Vehicle Routing Problem. *Algorithms* **2017**, *10*, 107. [CrossRef]
13. Ahmed, A.K.M.F.; Sun, J.U. Bilayer Local Search Enhanced Particle Swarm Optimization for the Capacitated Vehicle Routing Problem. *Algorithms* **2018**, *11*, 31. [CrossRef]
14. Desrochers, M.; Desrosiers, J.; Solomon, M. A New Optimization Algorithm for the Vehicle Routing Problem with Time Windows. *Oper. Res.* **1992**, *40*, 342–354. [CrossRef]
15. Tan, K.C.; Lee, L.H.; Zhu, Q.L.; Ou, K. Heuristic methods for vehicle routing problem with time windows. *Artif. Intell. Eng.* **2001**, *15*, 281–295. [CrossRef]
16. Yu, B.; Yang, Z.Z.; Yao, B.Z. A hybrid algorithm for vehicle routing problem with time windows. *Expert Syst. Appl.* **2011**, *38*, 435–441. [CrossRef]

17. Taner, F.; Galić, A.; Carić, T. Solving Practical Vehicle Routing Problem with Time Windows Using Metaheuristic Algorithms. *Promet-Traffic Transp.* **2012**, *24*, 343–351. [CrossRef]
18. Sripriya, J.; Ramalingam, A.; Rajeswari, K. A hybrid genetic algorithm for vehicle routing problem with time windows. In Proceedings of the 2015 International Conference on Innovations in Information, Embedded and Communication Systems (ICIIECS), Coimbatore, India, 19–20 March 2015; pp. 1–4.
19. Tadei, R.; Perboli, G.; Della, F. A Heuristic Algorithm for the Auto-Carrier Transportation Problem. *Transp. Sci.* **2002**, *36*, 55–62. [CrossRef]
20. Miller, B.M. Auto Carrier Transporter Loading and Unloading Improvement. Master's Thesis, Air Force Institute of Technology, Air University, Montgomery, AL, USA, March 2003. Available online: https://apps.dtic.mil/dtic/tr/fulltext/u2/a413017.pdf (accessed on 25 May 2018).
21. Dell'Amico, M.; Falavigna, S.; Iori, M. Optimization of a Real-World Auto-Carrier Transportation Problem. *Transp. Sci.* **2014**, *49*, 402–419. [CrossRef]
22. Tran, T.H.; Nagy, G.; Nguyen, T.B.T.; Wassan, N.A. An efficient heuristic algorithm for the alternative-fuel station location problem. *Eur. J. Op. Res.* **2018**, *269*, 159–170. [CrossRef]
23. Hosseinabadi, A.A.R.; Siar, H.; Shamshirband, S.; Shojafar, M.; Nasir, M.H.N.M. Using the gravitational emulation local search algorithm to solve the multi-objective flexible dynamic job shop scheduling problem in Small and Medium Enterprises. *Ann. Op. Res.* **2015**, *229*, 451–474. [CrossRef]
24. Instituto Nacional de Estadística y Geografía, Asociación Mexicana de la Industria Automotriz. *Estadísticas a Propósito de. . . la Industria Automotriz*; INEGI: México city, Mexico, 2016. Available online: http://www.dapesa.com.mx/dan-a-conocer-inegi-y-la-amia-libro-estadistico-de-la-industria-automotriz/ (accessed on 3 June 2018).
25. Morawska, L.; Moore, M.R.; Ristovski, Z.D. *Desktop Literature Review and Analysis: Health Impacts of Ultrafine Particles*; Technical Report; Department of the Environment and Heritage, Australian Government: Canberra, Australia, 2004. Available online: http://www.environment.gov.au/system/files/resources/00dbec61-f911-494b-bbc1-adc1038aa8c5/files/health-impacts.pdf (accessed on 17 July 2018).
26. Jiménez, M. Wastes to Reduce Emissions from Automotive Diesel Engines. *J. Waste Manag.* **2014**, *2014*, 1–5. [CrossRef]
27. Sajeevan, A.C.; Sajith, V. Diesel Engine Emission Reduction Using Catalytic Nanoparticles: An Experimental Investigation. *J. Eng.* **2013**, *2013*, 1–9, doi:10.1155/2013/589382. [CrossRef]
28. Cordeau, J.F.; Laporte, G.; Savelsbergh, M.W.P.; Vigo, D. Chapter 6. Vehicle Routing. In *Handbooks in Operations Research and Management Science*; Elsevier: Amsterdam, The Netherlands, 2007; Volume 14, pp. 367–428.
29. Bräysy, O.; Gendreau, M. Vehicle Routing Problem with Time Windows, Part I: Route Construction and Local Search Algorithms. *Transp. Sci.* **2005**, *39*, 104–118, doi:10.1287/trsc.1030.0056. [CrossRef]
30. Solomon, M.M. Algorithms for the Vehicle Routing and Scheduling Problems with Time Window Constraints. *Op. Res.* **1987**, *35*, 254–265. [CrossRef]
31. El-Sherbeny, N.A. Vehicle routing with time windows: An overview of exact, heuristic and metaheuristic methods. *J. King Saud Univ. Sci.* **2010**, *22*, 123–131. [CrossRef]

Article

On Neighborhood Structures and Repair Techniques for Blocking Job Shop Scheduling Problems [†]

Julia Lange [1,*] and Frank Werner [2]

[1] Division of Information Process Engineering, FZI Forschungszentrum Informatik Karlsruhe, 76131 Karlsruhe, Germany

[2] Fakultät für Mathematik, Otto-von-Guericke-Universität Magdeburg, D-39016 Magdeburg, Germany; frank.werner@ovgu.de

[*] Correspondence: lange@fzi.de

[†] This paper is an extended version of our paper published in the proceedings of the 9th IFAC Conference on Manufacturing Modeling, Management, and Control.

Received: 23 October 2019; Accepted: 11 November 2019; Published: 12 November 2019

Abstract: The job shop scheduling problem with blocking constraints and total tardiness minimization represents a challenging combinatorial optimization problem of high relevance in production planning and logistics. Since general-purpose solution approaches struggle with finding even feasible solutions, a permutation-based heuristic method is proposed here, and the applicability of basic scheduling-tailored mechanisms is discussed. The problem is tackled by a local search framework, which relies on interchange- and shift-based operators. Redundancy and feasibility issues require advanced transformation and repairing schemes. An analysis of the embedded neighborhoods shows beneficial modes of implementation on the one hand and structural difficulties caused by the blocking constraints on the other hand. The applied simulated annealing algorithm generates good solutions for a wide set of benchmark instances. The computational results especially highlight the capability of the permutation-based method in constructing feasible schedules of valuable quality for instances of critical size and support future research on hybrid solution techniques.

Keywords: job shop scheduling; blocking; total tardiness; permutations; repairing scheme; simulated annealing

---

1. Introduction

Complex scheduling problems appear in customer-oriented production environments, automated logistics systems, and railbound transportation as an everyday challenge. The corresponding job shop setting, where a set of jobs is to be processed by a set of machines according to individual technological routes, constitutes one of the non-trivial standard models in scheduling research. Even ·simple variants of this discrete optimization problem are proven to be NP-hard, see [1]. While the classical job shop scheduling problem with infinite buffers and makespan minimization has been a subject of extensive studies for many decades, see, for instance [2–4], solving instances with additional real-world conditions, such as the absence of intermediate buffers, given release and due dates of the jobs and recirculation, newly receives increasing attention.

The blocking job shop problem with total tardiness minimization (BJSPT) is regarded as an exemplary complex planning situation in this paper. Blocking constraints refer to a lack of intermediate buffers in the considered system. A job needs to wait on its current machine after processing, and thus blocks it, until the next required machine is idle. Such situations occur, for instance, in the manufacturing of huge items, in railbound or pipeline-based production and logistics as well as in train scheduling environments; see, for instance [5–8]. The consideration of a tardiness-based objective implements efficient economic goals like customer satisfaction and schedule reliability as they appear in most enterprises.

On the one hand, existing computational experiments indicate that exact general-purpose solution methods have significant difficulties in finding optimal and even feasible solutions for non-classical job shop instances of practically relevant size; see, for instance [9–11]. On the other hand, the application of special-purpose heuristics shows a necessity of complicated construction, repairing and guiding schemes to obtain good solutions; see [12–14]. This work is intended to analyze the capability of well-known scheduling-tailored heuristic search methods in determining high quality solutions for complex job shop scheduling problems. Structural reasons for the appearing complexity are detected and algorithms, which assure the applicability of basic strategies, are proposed.

As a natural foundation, permutations shall be used to represent feasible schedules of BJSPT instances. Widely applied interchange- and shift-based operators are chosen as transition schemes to set up neighboring solutions. Combining these ideas causes considerable redundancy and feasibility issues. A *Basic Repair Technique* (BRT) is proposed to construct a feasible schedule from any given permutation, cf. [15,16]. To fit the requirements in generating neighboring solutions, it is extended to an *Advanced Repair Technique* (ART), which defines a feasible neighboring schedule from an initial permutation and a desired interchange, see [15,16].

The resulting distances of solutions in a neighborhood are discussed to shed light onto the nature of the search space. In addition, different shifting strategies are analyzed with regard to their advantageousness in the search process. The presented neighborhood structures are embedded in a simulated annealing (SA) metaheuristic, which is applied to solve a diverse set of benchmark instances. Beneficial and critical aspects regarding the quality of the schedules found and the search guidance are pointed out by the computational results.

The remainder of the article is organized as follows. Section 2 summarizes existing work on complex job shop scheduling problems related to the BJSPT. A theoretical description of the problem and its notation are given in Section 3. Two variants of permutation-based encodings of schedules are discussed with regard to redundancy and feasibility in Section 4. Therein, the BRT is introduced and the distance of two schedules is defined. Section 5 incorporates explanations on the applied transition schemes, their implementation, and the operating principles of the ART. Furthermore, the neighborhoods are described and characteristics such as connectivity and solution distance are analyzed. Computational experiments on solving the BJSPT by an SA algorithm are reported in Section 6. Finally, Section 7 concludes with the main findings and further research perspectives.

2. Literature Review

A variety of exact and heuristic solution approaches to complex scheduling problems reported in the literature exist. This section will focus on observations and findings on job shop problems featuring constraints and optimization criteria similar to the BJSPT.

Exact solution methods are only sparsely applied to job shop problems with blocking constraints or tardiness-based objective functions. In 2002, Mascis and Pacciarelli [17] present a Branch & Bound procedure that is enhanced by scheduling-tailored bounds and a special branching technique. The approach is tested on complex instances with ten machines and ten jobs involving blocking constraints and makespan minimization. Obtaining proven optimal solutions for the benchmark problems takes between 20 min and four hours of computation time. Even if technical enhancements have been achieved and general-purpose mixed-integer programming solvers became more powerful, the job shop scheduling problem remains one of the hardest combinatorial optimization models. It is recently shown in [9,11,15,18] that even sophisticated mixed-integer programming solvers, such as IBM ILOG CPLEX and Gurobi, struggle with finding optimal and even feasible solutions to BJSPT instances with up to 15 machines processing up to 20 jobs in reasonable computation time.

Table 1 summarizes the heuristic approaches presented for job shop problems involving blocking constraints and tardiness-based objectives. A reference is stated in the first column, while the second column specifies whether a job shop problem with blocking constraints (BJSP) or without such restrictions (JSP) is examined. Column three contains the objective function regarded and the fourth column displays the maximal size (m, n) of the considered instances, where m denotes the number of machines and n defines the number of jobs. The last column presents the applied heuristic technique.

For reasons of comparison, the first two works [19,20] mentioned in Table 1 deal with a classical variant of the problem, namely the job shop problem with makespan minimization. The applied solution approaches constitute fundamental heuristic methods and the best-known algorithms to solve instances of the standard job shop setting until today. With the popular $(10, 10)$ instance of Fisher and Thompson having been open for decades, the size of standard job shop problems, for which good solutions can be obtained, has grown. However, most of the large instances have never been solved to optimality, which highlights the significant intricacy of the combinatorial optimization problems under study.

The following set of studies on JSPs with tardiness-based optimization criteria is intended to show the variety and the evolution of heuristic solution approaches applied together with the limitations in solvable problem size. A more comprehensive literature review including details on the types of instances solved can be found in [21].

In [22], a shifting bottleneck procedure is presented to generate schedules with minimal total tardiness for JSPs with release dates. The method is tested on a set of benchmark instances of size $(10, 10)$. A well-known critical path-oriented neighborhood, cf. [2], is discussed with regard to its applicability to pursue tardiness-based objectives in [23]. The authors tackle JSPs with total tardiness minimization by Simulated Annealing (SA) and show that a general neighborhood based on interchanges of adjacent operations on machines leads to better results. A hybrid genetic algorithm (GA) is proposed for JSPs with recirculation, release dates and various tardiness-based objective functions in [14]. Even if the procedure is enhanced by a flexible encoding scheme and a decomposition approach, the results do not significantly support GAs as a favorable solution method. The computational experiments are conducted on a set of twelve instances with maximum size $(8, 50)$. In [24], a generalized JSP consisting of a set of operations with general precedence constraints and required time lags is optimized with regard to total weighted tardiness. The authors apply a tabu search (TS) approach where neighboring solutions are also constructed by interchanges of adjacent operations, and the starting times of the operations are calculated based on a network flow. Here, the instance size of $(10, 10)$ is still critical.

A classical job shop setting involving release dates and minimizing the total weighted tardiness is considered in [25]. The authors combine a GA with an iterative improvement scheme and discuss the effect of various search parameters. Computational experiments are conducted on the set of benchmark instances with up to 15 machines and 30 jobs. The iterative improvement scheme seems to be a highly beneficial part of the solution approach, since it counteracts the occurring quality variance of consecutively constructed solutions. The same type of problems has also been tackled by a hybrid shifting bottleneck procedure including a preemption-allowing relaxation and a TS step in [26]. A more general type of optimization criteria, namely regular objective functions, are considered for JSPs in [27]. An enhanced local search heuristic with modified interchange-based neighborhoods is applied. The computational experiments are based on a large set of instances, where the most promising results are obtained for problems with up to 10 machines and 15 jobs. In [28], the critical path-based neighborhood introduced in [2] is used together with a block reversion operator in SA. Numerical results indicate that the combination of transition schemes is beneficial for finding job shop schedules with minimal total weighted tardiness. This work involves the largest test instances featuring 10 machines and 50 jobs as well as five machines and 100 jobs. However, these instances do not constitute a commonly used benchmark set so that no optimal solutions or lower bounds are known or presented to evaluate the capability of the method.

Table 1. Overview of existing heuristic solution approaches related to the BJSPT.

Reference	Problem	Objective *	Max. Size (m, n)	Solution Approach
Nowicki and Smutnicki 2005 [19]	JSP	C_{max}	$(10, 50), (20, 100)$	Tabu Search
Balas et al. 2008 [20]	JSP	C_{max}	$(22, 75)$	Shifting Bottleneck Procedure
Singer and Pinedo 1999 [22]	JSP	$\sum w_i T_i$	$(10, 10)$	Shifting Bottleneck Algorithm
Wang and Wu 2000 [23]	JSP	$\sum T_i$	$(30, 90)$	Simulated Annealing
Mattfeld and Bierwirth 2004 [14]	JSP	tardiness-based	$(8, 50)$	Genetic Algorithm
De Bontridder 2005 [24]	JSP	$\sum w_i T_i$	$(10, 10)$	Tabu Search
Essafi et al. 2008 [25]	JSP	$\sum w_i T_i$	$(10, 30), (15, 15)$	Hybrid Genetic Algorithm with Iterated Local Search
Bülbül 2011 [26]	JSP	$\sum w_i T_i$	$(10, 30), (15, 15)$	Hybrid Shifting Bottleneck Procedure with Tabu Search
Mati et al. 2011 [27]	JSP	regular	$(20, 30), (8, 50)$	Local Search Heuristic
Zhang and Wu 2011 [28]	JSP	$\sum w_i T_i$	$(15, 20), (10, 50), (5, 100)$	Simulated Annealing
Gonzalez et al. 2012 [29]	JSP	$\sum w_i T_i$	$(10, 30), (15, 20)$	Hybrid Genetic Algorithm with Tabu Search
Kuhpfahl and Bierwirth 2016 [30]	JSP	$\sum w_i T_i$	$(10, 30), (15, 15)$	Local Descent Scheme, Simulated Annealing
Bierwirth and Kuhpfahl 2017 [21]	JSP	$\sum w_i T_i$	$(10, 30), (15, 15)$	Greedy Randomized Adaptive Search Procedure
Brizuela et al. 2001 [12]	BJSP	C_{max}	$(20, 20)$	Genetic Algorithm
Mati et al. 2001 [8]	BJSP	C_{max}	$(10, 30)$	Tabu Search
Mascis and Pacciarelli 2002 [17]	BJSP	C_{max}	$(10, 30), (15, 20)$	Greedy Heuristics
Meloni et al. 2004 [31]	BJSP	C_{max}	$(10, 10)$	Rollout Metaheuristic
Gröflin and Klinkert 2009 [13]	BJSP	C_{max}	$(10, 50), (15, 20), (20, 20)$	Tabu Search
Oddi et al. 2012 [32]	BJSP	C_{max}	$(10, 30), (15, 15)$	Iterative Improvement Scheme
AitZai and Boudhar 2013 [33]	BJSP	C_{max}	$(10, 30), (15, 15)$	Particle Swarm Optimization
Pranzo and Pacciarelli 2016 [34]	BJSP	C_{max}	$(10, 30), (15, 20)$	Iterative Greedy Algorithm
Bürgy 2017 [9]	BJSP	regular	$(10, 30), (15, 20), (20, 30)$	Tabu Search
Dabah et al. 2019 [35]	BJSP	C_{max}	$(10, 30), (15, 15)$	Parallel Tabu Search

* Objective Functions: makespan C_{max}, total tardiness $\sum T_i$, total weighted tardiness $\sum w_i T_i$, various *regular* or *tardiness-based* objectives.

A JSP with setup times and total weighted tardiness minimization is tackled by a hybrid heuristic technique in [29]. A TS method is integrated into a GA to balance intensification and diversification in the search process. Furthermore, an improvement potential evaluation is applied to guide the selection of neighboring solutions in the TS. Promising results are found on a widely used set of benchmark instances. Different neighborhood structures are discussed and analyzed according to their capability of constructing schedules for a JSP with release dates and total weighted tardiness minimization in [30]. The experimental results show that the choice of the main metaheuristic method and the initial solution influence the performance significantly. Complex neighborhood structures involving several partially critical path-based components yield convincing results for instances with up to 15 machines and 30 jobs. In [21], an enhanced Greedy Randomized Adaptive Search Procedure (GRASP) is proposed and tested on the same set of benchmark instances. The applied method involves a neighborhood structure based on a critical tree, a move evaluation scheme as well as an amplifying and a path relinking strategy. The comprehensive computational study of Bierwirth and Kuhpfahl [21] shows that the presented GRASP is able to compete with the most powerful heuristic techniques tackling JSP instances with total tardiness minimization, namely the GA-based schemes proposed by Essafi et al. [25] and Gonzalez et al. [29]. Overall, the complexity of the applied methods, which is required to obtain satisfactory results for instances of still limited size, highlights the occurring difficulties in guiding a heuristic search scheme based on tardiness-related objective functions.

Considering the second set of studies on BJSPs given in Table 1, an additional feasibility issue arises and repairing or rescheduling schemes become necessary. The inclusion or exclusion of swaps of jobs on machines constitutes a significant structural difference with regard to real-world applications and the applied solution approach, see Section 4.2 for further explanations. Note that almost all existing solution approaches are dedicated to makespan minimization, even if this does not constitute the most practically driven objective.

In [12], a BJSP involving up to 20 machines and 20 jobs with swaps is tackled by a GA based on a permutation encoding. The well-known critical path-oriented transition scheme, cf. [2], is applied together with a job insertion-based rescheduling method in a TS algorithm in [8]. The authors consider a real-world application where swaps are not allowed and test their approach on instances with up to 10 machines and 30 jobs. Different greedy construction heuristics are compared in solving BJSP instances with and without swaps in [17]. Even for small instances, the considered methods have significant difficulties in constructing feasible schedules, since the completion of an arbitrary partial schedule is not always possible. The same issue occurs in [31], where a rollout metaheuristic involving a scoring function for potential components is applied to BJSP instances of rather small size with and without swaps.

A connected neighborhood relying on interchanges of adjacent operations and job reinsertion is presented for the BJSP in [13]. Instances involving setup and transfer times, and thus excluding swaps, are solved by a TS algorithm with elite solutions storage. Computational experiments are conducted on a large set of benchmark instances with up to 20 machines and 50 jobs. In [32], an iterative improvement algorithm incorporating a constraint-based search procedure with relaxation and reconstruction steps is proposed for the BJSP with swaps. A parallel particle swarm optimization is tested on instances of the BJSP without swaps in [33] but turns out not to be competitive with the method proposed in [13] and the following one. In [34], an iterated greedy algorithm, which loops deconstruction and construction phases, is applied to problems with and without swaps. Computational experiments on well-known benchmark instances with up to 15 machines and 30 jobs imply that forced diversification of considered solutions is favorable to solve the BJSP. A study tackling instances with up to 20 machines and 30 jobs and approaching a wider range of regular objective functions including total tardiness is reported in [9]. The authors embed a job reinsertion technique initially proposed in [36] in a TS and test their method on the BJSP with swaps. A parallel TS including the critical path-oriented neighborhood, cf. [2], and construction heuristics to recover feasibility is presented in [35]. Parallel search trajectories without communication are set up to increase the number of considered solutions.

Overall, the most promising approaches to solve BJSP instances proposed by Bürgy [9], Dabah et al. [35], and Pranzo and Pacciarelli [34] give evidence for focusing on the application of sophisticated neighborhood and rescheduling structures instead of increasing the complexity of the search procedure itself. This motivates the following work on evaluating the capability of basic scheduling-tailored techniques. Furthermore, a study on the interaction of blocking constraints and tardiness-based optimization criteria will be provided.

3. Problem Description and Benchmark Instances

The BJSPT is defined by a set of machines $\mathcal{M} = \{M_k \mid k = 1,\ldots,m\}$ which are required to process a set of jobs $\mathcal{J} = \{J_i \mid i = 1,\ldots,n\}$ with individual technological routes. Each job consists of a set of operations $\mathcal{O}^i = \{O_{i,j} \mid j = 1,\ldots,n_i\}$, where operation $O_{i,j}$ describes the j-th non-preemptive processing step of job J_i. The overall set of operations is defined by $\mathcal{O} = \bigcup_{J_i \in \mathcal{J}} \mathcal{O}^i$ containing n_{op} elements. Each operation $O_{i,j}$ requires a specific machine $Ma(O_{i,j})$ for a fixed processing time $p_{i,j} \in \mathbb{Z}_{>0}$. The recirculation of jobs is allowed. Furthermore, a release date $r_i \in \mathbb{Z}_{\geq 0}$ and a due date $d_i \in \mathbb{Z}_{>0}$ are given for every job $J_i \in \mathcal{J}$.

Blocking constraints are introduced for every pair of operations $O_{i,j}$ and $O_{i',j'}$ of different jobs requiring the same machine. Given that $O_{i,j} \rightarrow O_{i',j'}$ determines the operation sequence on the corresponding machine M_k and $j \neq n_i$ holds, the processing of operation $O_{i',j'}$ cannot start before job J_i has left machine M_k, in other words, the processing of operation $O_{i,j+1}$ has started. To account for the optimization criterion, a tardiness value is determined for every job with $T_i = max\{0, C_i - d_i\}$, where C_i describes the completion time of the job.

There exist different mathematical formulations of the described problem as a mixed-integer optimization program. For an overview of applicable sequence-defining variables and comprehensive studies on advantages and disadvantages of the corresponding models, the reader is referred to [11,15]. According to the well-known three-field notation, cf. for instance [37,38], the BJSPT can be described by

$$Jm \mid block, recrc, r_i \mid \sum T_i.$$

A feasible schedule is defined by the starting times $s_{i,j}$ of all operations $O_{i,j} \in \mathcal{O}$, which fulfill the processing sequences, the technological routes and the release dates of all jobs as well as the blocking constraints. Since the minimization of total tardiness constitutes a regular optimization criterion, it is sufficient to consider semi-active schedules where no operation can be finished earlier without modifying the order of processing of the operations on the machines, see e.g., [38,39]. Thus, the starting times of the operations and the operation sequences on the machines constitute uniquely transformable descriptions of a schedule. If a minimal value of the total tardiness of all jobs $\sum_{J_i \in \mathcal{J}} T_i$ is realized, a feasible schedule is denoted as optimal. Note that, regarding the complexity hierarchies of shop scheduling problems, see for instance [38,40] for detailed explanations, the BJSPT is harder than the BJSP with the minimization of the makespan C_{max}.

To discuss the characteristics of neighborhood structures and to evaluate their performance, a diverse set of benchmark instances is used. It is intended to involve instances of different sizes (m, n) featuring different degrees of inner structure. The set of problems contains train scheduling-inspired (ts) instances that are generated based on a railway network, distinct train types, routes and speeds, cf. [11,15], as well as the Lawrence (la) instances which are set up entirely random with $n_i = m$ for $J_i \in \mathcal{J}$, cf. [41]. The problems include 5 to 15 machines and 10 to 30 jobs. The precise instance sizes can be found in Tables 2–4.

For all instances, job release dates and due dates are generated according to the following terms in order to create computationally challenging problems. The release dates are restricted to a time interval which forces jobs to overlap and the due dates are determined with a tight due date factor, see [9,11,30]:

$$r_i \in \left[0, 2 \cdot \min_{J_i \in \mathcal{J}} \left\{\sum_{j=1}^{n_i} p_{ij}\right\}\right] \quad \text{and} \quad d_i = \left\lceil r_i + \left(1.2 \cdot \sum_{j=1}^{n_i} p_{ij}\right)\right\rceil \qquad \text{for all } J_i \in \mathcal{J}. \tag{1}$$

4. Representations of a Schedule

The encoding of a schedule is basic to every heuristic solution approach. In contrast to most of the existing work on BJSPs, the well-known concept of permutation-based representations is used here. In the following, redundancy and feasibility issues will be discussed and overcome, and a distance measure for two permutation-based schedules is presented.

4.1. Permutation-Based Encodings

An *operation-based representation* s^{op} of a schedule, also called permutation, is given as a single list of all operations. Consider exemplarily

$$s^{op} = [O_{i,1}, O_{i',1}, O_{i,2}, O_{i'',1}, O_{i,3}, O_{i',2}, \ldots].$$

The permutation defines the operation sequences on the machines, and the corresponding starting times of all operation can be determined by a list scheduling algorithm. Note that the processing sequences of the jobs are easily satisfiable with every operation $O_{i,j}$ having a higher list index than its job predecessor $O_{i,j-1}$. Furthermore, blocking restrictions can be implemented by list index relations so that the feasibility of a schedule can be assured with the operation-based representation. However, when applying the permutation encoding in a heuristic search procedure, redundancy issues need to be taken into account. Regarding the list s^{op} shown above and assuming that the first two operations $O_{i,1}$ and $O_{i',1}$ require different machines, the given ordering $O_{i,1} \rightarrow O_{i',1}$ and the reverse ordering $O_{i',1} \rightarrow O_{i,1}$ imply exactly the same schedule. Generally, the following conditions can be identified for two adjacent operations in the permutation being interchangeable without any effects on the schedule encoded, cf. [15]:

- The operations belong to different jobs.
- The operations require different machines.
- The operations are not connected by a blocking constraint.
- None of the operations is involved in a swap.

Details on the relation of two operations due to a blocking constraint and the implementation of swaps are given in the subsequent Sections 4.2 and 4.3. To avoid unnecessary computational effort caused by treating redundant permutations as different schedules, the application of a unique representation is desirable.

A second permutation-based encoding of a schedule, namely the *machine-based representation* s^{ma}, describes the operation sequences on the machines as a nested list of all operations. Consider

$$s^{ma} = [[O_{i,1}, O_{i'',1}, \ldots], [O_{i',1}, O_{i,2}], \ldots], [O_{i,3}, \ldots], [O_{i',2}, \ldots], \ldots]$$

as a general example, where the k-th sublist indicates the operation sequence on machine M_k. It can be observed that the machine-based representation uniquely encodes these operation sequences and any modification leads to the creation of a different schedule. However, since the machine-based encoding does not incorporate any ordering of operations requiring different machines, the given schedule may be infeasible with regard to blocking constraints. Preliminary computational experiments have shown that this blocking-related feasibility issue frequently appears when constructing BJSP schedules in heuristic search methods. Therefore, both representations are simultaneously used here to assure the uniqueness and the feasibility of the considered schedules.

As a consequence, the applied permutation-based encodings need to be transformed efficiently into one another. Taking the general representations given above as examples, the operation-based encoding features list indices $lidx(O_{i,j})$ and required machines $Ma(O_{i,j})$ as follows:

$$s^{op} = [\quad O_{i,1}, \quad O_{i',1}, \quad O_{i,2}, \quad O_{i'',1}, \quad O_{i,3}, \quad O_{i',2}, \quad \ldots]$$

$$
\begin{array}{lccccccc}
lidx(O_{i,j}): & 1 & 2 & 3 & 4 & 5 & 6 & \ldots \\
Ma(O_{i,j}): & M_1 & M_2 & M_2 & M_1 & M_3 & M_4 & \ldots
\end{array}
$$

The transformation $s^{op} \rightarrow s^{ma}$ can be performed by considering the operations one by one with increasing list indices in s^{op} and assigning them to the next idle position in the operation sequence of the required machine in s^{ma}, see [15,16]. As an example, after transferring the first two operations $O_{i,1}$ and $O_{i',1}$ from s^{op} to s^{ma}, the machine-based representation turns out as $s^{ma} = [[O_{i,1}],[O_{i',1}],[],[],\ldots]$. After transferring all operations given in the permutation s^{op}, the machine-based encoding exactly corresponds to the nested list shown above.

While performing the transformation $s^{ma} \rightarrow s^{op}$, the redundancy of operation-based encodings needs to be taken into account. If the machine-based representation s^{ma} is constructed from an operation-based representation s^{op} of a specific schedule, it will be desirable that the reverse transformation yields exactly the initially given permutation s^{op} instead of a redundant equivalent. To assure that the resulting list of operations is equivalent or closest possible to an initially given operation-based representation, the following method is proposed.

Priority-Guided Transformation Scheme $s^{ma} \rightarrow s^{op}$, **cf. [15]:** In transferring a machine-based representation s^{ma} to a permutation $s^{op'}$, the set of candidate operations to be added to the permutation $s^{op'}$ next consists of all operations $O_{i,j}$ in s^{ma}, for which the job predecessor $O_{i,j-1}$ and the machine predecessor given in s^{ma} either do not exist or are already present in $s^{op'}$. Considering the machine-based representation s^{ma} given above and an empty permutation $s^{op'}$, the set of candidate operations to be assigned to the first list index in $s^{op'}$ contains the operations $O_{i,1}$ and $O_{i',1}$. To guarantee the recreation of the initially given permutation s^{op}, the operation $O_{i,j}$ with the maximum priority $prio(O_{i,j})$ is chosen among all candidate solutions, whereby

$$
prio(O_{i,j}) = \begin{cases}
\frac{1}{lidx(O_{i,j})-lidx'(*)} & \text{if } lidx'(*) < lidx(O_{i,j}), \\
2 & \text{if } lidx'(*) = lidx(O_{i,j}), \\
lidx'(*) - lidx(O_{i,j}) + 2 & \text{if } lidx(O_{i,j}) < lidx'(*),
\end{cases}
\tag{2}
$$

with $lidx(O_{i,j})$ being the list index of the operation in the initially given permutation s^{op} and $lidx'(*)$ being the currently considered, idle list index in the newly created list $s^{op'}$. Recalling the example described above, the currently considered index features $lidx'(*) = 1$, while $prio(O_{i,1}) = 2$ due to $lidx'(*) = lidx(O_{i,1}) = 1$ and $prio(O_{i',1}) = \frac{1}{2-1} = 1$ due to $lidx'(*) = 1 < 2 = lidx(O_{i',1})$. Thus, $s^{op'} = [O_{i,1}]$ holds after the first iteration, and the set of candidate operations to be assigned to the next idle list index $lidx'(*) = 2$ consists of the operations $O_{i'',1}$ and $O_{i',1}$. Following this method iteratively, the newly constructed permutation $s^{op'}$ will be equivalent to the initially given permutation s^{op} from which s^{ma} has been derived.

Given that the considered machine-based representation is feasible with regard to the processing sequences and the technological routes of the jobs, the priority-guided transformation scheme will assign exactly one operation to the next idle list index of the new permutation in every iteration and can never treat an operation more than once. Thus, the method constructs a unique operation-based representation from a given machine-based representation in a finite number of $O(n_{op} \cdot m)$ steps, see [15] for detailed explanations.

4.2. Involving Swaps

When considering a lack of intermediate buffers in the job shop setting, the moments in which jobs are transferred from one machine to the next require special attention. The situation where two or more jobs interchange their occupied machines at the same point in time is called a swap of jobs. Figure 1 shows an excerpt of a general BJSP instance involving operations of three jobs with unit processing times on two machines. The Gantt chart in part (a) of the figure illustrates a swap of the jobs J_i and $J_{i'}$ on the machines M_k and $M_{k'}$ at time point $\bar{t}$.

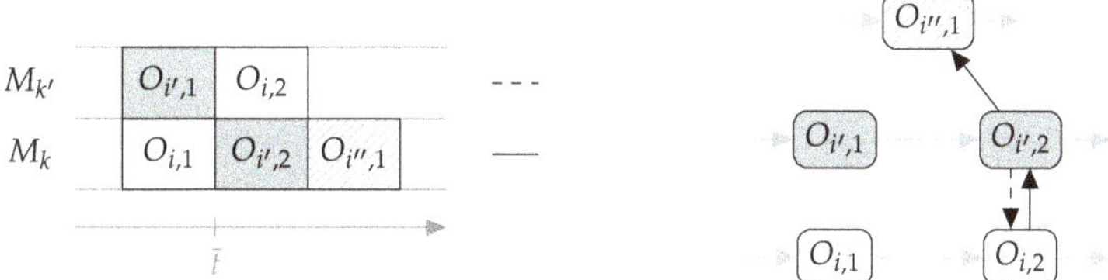

(**a**) A swap of jobs as a feasible cycle in the BSJP schedule

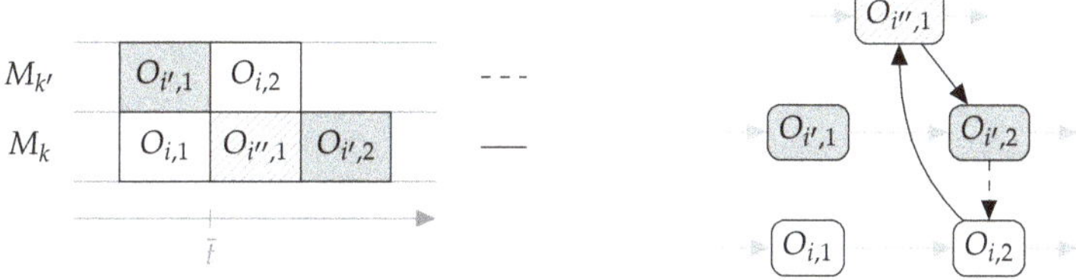

(**b**) A blocking-infeasible cycle in the BJSP schedule

Figure 1. Illustration of feasible and infeasible cycles in BJSP schedules.

Dependent on the type of real-world application, swaps are treated as feasible or infeasible in a BJSP schedule; see, for instance [8,17]. In particular, the implementation of rail-bound systems and the existence of setup times require their exclusion, cf. [5,6,13,34]. In this work, it is assumed that, even if jobs cannot be stored between consecutive processing steps, it is possible to move several jobs simultaneously on the production site. Thus, swaps are treated as feasible here.

Considering the alternative graph representation of a BJSP, initially proposed in [42], reveals an upcoming issue related to swaps and feasibility. In Figure 1, the corresponding excerpt of the alternative graph implementing the given operations as nodes and the implied ordering constraints as arcs is shown on the right next to the Gantt chart. The gray arcs represent the processing sequences of the jobs, while the black arcs indicate the existing operation sequence and blocking constraints. Taking the swap situation in part (a) of Figure 1 as an example again and assuming that the operation $O_{i',2}$ is the last processing step of job $J_{i'}$, the operation sequence $O_{i,1} \rightarrow O_{i',2} \rightarrow O_{i'',1}$ on machine M_k implies the solid arcs $(O_{i,2}, O_{i',2})$ and $(O_{i',2}, O_{i'',1})$ in the alternative graph. Equivalently, the operation sequence $O_{i',1} \rightarrow O_{i,2}$ on machine $M_{k'}$ causes a blocking constraint, which is represented by the dashed arc $(O_{i',2}, O_{i,2})$. The resulting structure of arcs shows that swaps appear as cycles in the alternative graph representation of the schedule. These cycles refer to feasible situations, since the underlying blocking inequalities can simultaneously be fulfilled by an equivalent starting time of all involved operations.

In part (b) of Figure 1, a Gantt chart and the corresponding graph-based representation of infeasible operation sequences are shown as a contrasting example. When trying to determine the starting times of the operations according to the operation sequences on the machines, an infeasible cyclic dependency of ordering and blocking constraints occurs at point $\bar{t}$ as follows:

$$s_{i'',1} + p_{i'',1} \leq s_{i',2} \leq s_{i,2} \leq s_{i'',1}.$$

It can be observed that such infeasible operation sequences similarly appear as cycles in the alternative graph representation. Thus, treating swaps as feasible results in a need to differentiate feasible and infeasible cycles when encoding BJSP schedules by an alternative graph. Following findings presented in [39] on a weighted graph representation, a simple structural property to contrast feasible swap cycles from infeasible sequencing cycles can be proposed. An alternative graph represents a feasible schedule for the BJSP, if all cycles involved do only consist of operations of different jobs requiring different machines, cf. [15]. The operations forming the cycle and featuring their start of processing at the same point in time are called a swap group. The given property facilitates the interchange of two or more jobs on a subset of machines, since it assures that every machine required by an operation of the swap group is currently occupied by the job predecessor of another operation of the group. Comparing the cycles in Figure 1, it can be seen that the arcs involved in the feasible swap cycle in part (a) feature different patterns, since the operations at their heads require different machines. On the contrary, two of the arcs forming the infeasible cycle in part (b) are solid arcs indicating that the two involved operations $O_{i'',1}$ and $O_{i',2}$ require the same machine.

Since this work relies on permutation-based encodings of schedules and corresponding feasibility checking procedures, the concept of swap groups is used to handle feasible cyclic dependencies. In the previous section, it is already mentioned that relations between operations on different machines can only be included in the operation-based representation of a schedule. Thus, the appearance of a swap is implemented in a single list by forming a swap group of operations which is assigned to one single list index. This list index fulfills the existing processing sequence and blocking constraints of all involved operations, and indicates that these operations will also feature a common starting time in the schedule. Considering the small general example given in part (a) of Figure 1, an operation-based representation of this partial schedule may result in $s^{op} = [\ldots, O_{i,1}, O_{i',1}, (O_{i,2}, O_{i',2}), O_{i'',1}, \ldots]$.

4.3. Feasibility Guarantee

In the following, the feasibility of a schedule given by its operation-based representation shall be examined more closely. As mentioned before, the processing sequences of the jobs and the blocking constraints can be translated to required list index relations of pairs of operations in the permutation.

For two consecutive operations $O_{1,j}$ and $O_{1,j'}$ of a job J_1 with $j < j'$, the starting time constraint $s_{1,j} + p_{1,j} \leq s_{1,j'}$ has to be fulfilled by a feasible schedule. Since these starting times are derived from the ordering of the operations in the permutation-based encoding, the required processing sequence can easily be implemented by assuring $lidx(O_{1,j}) + 1 \leq lidx(O_{1,j'})$. Blocking constraints can be described using list indices following the same pattern. Assume that, besides the operations $O_{1,j}$ and $O_{1,j'}$ requiring two machines M_k and $M_{k'}$, respectively, there is another operation $O_{2,j''}$ requiring machine M_k. If $O_{1,j} \rightarrow O_{2,j''}$ is determined as the operation sequence on this machine, the absence of intermediate buffers causes the starting time constraint $s_{1,j+1} \leq s_{2,j''}$. Translating this blocking restriction to a list index constraint implies that the list index of the job successor of the machine predecessor of an operation needs to be smaller than the list index of the operation. Formally, for two operations $O_{i,j}$ and $O_{i',j'}$ of different jobs requiring the same machine and a given operation sequence $O_{i,j} \rightarrow O_{i',j'}$, the following list index relation has to be fulfilled by a feasible permutation:

$$lidx(O_{i,j+1}) + 1 \leq lidx(O_{i',j'}), \tag{3}$$

provided that the operation $O_{i,j+1}$ exists.

This type of list index constraints constitutes the basis of checking and retrieving the feasibility of a BJSP schedule given by a single list of operations. The proposed method, called the *Basic Repair Technique (BRT)*, takes any permutation *perm*, which is feasible with regard to the processing sequences of the jobs, as input and constructs the operation-based representation s^{op} of a feasible schedule for the BJSP, cf. [15,16]. Note that the different terms *perm* and s^{op} both describing an operation-based encoding of the schedule are only used for reasons of clarification here. Figure 2 outlines the algorithm.

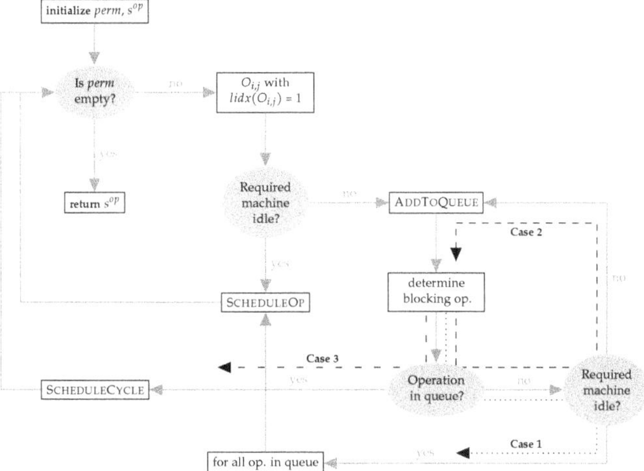

Figure 2. Schematic outline of the Basic Repair Technique (BRT), cf. [15].

A permutation *perm*, from which a feasible schedule is to be constructed, is initially given. The return permutation s^{op} is initialized with an empty list. The basic strategy of the BRT is to iterate over the list *perm*, take at least one operation from this list in each iteration and place it in the list s^{op}, so that all BJSP constraints are satisfied. As long as *perm* is not empty, the operation $O_{i,j}$ at the first list index is considered in the current iteration. If the required machine $Ma(O_{i,j})$ is idle, meaning that there is no other operation blocking it, the function SCHEDULEOP is called on operation $O_{i,j}$. This function

- determines and stores the earliest possible starting time of the considered operation,
- removes the operation from *perm*,
- adds the operation to the next idle list index in s^{op}, and
- sets the status of $Ma(O_{i,j})$ to blocked provided that a job successor $O_{i,j+1}$ exists.

With this, operation $O_{i,j}$ is said to be scheduled and the algorithm continues with the next iteration.

In case the required machine $Ma(O_{i,j})$ is not idle, meaning that it is blocked by another operation, the currently considered operation is involved on the right-hand side of a blocking constraint as given in (3). The operation $O_{i,j}$ is added to a queue and the operation required at a smaller list index to satisfy the blocking constraint is determined. The required operation is denoted as $O_{a,b}$ in the following. At this point, the BRT proceeds according to one of three different paths indicated in Figure 2.

Case 1: If the operation $O_{a,b}$ is not involved in the queue and its required machine $Ma(O_{a,b})$ is idle, operation $O_{a,b}$ and all operations in the queue are scheduled following a last in-first out strategy. Note that, when Case 1 is singly executed, exactly two operations are transferred from *perm* to the new operation-based representation s^{op}.

Case 2 → Case 1: If operation $O_{a,b}$ is not involved in the queue but its required machine $Ma(O_{a,b})$ is not idle, operation $O_{a,b}$ is added to the queue and the next required operation to fulfill the occurring blocking constraint is determined. Operations are added to the queue according to Case 2 until a required operation with an idle machine is found. Then, Case 1 is executed and all queuing operations are scheduled. Note that following this path, at least three operations are transferred from *perm* to s^{op}.

Case 2 → Case 3: Equivalent to the previous path, the operation $O_{a,b}$ is added to the queue and the next required operation is determined. Case 2 is executed until an operation already present in the queue is found. This refers to the situation where a cyclic dependency of blocking constraints exists and a swap needs to be performed in the schedule. The swap group is defined by all operations in the queue added in between the two occurrences of the currently considered operation. Following case 3, all operations of the swap group are scheduled with equivalent starting times and potentially remaining operations in the queue are scheduled correspondingly after. Since the smallest possible swap cycle is formed by two operations, at least two operations are transferred from *perm* to s^{op} when this path is executed.

With this, the BRT captures all occurring dependencies in arbitrary permutation-based encodings of BJSP schedules. The method assures that all blocking constraints are fulfilled by shifting required operations to positions with smaller list indices and by forming and modifying swap groups. Given that the number of operations in the problem is finite and the initially given permutation is feasible with respect to the processing sequences of the jobs, the following proposition holds, cf. [15].

Proposition 1. *The Basic Repair Technique (BRT) terminates and constructs an operation-based representation s^{op} of a feasible schedule for the BJSP.*

Proof. It has to be shown that

(1) the resulting permutation s^{op} is feasible with regard to the processing sequences of all jobs $J_i \in \mathcal{J}$,

(2) the resulting permutation s^{op} is feasible with regard to blocking constraints and

(3) every operation $O_{i,j} \in \mathcal{O}$ is assigned to a position in the feasible permutation s^{op} exactly once.

An unsatisfied blocking constraint $s_{i',j'} \leq s_{i,j}$ is detected in the BRT while an operation $O_{i',j'-1}$ is already scheduled in the feasible partial permutation s^{op} and operation $O_{i,j}$ is the currently considered operation, for which $Ma(O_{i,j})$ is not idle. The BRT shifts required operation(s), here only operation $O_{i',j'}$, to the next idle position $lidx'(*) > lidx'(O_{i',j'-1})$ and will never affect list indices prior to or equal to $lidx'(O_{i',j'-1})$. Hence, a given feasible ordering accounting for processing sequences and technological routes, such as $lidx(O_{i',j'-1}) < lidx(O_{i',j'})$, can never be violated by changes in the operation sequences made to fulfill blocking constraints. (1) is true.

The initially empty permutation s^{op} is expanded iteratively in the BRT. Every time an operation $O_{i,j}$ is considered to be assigned to the next idle list index $lidx'(*)$, unsatisfied blocking constraints are detected and fulfilled. Accordingly assigning an operation $O_{i',j'}$ to the list index $lidx'(*)$ in s^{op} prior to its initially given index $lidx(O_{i,j})$ in *perm* may implement a change in the operation sequence on the concerned machine. This may only cause new blocking constraints referring to the positions of the job successor $O_{i',j'+1}$ and the machine successor of operation $O_{i',j'}$. Due to given feasible processing sequences, affected operations cannot be part of the current partial permutation s^{op} and unsatisfied blocking constraints do only arise in the remainder of the permutation *perm*. Thus, it is assured that the existing partial permutation s^{op} is feasible with regard to blocking constraints in every iteration. Since this remains true until the BRT terminates, (2) is shown.

The consideration of operations in the BRT follows the ordering given in the initial list *perm* starting from the first position. Since the assignment of an operation $O_{i,j}$ to the next idle list index $lidx'(*)$ in s^{op} may only affect constraints that are related to succeeding operations in the initial list *perm*, the necessity of a repeated consideration of an operation can never occur, once it is added to the feasible ordering s^{op}. Therefore, (3) is true. □

Considering the remarks on the numbers of operations scheduled in every iteration of the BRT, it can already be expected that a feasible schedule is determined by the BRT in polynomial time. In [15], it is shown in detail that the schedule construction takes $O(n_{op} \cdot m)$ steps. Thus, the BRT is an appropriate basic tool to be applied in heuristic search schemes.

4.4. Distance of Schedules

The distance of feasible solutions is an important measure in analyzing search spaces and neighborhood structures of discrete optimization problems, cf. [43,44]. When a heuristic search method is applied, the distance of two consecutively visited solutions refers to the size of the search step. In such a procedure, the step size may act as a control parameter or observed key measure to guide the search. Intensification and diversification are strategically implemented by conducting smaller or bigger steps to avoid an early entrapment in locally optimal solutions.

In scheduling research, the distance $\delta(s, s')$ of two feasible schedules s and s' is commonly defined by the minimum number of basic operators required to construct one schedule from the other, cf. [44,45]. Here, the adjacent pairwise interchange (API) of two neighboring operations in the machine-based representation of a schedule is used as the basic operator. Formally, it can be described by the introduction of an indicator variable for all pairs of operations $O_{i,j}$ and $O_{i',j'}$ with $i < i'$ requiring the same machine as follows:

$$
h_{i,j,i',j'} = \begin{cases} 1, & \text{if an ordering } O_{i,j} \to O_{i',j'} \text{ or } O_{i',j'} \to O_{i,j} \text{ in } s \text{ is reversed in } s', \\ 0, & \text{else.} \end{cases}
\tag{4}
$$

Consequently, the distance of two schedules is determined by

$$
\delta(s, s') = \sum_{\substack{O_{i,j}, O_{i',j'} \in \mathcal{O} \text{ with} \\ Ma(O_{i,j}) = Ma(O_{i',j'}),\ i<i'}} h_{i,j,i',j'} \, .
\tag{5}
$$

Note that, when describing the BJSP with a mixed-integer program and implementing the pairwise ordering of operations with binary variables, the given distance measure is highly related to the well-known Hamming distance of binary strings, see [15] for further explanations.

5. Neighborhood Structures

In the following, neighborhood structures, which apply interchanges and shifts to the permutation-based representations of a schedule, are defined. The generation of feasible neighbors receives special attention, and the connectivity of the neighborhoods when dealing with complex BJSPT instances is discussed. A statistical analysis of a large set of generated neighboring solutions is reported to detect critical characteristics of the repairing scheme and the search space in general.

5.1. Introducing Interchange- and Shift-Based Neighborhoods

5.1.1. Transition Schemes and Their Implementation

In line with the findings presented in the literature, intensification and diversification shall both be realized in a heuristic search procedure by appropriate moves. When solving general sequencing problems, interchanges and shifts of elements in permutations constitute generic operators which are widely used, cf. [45,46]. The interchange-based moves applied here to the BJSPT and their implementation in the permutation-based encodings are defined as follows, see [15,16].

Definition 1. *An **API move** denotes the interchange of two adjacent operations $O_{i,j}$ and $O_{i',j'}$ of different jobs requiring the same machine $M_k \in \mathcal{M}$ in the machine-based representation of the schedule. Adjacency is defined in a strict sense. A pair of operations $O_{i,j}$ and $O_{i',j'}$ is called **adjacent** if there is no idle time on machine M_k between the preceding operation leaving the machine and the start of the processing of the succeeding operation.*

Definition 2. *A **TAPI move** denotes an interchange of two adjacent operations $O_{i,j}$ and $O_{i',j'}$ of different jobs requiring the same machine $M_k \in \mathcal{M}$ with $O_{i,j} \to O_{i',j'}$ in the machine-based representation of the schedule, where strict adjacency is given and the job $J_{i'}$ is currently tardy.*

The limitation to pairs of operations, which are strictly adjacent in a schedule, can be made without loss of search capability, cf. [15]. An idle time between two consecutively processed operations of different jobs on a machine may only occur due to

1. the technological routes of the jobs and the corresponding processing sequences on other machines or

2. the release date of the job of the succeeding operation.

In the first case, there always exists a sequence of applicable API moves that eliminates the idle time and enables an interchange of the considered pair of operations. In the second case, an interchange of the considered pair of operations will only result in postponing the starting time of the initially preceding operation, since the succeeding operation cannot be processed earlier. If such a postponement is beneficial, this will also be indicated by an applicable API at another point in the schedule. Otherwise, postponing the preceding operation can never be advantageous with regard to total tardiness.

These operators are intended to construct close neighboring solutions with a desired distance $\delta(s, s') = 1$. Small steps are supposed to intensify the search and make a heuristic search procedure nicely tractable towards locally optimal schedules. The advantageousness of restricting the set of potential API moves based on the objective function value, namely considering only TAPI moves, shall be closely investigated in the computational experiments. Figure 3 shows all applicable API moves (solid arrows) and TAPI moves (dashed arrows) for a small BJSPT instance with three machines and three jobs. It can be observed that, referring to the same schedule, the set of TAPI moves is a subset of the set of API moves. Note that there is an idle time occurring between three pairs of consecutively processed operations on the machines M_2 and M_3, while there is no blocking time on any machine in the given schedule.

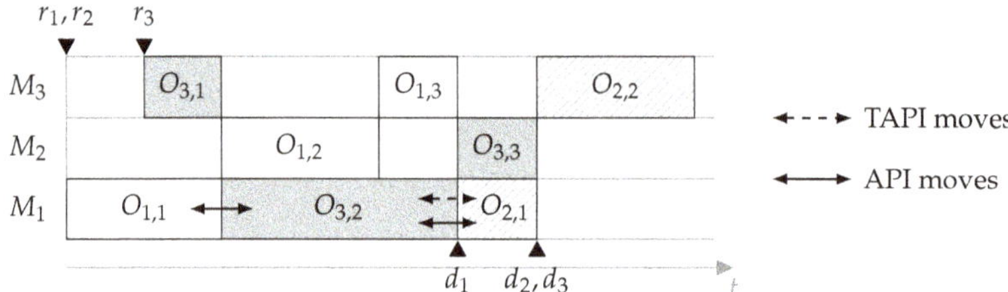

Figure 3. Illustration of applicable API and TAPI moves in a given schedule for the BJSPT

To avoid redundancy in the neighborhood structure on the one hand, API moves are applied to the machine-based representation of a schedule. To check the feasibility of the constructed neighboring schedule on the other hand, the applied API move needs to transferred to the operation-based representation of the schedule, cf. [16]. Since the interchanged operations are not necessarily directly adjacent in the permutation, this can be done by a left shift or a right shift transformation, respectively. First, consider the API move $O_{1,1} \leftrightarrow O_{3,2}$ in the schedule given in Figure 3 and the following permutation encoding this schedule:

$$s^{op} = [O_{1,1}, O_{3,1}, O_{1,2}, O_{3,2}, O_{1,3}, O_{3,3}, O_{2,1}, O_{2,2}].$$

The API move can be implemented either by shifting operation $O_{1,1}$ to the right together with its job successor $O_{1,2}$ to preserve the processing sequence or by shifting operation $O_{3,2}$ to the left together with its job predecessor $O_{3,1}$. In both cases, the following permutation *perm* is generated:

$$perm = [O_{3,1}, O_{3,2}, O_{1,1}, O_{1,2}, O_{1,3}, O_{3,3}, O_{2,1}, O_{2,2}].$$

It appears that the permutations generated by implementing API moves are infeasible with regard to blocking constraints. Here, operation $O_{1,1}$ cannot be scheduled on machine M_2, since this machine is blocked by operation $O_{3,2}$. Thus, the given list needs to be repaired. After applying the BRT to *perm*, the following feasible neighboring schedule $s^{op'}$, which incidentally turns out as a permutation schedule, is constructed:

$$s^{op'} = [O_{3,1}, \underline{O_{3,2}}, O_{3,3}, \underline{O_{1,1}}, O_{1,2}, O_{1,3}, O_{2,1}, O_{2,2}].$$

Considering the second applicable API move $O_{3,2} \leftrightarrow O_{2,1}$ in the schedule of the $(3,3)$-instance in Figure 3, the left shift of operation $O_{2,1}$ and the right shift of operation $O_{3,2}$ in s^{op} generate two different permutations $perm_1$ and $perm_2$, respectively:

$$perm_1 = [O_{1,1}, O_{3,1}, O_{1,2}, \underline{O_{2,1}}, \underline{O_{3,2}}, O_{1,3}, O_{3,3}, O_{2,2}],$$
$$perm_2 = [O_{1,1}, O_{3,1}, O_{1,2}, O_{1,3}, \underline{O_{2,1}}, \underline{O_{3,2}}, O_{3,3}, O_{2,2}].$$

Applying the BRT to $perm_1$ constructs a feasible neighboring schedule

$$s_1^{op'} = [O_{1,1}, O_{3,1}, O_{1,2}, \underline{O_{2,1}}, (O_{2,2}, \underline{O_{3,2}}), O_{1,3}, O_{3,3}],$$

which is displayed in Figure 4. This schedule features a swap of the jobs J_2 and J_3 on the machine M_1 and M_3 and two periods of blocking time on the machines M_2 and M_3 indicated by the curved lines.

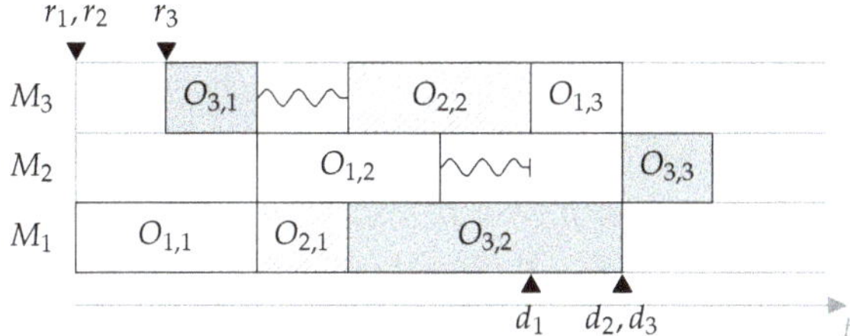

Figure 4. Illustration of the feasible neighboring schedule $s_1^{op'}$ resulting from an API move.

Applying the BRT to $perm_2$ reveals a major difficulty of using permutations, interchange-based operators, and repairing schemes in solution approaches for BJSPs. After the first three operations have been added to the partial permutation $s_2^{op'} = [O_{1,1}, O_{3,1}, O_{1,2}]$, the operation $O_{1,3}$ is considered. It requires machine M_3, which is blocked by operation $O_{3,1}$. Thus, the job successor $O_{3,2}$ must be scheduled prior to operation $O_{1,3}$, and the BRT reverts the given API to regain the feasibility of the schedule. A graphical representation of the critical step is given in Figure 5.

$$perm_2 = [O_{1,1}, O_{3,1}, O_{1,2}, O_{1,3}, \underline{O_{2,1}, O_{3,2}}, O_{3,3}, O_{2,2}]$$

Figure 5. Schematic presentation of an API reverted by applying the BRT.

It can easily be seen that the operation sequence given in the first part of the permutation $perm_2$ and the operation sequence resulting from the API cannot be implemented together. Additional changes in the schedule are necessary to construct a feasible solution involving the desired ordering $O_{2,1} \rightarrow O_{3,2}$. Preliminary experiments have shown that a reversion occurs in 80 to 90% of all generated and repaired neighboring schedules. Therefore, an enhanced repairing scheme is required to find feasible solutions that contain given orderings while featuring as few changes as possible compared to the initially given ones.

The desired operation sequence can be interpreted as a partial schedule with exactly two elements. While the decision problem on the existence of a completion of an arbitrarily large partial schedule is NP-complete for the BJSP, cf. [17], the generation of a feasible schedule with a given ordering of exactly two operations is always possible. Nonetheless, the challenging task is to find a structured and commonly applicable procedure, which returns a feasible neighbor from an initially given schedule and a desired operation sequence resulting from an API. The subsequent section deals with this issue in detail.

It is indicated by previous studies on BJSPs that a diversification strategy is beneficial to reach promising regions of the search space, see [32,34]. Therefore, a randomized and objective function-oriented transition scheme is defined. It is applied to the operation-based representation of a schedule and relies on shifts of operations in the permutation as generic operators, see [15,16].

Definition 3. *A **TJ move** is defined by applying random leftward shifts to all operations of a tardy job J_i in the permutation-based representation of a schedule, while preserving the processing sequence $O_{i,1} \rightarrow O_{i,2} \rightarrow \cdots \rightarrow O_{i,n_i}$ of the job.*

The resulting permutation might be infeasible with regard to blocking constraints, and the BRT is used to construct a feasible neighboring schedule. Since a TJ move creates desired partial sequences for every shifted operation, it is not guaranteed that a solution involving all of these orderings simultaneously exists. Thus, no fixation can be applied and the BRT is potentially able to revert all shifts. To avoid neighboring schedules which are equivalent to the initially given ones, sufficiently large shifts are executed.

5.1.2. Generating Feasible API-Based Neighbors

As mentioned in the previous section, the generation of feasible neighboring schedules for BJSPs involving a given API-based ordering is a critical issue. Since potentially required additional changes in the schedule are not contained in the BRT, an *Advanced Repair Technique (ART)* is proposed, cf. [15,16]. This method takes the operation-based representation of a schedule s, named *perm*, and a desired sequence of two operations $O_{a,b} \rightarrow O_{i',j'}$ and returns the operation-based representation s^{op} of a neighboring schedule s', which involves the given ordering. All additional APIs necessary to transform the schedule s into s' follow a basic rule. Instead of reverting a given pairwise sequence $O_{a,b} \rightarrow O_{i',j'}$, the initial permutation is adapted by leftward interchanges of an operation of the job J_a and the repairing scheme is restarted. Figure 6 gives a schematic illustration of the ART in total.

It can be observed in the left part of the chart that the BRT constitutes the foundation of the ART. The operations are iteratively taken from the list *perm*, requiring machines are checked for idleness, and blocking operations are determined, if necessary. The first important difference is indicated by the ellipsoid node printed in bold face. An operation is defined to be fixed, if it acts as the successor operation in a given pairwise sequence. The corresponding predecessor is denoted as the associated operation. In the general example stated above, operation $O_{i',j'}$ is fixed with the associated operation $O_{a,b}$. In case a fixed operation shall be added to the queue and its associated operation is already scheduled in the feasible permutation s^{op}, no reversion occurs and the procedure continues in the basic version. In case the associated operation $O_{a,b}$ is not yet scheduled, the given ordering $O_{a,b} \rightarrow O_{i',j'}$ would be reverted by adding operation $O_{i',j'}$ to the queue. To avoid this, the ART follows one of four modification paths in the gray box, the initial list *perm* is adapted, and the whole procedure is restarted.

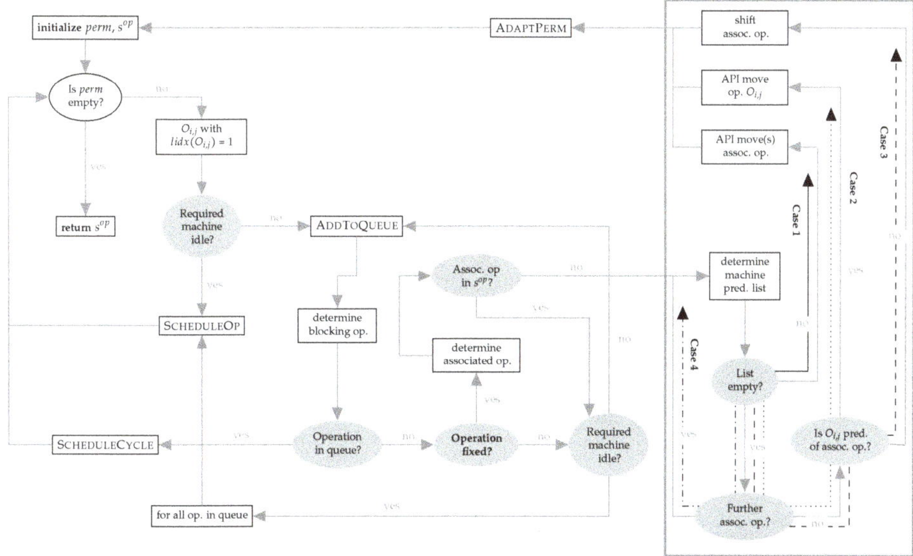

Figure 6. Schematic outline of the Advanced Repair Technique (ART), cf. [15].

Figure 7 illustrates the four possible cases of adaptation by means of Gantt charts. Assume that $O_{a,b} \to O_{i',j'}$ is a given ordering and operation $O_{i,j}$, shown in striped pattern, is the operation currently considered by the ART. Irreversible pairwise sequences are given by bold rightward arrows connecting adjacent operations on a machine, and the required additional APIs are indicated by leftward arrows with case-corresponding line patterns, see Figure 6. Let the feasible partial schedule s^{op} already contain operation $O_{i',j'-1}$ in the cases 1, 2, and 3, and operation $O_{i',j'-2}$ in case 4. The consideration of operation $O_{i,j}$ to be scheduled next requires the following blocking constraint to be fulfilled: $lidx(O_{i',j'}) < lidx(O_{i,j})$. Since the associated operation $O_{a,b}$ is not yet included in the list s^{op}, the fixed operation $O_{i',j'}$ cannot be positioned at the next idle list index prior to operation $O_{i,j}$. To resolve the situation, the basic strategy is to shift or interchange the associated operation $O_{a,b}$ further to the left on its machine, so that it will be scheduled before the required repairing shift of operation $O_{i',j'}$ occurs in the next run of the procedure. Therefore, the machine predecessor list of $O_{a,b}$ excluding operations of job J_a is determined, see Figure 6, and the adaptation of the initial permutation is conducted according to one the following cases:

Case 1: If there exists a machine predecessor $\alpha(O_{a,b})$, an additional API move is performed and the ordering $O_{a,b} \to \alpha(O_{a,b})$ is defined to be fixed additionally, see part (a) of Figure 7. The API is implemented in the list *perm* by a left shift of operation $O_{a,b}$.

Case 2: If there exists no machine predecessor of operation $O_{a,b}$ and there exists no other operation associated with operation $O_{i',j'}$, the currently considered operation $O_{i,j}$ might itself be a job predecessor $O_{a,b'}$ of the associated operation $O_{a,b}$. If this is true, an API move is performed with its machine predecessor $O_{i',j'-1}$, see part (b) of Figure 7. Note that, for this situation to occur, the machine predecessor necessarily needs to exist and belong to job $J_{i'}$.

Case 3: Assume that there exists no machine predecessor of operation $O_{a,b}$ and no other operation associated with operation $O_{i',j'}$, and, furthermore, the currently considered operation is not a job predecessor of operation $O_{a,b}$. Then, the associated operation is shifted leftward in the permutation *perm* to the position prior to the currently considered operation $O_{i,j}$, see part (c) of Figure 7. This shift does not implement an API move but is sufficient to satisfy the given blocking constraint.

Case 4: This situation differs structurally from the other three cases. It involves at least three machines, and it can only appear with operations of recirculating jobs after one or more additional APIs have already been performed. In part (d) of Figure 7, besides the initially given ordering, the pairwise sequences $O_{a,b'} \rightarrow O_{i',j'-1}$ and $O_{a,b''} \rightarrow O_{i',j'-1}$ are exemplarily fixed. After the machine predecessor list of the associated operation $O_{a,b'}$ has been determined as empty, a second operation associated with the fixed operation $O_{i',j'-1}$ can be found, namely operation $O_{a,b''}$. Dependent on the existence of a machine predecessor $\alpha(O_{a,b''})$, the ART proceeds according to Cases 1, 2, or 3 with an adaptation of the list *perm*. In the depicted Gantt chart, a shift following Case 1 is shown as an example.

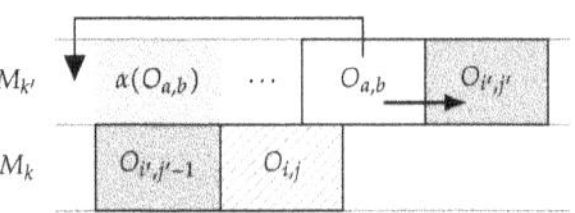

(a) Case 1: Shifting the associated operation $O_{a,b}$ in the operation sequence on $Ma(O_{a,b})$

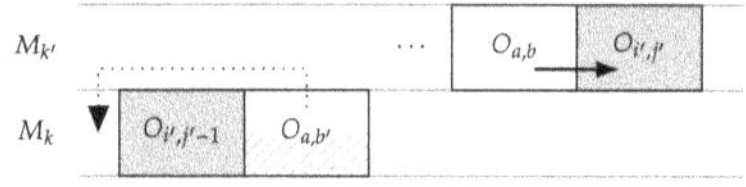

(b) Case 2: Shifting the currently considered operation $O_{i,j} = O_{a,b'}$, $(b' < b)$ in the operation sequence on $Ma(O_{a,b'})$

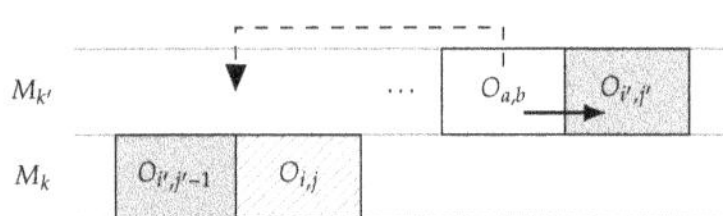

(c) Case 3: Shifting the associated operation $O_{a,b}$ in *perm*

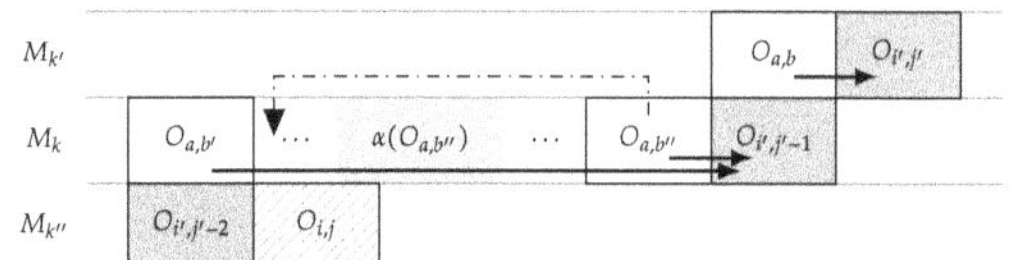

(d) Case 4: Shifting further associated operations like operation $O_{a,b''}$, $(b' < b'' < b)$ of the fixed blocking operation $O_{i',j'-1}$ in the operation sequence on $Ma(O_{a,b''})$

Figure 7. Adapting the permutation *perm* in the ART, cf. [15].

After the permutation is adapted according to one of the four cases, the ART is restarted involving one additionally fixed pairwise sequence. The following observation can be made regarding the operations moved during adaptation.

Observation 1. *While executing the ART, the operation interchanged or shifted leftwards when adapting the permutation is always the associated operation $O_{a,b}$ defined by the initially given fixed sequence $O_{a,b} \rightarrow O_{i',j'}$ or one of its job predecessors.*

Based on this, arguments indicating the correctness of the ART can be derived as follows, cf. [15]:

Proposition 2. *The ART terminates and returns a permutation s^{op} encoding a feasible schedule for the BJSP involving a predefined ordering $O_{a,b} \rightarrow O_{i',j'}$ of two operations of different jobs requiring the same machine.*

Proof. It is equivalently assumed here that the initially given list *perm* is feasible with regard to the processing sequences of all jobs $J_i \in \mathcal{J}$. The ART proceeds like the BRT until there is a fixed operation $O_{i',j'}$ to be scheduled prior to its associated operation $O_{a,b}$. According to Proposition 1, the BRT terminates and returns a feasible encoding s^{op} of a BJSP schedule from a given permutation. Consequently, the adaptation of the permutation and the restart of the ART are the only critical aspects to regard **here** in detail.

It needs to be shown that

(1) an adaptation does not violate the processing sequences of the jobs,
(2) an adaptation can never be reverted,
(3) the number of possible adaptations is finite and
(4) there exists a sequence of adaptations leading to a feasible schedule for the BJSP including the predefined pairwise sequence $O_{a,b} \rightarrow O_{i',j'}$.

When an adaptation is performed, an operation of job J_a is shifted leftwards in the permutation. The processing sequence of this job is the only one potentially affected and it may only get violated, if the operation is moved prior to one or more of its job predecessors. This situation is checked during the adaptation and job predecessors are additionally shifted, if necessary. Thus, (1) is always true.

The set of irreversible orderings is extended by one pairwise sequence in every execution of the adaptation procedure. Thus, the incorporated BRT mechanisms can never reverse an adaptation. A consecutively required API or shift can only result from a fixed ordering $O_{a,b'} \rightarrow O_{c,d}$ with $b' \in \{1, \dots, b\}$, where the currently regarded operation $O_{i,j}$ features a list index prior to $lidx(O_{a,b'})$ in *perm*. This means that a consecutively required adaptation does always appear at a position prior to the previous adaptation causing the fixed sequence $O_{a,b'} \rightarrow O_{c,d}$. As a consequence, the operation $O_{a,b'}$ or one of its job predecessors is moved to a list index smaller than $lidx(O_{i,j})$, for which $lidx(O_{i,j}) < lidx(O_{a,b'}) < lidx(O_{c,d})$ holds. Thus, an implemented adaptation can never be reverted by an ART mechanism. (2) is true.

Since the list *perm* contains a finite number of elements, and the set of shifted operations is restricted to all operations of a job $J_a \in \mathcal{J}$, see Observation 1, and (2) is true, the number of possible adaptations is finite. (3) holds.

The strategy of the ART can be summarized as shifting the operations of a job J_a iteratively leftwards in the operation sequences on the required machines. This is repeatedly applied until the given pairwise sequence is realized in a feasible schedule for the BJSP constructed by BRT mechanisms only. Following from Observation 1 and statement (2), all moved operations may end up at the first positions in the operation sequences on their machines in the extreme case. Thus, the job J_a involving operation $O_{a,b}$ is scheduled prior to all other jobs involved in the problem and $O_{a,b} \rightarrow O_{i',j'}$ is guaranteed. (4) is shown. $\square$

Considering that the total number of operations is given by n_{op} and, furthermore, taking this measure as a worst case estimate for the number of operations requiring a certain machine, the ART determines a feasible schedule involving a given pairwise sequence in $O\left((n_{op})^4\right)$ steps, cf. [15]. This method enables the usage of APIs as generic operators in neighborhood structures for BJSPs.

5.1.3. Definition of the Neighborhoods

In line with the transition schemes described in the previous section, the examined neighborhoods are defined as follows:

Definition 4. *The **API neighborhood** of a schedule s is defined as the set of schedules s', where s' is a feasible schedule involving a given API move implemented by a left shift or a right shift.*

Definition 5. *The **TAPI neighborhood** of a schedule s is defined as the set of schedules s', where s' is a feasible schedule involving a given TAPI move implemented by left shift or right shift.*

Definition 6. *The **TJ neighborhood** of a schedule s is defined as the set of schedules s', where s' is a feasible schedule resulting from a TJ move.*

In the following, all neighboring solutions constructible through a given API or TAPI are generally denoted as API-based neighbors of a schedule. Note that, due to the required repairing schemes, the actual distances of neighboring schedules are not precisely determined, cf. [15]. The minimum distance of a schedule and its API-based neighbor is given by 1, and the maximum distance is theoretically bounded by $\sum_{M_k \in \mathcal{M}} \binom{|\Omega^k|}{2}$, where Ω^k defines the set of operations requiring machine M_k. As mentioned in the previous section, schedules in the TJ neighborhood might have a minimum distance of 0 to the initially given one, while the maximum distance is equivalently restricted by the structural upper bound. While the leftward shifting strategy applied by the ART in the adaptation of the permutation is required for the termination of the method, it is not guaranteed that the smallest number of necessary changes is implemented. To the best of the authors' knowledge, there does not yet exist a general neighborhood structure or repairing scheme for BJSP schedules capable of certainly constructing the closest possible neighbor for an initial solution and a given change. An empirical study on the distances resulting for the proposed neighborhoods together with the ART is reported in the next section.

5.2. Characteristics and Evaluation

5.2.1. Connectivity of the Neighborhoods

A neighborhood is said to be connected, if every existing feasible schedule can be transformed into every other existing feasible schedule by (repeatedly) applying a given neighbor-defining operator, see [13,39,45]. Here, the neighbor-defining operators consist of a move and a repairing scheme. The connectivity of the neighborhood is of significant importance in the application of search procedures, since it guarantees that the methods are capable of finding optimal solutions. However, such a structural result can only be interpreted as an indication for the actual performance of a neighborhood-based heuristic solution approach on practically relevant instances.

Proposition 3. *Given general release dates $r_i \in \mathbb{Z}_{\geq 0}$ for $J_i \in \mathcal{J}$ and the minimization of total tardiness as the optimization criterion, the proposed neighborhoods, namely the API neighborhood, the TAPI neighborhood and the TJ neighborhood, are* **not connected**.

Proof. Consider API and TAPI moves first. As described in Section 5.1, in a feasible schedule, there may exist two subsequent operations $O_{i,j}$ and $O_{i',j'}$ on a machine in the schedule which are considered as non-adjacent due to an idle time caused by the release date of the succeeding job $J_{i'}$. Such a pair of operations can never be chosen for an API or a TAPI move in constructing neighboring schedules. Thus, a schedule involving the ordering $O_{i',j'} \rightarrow O_{i,j}$ cannot be reached by applying the neighbor-defining operators even if it is feasible for the BJSP. Thus, the API and the TAPI neighborhoods are not connected.

Furthermore, regarding the TJ move which shifts all operations of a currently tardy job in the permutation, the limitation to choosing a job with a strictly positive tardiness value implies that the neighborhood of feasible schedules with a total tardiness of 0 is empty. Even if optimal solutions for the BJSPT are found in this case, these schedules are isolated by definition and no other feasible schedule can be constructed subsequently. Thus, the TJ neighborhood is not connected. $\square$

Despite these negative findings on the connectivity of the neighborhoods, the proposed structures are still supposed to be successfully applicable in a metaheuristic search method for the BJSPT. It can be expected that extraordinarily widespread release dates, which cause a disjoint partitioning of the search space, do not occur in practically relevant problems. Furthermore, in most of the cases, it is not necessary to continue the search once an optimal solution is found.

However, the questions on whether the described API neighborhood is connected for the special case $r_i = 0$, $J_i \in \mathcal{J}$ or for a specific combination of release date and processing time ranges remain open. It is conjectured that the API neighborhood together with the ART feature the connectivity property for the BJSP without release dates.

5.2.2. Observations on the Interchange-Based Transition Scheme

Besides the general problem solving capability of a metaheuristic involving the proposed neighborhood structures, the API-based transition schemes shall be evaluated with regard to their ability to generate small distance and high quality neighbors. Special attention is given to the differences appearing in using a left shift or a right shift transformation to implement an API in the operation-based encoding of a schedule. For all API and TAPI neighbors constructed during the computational experiments, a specific interchange is chosen, left shift and right shift transformation are performed, and both resulting solutions are evaluated with regard to their distances from the initially given schedule and their total tardiness values.

Figure 8 displays the distributions of the distances of API-based neighbors with left shift (LS) and right shift (RS) transformation for the benchmark instances by boxplots. The range in which 50% of the distance measures of the neighbors can be found, the so-called interquartile range, is represented by the box. The black horizontal line indicates the median of the sample. The whiskers plot the minimal and the maximal distance value which are not more than one and a half interquartile ranges away from the box.

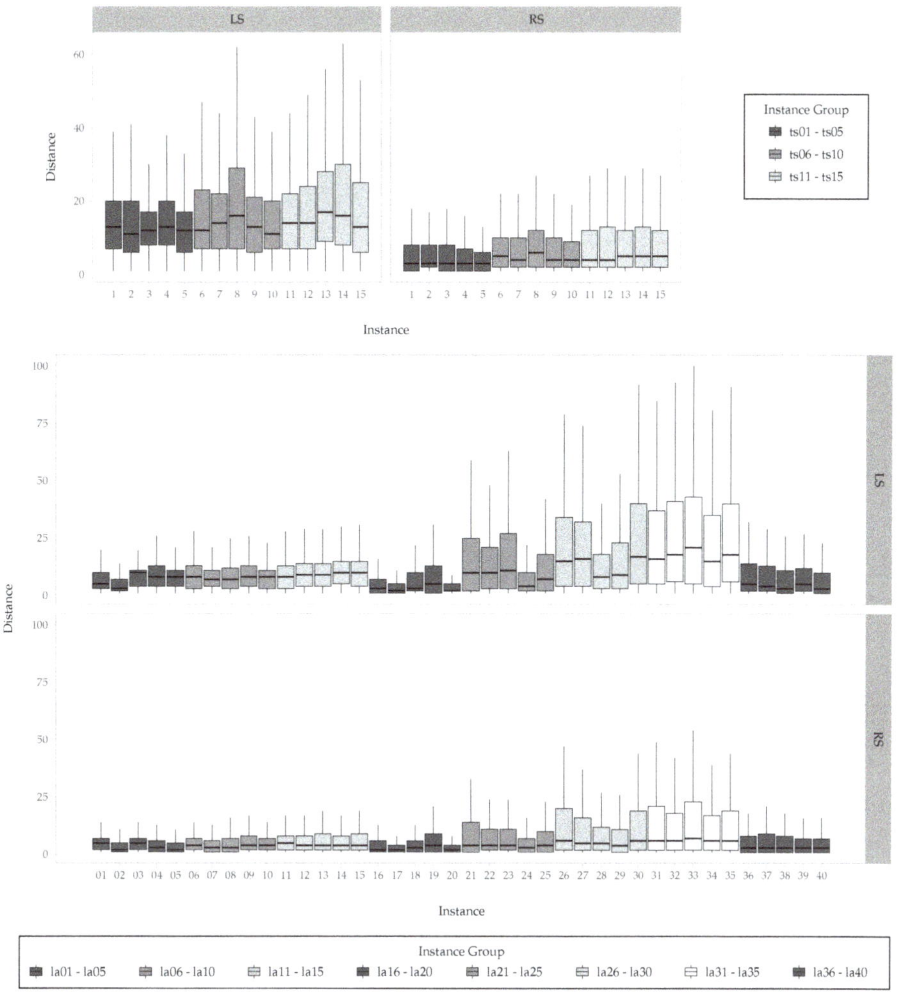

Figure 8. Boxplots representing the distribution of the distance measure among neighbors based on API-moves for the benchmark instances of the BJSPT, cf. [15].

Considering all transitions independent of the direction of implementation, it can be stated that the distance of neighboring schedules based on a single API is remarkably large for the BJSP. Evidently, a significant amount of additional adaptations is required to fit a given pairwise sequence to a feasible schedule. Even if it is not guaranteed that the closest neighbor is generated by the ART, these results highlight the complexity of the search space caused by blocking constraints. This may lead to difficulties in the effectiveness and the control of a heuristic search method, since an iterative execution of small changes is desirable to systematically explore the set of feasible schedules. Comparing the directed implementations of APIs, the distances of neighbors constructed using a right shift transformation are significantly smaller than the measures of neighboring schedules generated by a left shift transformation. Thus, it is recommendable to implement APIs by a right shift in the operation-based encoding to support the execution of smaller search steps, cf. [15].

The chart in Figure 9 shows the proportion of APIs for which the neighbors resulting from a left shift and a right shift transformation are equivalent (EQ). Given that two different schedules arise, it is displayed to which extent the schedule with a smaller total tardiness value is generated by a left shift or by a right shift implementation of the API in the operation-based encoding. It can be observed that the majority of neighboring schedules based on the same API end up to be equivalent after applying the ART. Regarding the cases where different schedules are constructed, the right shift transformation clearly outperforms the left shift transformation by means of total tardiness. Since heuristic methods are intended to require limited computational effort, it is reasonable to implement APIs by right shift transformation only, cf. [15]. The analysis indicates that the proportion of cases in which the best possible neighboring schedule is not generated is less than 13%.

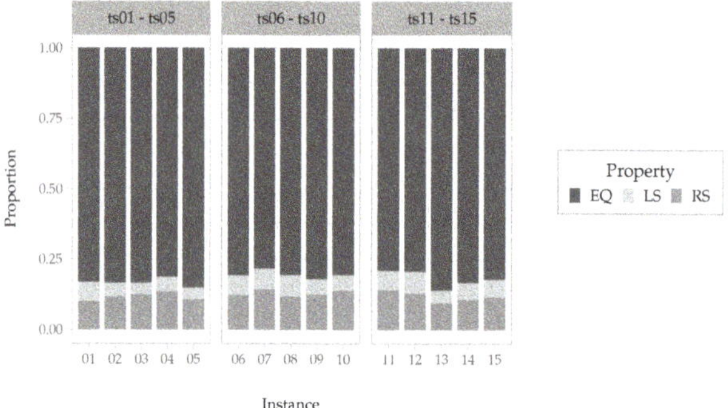

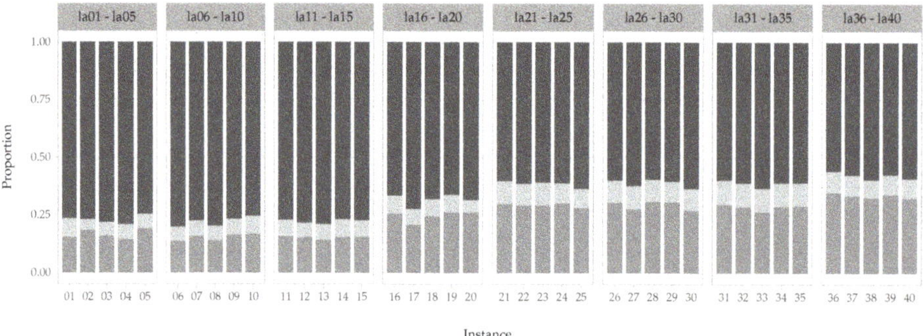

Figure 9. Proportion of equivalent solutions and performance of right shift and left shift transformation among all API-based neighbors for the benchmark instances of the BJSPT, cf. [15].

6. Computational Experiments and Results

Finally, the proposed neighborhood structures and repairing schemes are used to solve the benchmark instances of the BJSPT introduced in Section 3. In line with the findings in the literature, SA is chosen as a simple and generic local search scheme. The neighborhoods and repairing schemes can easily be embedded and the method facilitates moves to inferior neighboring solutions. The latter aspect seems especially promising with regard to the observed ruggedness of the search space of the

BJSPT. The following computational results give insight to the general capability of permutation-based procedures in solving the problem under study. Furthermore, the potential change in the performance when guiding the search scheme by objective function values is observed.

6.1. A Simulated Annealing Algorithm

The metaheuristic framework is implemented in the standard variant, see, for instance [2,38,39], with a geometric cooling scheme $t_{\tau+1} = c \cdot t_\tau$. Correspondingly, the initial temperature t_0, the terminal temperature T, and the cooling factor c act as the control parameters of the procedure. With regard to the asymptotic convergence of SA, $n_{op} - m$ neighboring solutions are evaluated for every temperature level.

Since the probability for a generated neighbor to be accepted as the new current solution depends on the objective function values of the considered schedules next to the temperature level, the parameter setting needs to be adjusted according to the magnitude of the total tardiness. Preliminary experiments indicated the following settings (t_0, T, c) as beneficial for the benchmark problems: (20, 0.5, 0.9925) and (20, 10, 0.999) for the train scheduling-inspired instances and (200, 50, 0.995) for the Lawrence instances, cf. [15,47]. Dependent on the size of the instances, 11,000 to 84,000 iterations are performed, cf. [15].

Furthermore, the extent to which the API-based and the randomized shift-based neighborhood structures are used to include intensification and diversification advantageously has been part of an initial study. The proposed algorithm applies either the API or the TAPI neighborhood combined with the TJ neighborhood, respectively. This implies that the effectiveness of an objective function-oriented guidance can be analyzed, while a random component is always involved. For every generation of a neighbor, the API-based neighborhood is chosen with a probability of 0.9 and a TJ neighbor is constructed with a probability of 0.1, cf. [15,47]. If an API is performed, both schedules resulting from a left shift and a right shift transformation are evaluated, and the superior one becomes the candidate to represent the next incumbent solution. The created algorithm is called permutation-based simulated annealing (PSA).

6.2. Numerical Results

The computational experiments are conducted on a notebook featuring an Intel Dual Core i5 processor (2.20 GHz) with 8 GB RAM. Algorithm PSA is implemented in Python 3. Tables 2–4 summarize the numerical results of five independent runs operated for each instance, parameter setting, and neighborhood structure. The first two columns of each table display the instance and the corresponding size (m, n). For reasons of comparison, the third column contains the best total tardiness value obtained by solving the considered problem with the help of IBM ILOG CPLEX 12.8 using a mixed-integer programming (MIP) formulation with pairwise precedence variables, see [15] for detailed explanations on the model. Objective function values with proven optimality are denoted by an asterisk. The next pairs of columns show the average total tardiness $\overline{\sum T_i}$ and the minimal total tardiness $\min(\sum T_i)$ obtained for each instance by Algorithm PSA based on the API and the TAPI neighborhood, respectively. Contrasting the mean total tardiness values reached, the smaller measure is highlighted by boldface printing.

Generally, it can be stated that Algorithm PSA yields satisfactory results for instances with and without inner structure especially when being compared to the MIP approach. For instances of small size and an equivalent number of jobs and machines, such as ts01 to ts05, la02 to la05, la07 and la08, la17, la19 and la20, the method is capable of finding an optimal solution. Even more important, the algorithm is able to generate medium quality solutions for large problems like la31 to la35, for which a general-purpose method might even struggle in generating feasible schedules. This gives evidence for the advantageousness of the proposed heuristic approach in solving real-world production planning instances of critical size.

Nonetheless, the complex repairing schemes constitute a drawback of the heuristic algorithm with regard to computation time. PSA requires 2 to 70 min of runtime dependent on the size of the instances, while the MIP technique is able to solve small instances to optimality in a few seconds, cf. [15]. As indicated by the statistical analysis of the neighborhoods in the previous section, the runtime of PSA can be improved by implementing APIs by a right shift transformation only. To further overcome these difficulties, a hybrid method combining heuristic and MIP mechanisms seems promising. Heuristic methods can be used to generate feasible schedules for large instances quickly, while solving smaller subproblems by MIP may be a superior improvement strategy towards locally optimal solutions. First, results following this research direction are presented in [15,18].

Table 2. Computational results of Algorithm PSA with $(20, 0.5, 0.9925)$ applied to the train scheduling-inspired instances, cf. [15].

Inst.	(m, n)	MIP	API		TAPI	
			$\overline{\sum T_i}$	$\min(\sum T_i)$	$\overline{\sum T_i}$	$\min(\sum T_i)$
ts01	$(11, 10)$	138 *	**140.0**	138 *	142.6	138 *
ts02	$(11, 10)$	90 *	**95.0**	91	96.6	90 *
ts03	$(11, 10)$	72 *	**78.8**	72 *	84.8	76
ts04	$(11, 10)$	41 *	41.4	41 *	**41.2**	41 *
ts05	$(11, 10)$	71 *	**71.2**	71 *	71.6	71 *
ts06	$(11, 15)$	88 *	125.0	108	**119.4**	109
ts07	$(11, 15)$	172 *	**196.0**	184	201.0	192
ts08	$(11, 15)$	163 *	185.6	163 *	185.6	181
ts09	$(11, 15)$	153	**174.0**	160	175.2	161
ts10	$(11, 15)$	97 *	116.6	107	**112.6**	108
ts11	$(11, 20)$	366	**409.4**	387	411.8	392
ts12	$(11, 20)$	419	**429.2**	412	442.4	419
ts13	$(11, 20)$	452	492.2	472	**478.2**	445
ts14	$(11, 20)$	459	**500.6**	473	508.8	492
ts15	$(11, 20)$	418	433.2	413	**428.2**	387

Table 3. Computational results of Algorithm PSA with $(20, 10, 0.999)$ applied to the train scheduling-inspired instances, cf. [15].

Inst.	(m, n)	MIP	API		TAPI	
			$\overline{\sum T_i}$	$\min(\sum T_i)$	$\overline{\sum T_i}$	$\min(\sum T_i)$
ts01	$(11, 10)$	138 *	140.2	138 *	**140.0**	138 *
ts02	$(11, 10)$	90 *	**94.6**	91	95.2	91
ts03	$(11, 10)$	72 *	**74.2**	72 *	74.4	72 *
ts04	$(11, 10)$	41 *	41.8	41 *	**41.0 ***	41 *
ts05	$(11, 10)$	71 *	71.4	71 *	**71.0 ***	71 *
ts06	$(11, 15)$	88 *	121.6	107	**119.8**	111
ts07	$(11, 15)$	172 *	195.4	189	**192.8**	185
ts08	$(11, 15)$	163 *	**184.2**	179	185.0	181
ts09	$(11, 15)$	153	178.8	168	**177.4**	174
ts10	$(11, 15)$	97 *	114.8	97 *	**112.0**	105
ts11	$(11, 20)$	366	406.4	390	**401.6**	387
ts12	$(11, 20)$	419	428.2	412	**424.6**	405
ts13	$(11, 20)$	452	462.6	448	**460.6**	447
ts14	$(11, 20)$	459	**462.8**	418	495.0	466
ts15	$(11, 20)$	418	**419.4**	401	435.0	414

Comparing the API and TAPI neighborhood with regard to solution quality over all instances, no transition scheme clearly dominates. An advantageousness of guiding the search by current total tardiness values cannot be observed. A preliminary performance testing might be beneficial for every individual application of the API-based neighborhood structures to other BJSPT instances, since the solution quality reached may dependent on the problems size and structure as well as on the setting of the metaheuristic framework. It can be remarked that, based on the experiments on the ts instances with two different parameter settings, the API neighborhood performs better with lower temperature levels between 20 and 0.5, while the TAPI neighborhood is favorable combined with higher temperature levels between 20 and 10. This implies that simultaneously applying a strict limitation of the acceptance of inferior schedules in the search procedure and a restriction of the possible interchanges based on the objective function value is not reasonable.

Table 4. Computational results of Algorithm PSA with $(200, 50, 0.995)$ applied to the Lawrence instances, cf. [15].

Inst.	(m, n)	MIP	API		TAPI	
			$\overline{\sum T_i}$	$\min(\sum T_i)$	$\overline{\sum T_i}$	$\min(\sum T_i)$
la01	$(5, 10)$	762 *	787.4	773	**783.8**	773
la02	$(5, 10)$	266 *	283.4	266 *	**277.6**	266 *
la03	$(5, 10)$	357 *	357.0 *	357 *	357.0 *	357 *
la04	$(5, 10)$	1165 *	**1217.2**	1165 *	1284.2	1165 *
la05	$(5, 10)$	557 *	557.0 *	557 *	557.0 *	557 *
la06	$(5, 15)$	2516	**2790.0**	2616	2912.4	2847
la07	$(5, 15)$	1677 *	1942.2	1869	**1904.2**	1677 *
la08	$(5, 15)$	1829 *	2335.0	1905	**2129.6**	1829 *
la09	$(5, 15)$	2851	3275.2	3161	**3226.6**	3131
la10	$(5, 15)$	1841 *	2178.2	2069	**2119.4**	2046
la11	$(5, 20)$	6534	6186.2	5704	**5846.4**	5253
la12	$(5, 20)$	5286	5070.0	4859	**4997.8**	4809
la13	$(5, 20)$	7737	7850.6	7614	**7611.8**	7342
la14	$(5, 20)$	6038	**6616.8**	5714	6872.4	6459
la15	$(5, 20)$	7082	**7088.6**	5626	7153.6	6330
la16	$(10, 10)$	330 *	395.8	335	**360.8**	335
la17	$(10, 10)$	118 *	144.2	120	**118.8**	118 *
la18	$(10, 10)$	159 *	**229.4**	159 *	264.0	235
la19	$(10, 10)$	243 *	306.6	243 *	**301.0**	243 *
la20	$(10, 10)$	42 *	55.6	42 *	**42.0 ***	42 *
la21	$(10, 15)$	1956	**2847.2**	2101	2961.8	2680
la22	$(10, 15)$	1455	**2052.8**	1773	2123.0	1988
la23	$(10, 15)$	3436	**3692.6**	3506	3746.8	3424
la24	$(10, 15)$	560 *	966.8	761	**724.0**	644
la25	$(10, 15)$	1002	**1557.4**	1289	1583.0	1390
la26	$(10, 20)$	7961	9275.8	8475	**8600.8**	7858
la27	$(10, 20)$	8915	**7588.0**	6596	7641.8	6457
la28	$(10, 20)$	2226	3430.8	2876	**3367.6**	2849
la29	$(10, 20)$	2018	**2948.0**	2432	3099.0	2626
la30	$(10, 20)$	6655	7621.6	6775	**7372.8**	6395
la31	$(10, 30)$	20,957	18,921.8	17,984	**18,409.6**	17,751
la32	$(10, 30)$	23150	21,991.4	20,401	**21,632.2**	20,546
la33	$(10, 30)$	none	**22,494.2**	19,750	22,913.2	20,553
la34	$(10, 30)$	none	**20,282.8**	18,633	21,911.8	19,577
la35	$(10, 30)$	none	21,895.0	18,778	**21,384.4**	20,537
la36	$(15, 15)$	675	1856.0	1711	**1839.0**	1599
la37	$(15, 15)$	1070	**1774.2**	1621	1835.8	1594
la38	$(15, 15)$	489 *	760.4	645	**745.4**	676
la39	$(15, 15)$	754	**1573.0**	1391	1850.2	1551
la40	$(15, 15)$	407 *	**1008.6**	613	1187.6	912

Moreover, it can be observed that the mean and the minimal total tardiness values differ significantly for most of the instances. Especially for problems of larger size, the mean objective function value often exceeds the minimal one by more than 10%. This aspect numerically emphasizes the ruggedness of the search space of the BJSPT, which leads to difficulties in the guidance of any heuristic search method. There seems to be a necessity of developing tailored neighborhood structures to more efficiently solve job shop problems with practically relevant constraints and objective functions. Based on these results, the involvement of random and diversifying components in a solution approach is recommendable together with the performance of several independent runs when using standard scheduling-tailored mechanisms.

7. Conclusions

In this paper, instances of a complex job shop scheduling problem are solved by a permutation-based heuristic search method. Two repairing schemes are proposed to facilitate the usage of well-known list encodings and generic operators for job shop problems with blocking constraints. In applying interchange- and shifts-based transition schemes, three neighborhoods are defined and analyzed with regard to structural issues and performance in an SA algorithm.

The computational experiments indicate that the proposed heuristic method using basic scheduling-tailored operators is capable of finding optimal and near-optimal schedules for small and medium size instances. Furthermore, it outperforms general-purpose techniques in generating feasible schedules for problems of large size. This gives evidence to its applicability in decision support systems for solving problems of practical relevance in production planning and logistics.

It turns out that the implementation of APIs by right shifts in the operation-based representation of a schedule is favorable compared to other mechanisms with respect to small search steps and solution quality. This narrows the required computational effort for heuristic search schemes using these types of operators. The complexity of the problem under study becomes clearly visible in the necessary enhancements of neighbor-defining moves and the resulting large distances of feasible schedules in the API-based neighborhoods. This work shows that existing generic scheduling-tailored operators have limits in their applicability to job shop problems with blocking constraints and tardiness-based objectives. The development of dedicated heuristic solution approaches, which allow more controllable search patterns, can be named as an important aspect of future research.

An advantage of guiding the choice of the executed interchanges by the objective function value is not substantiated by the numerical results. Furthermore, considering the total tardiness values obtained in several independent runs of the metaheuristic on the same instances, a high variance in quality of the best schedules found is observed. Thus, the ruggedness of the search space of the BJSPT and remarkable feasibility issues in the generation of neighboring schedules can be named as reasons for the ongoing difficulties in solving instances of practically relevant size. However, the computational results give evidence for hybrid solution approaches as a promising future research direction to overcome such issues. The combination of a heuristic technique to find feasible schedules for large instances and a general-purpose MIP method to quickly generate superior neighboring solutions is expected to be beneficial.

Overall, the proposed permutation-based heuristic can enhance solving capability of complex job shop scheduling problems. Important insights are gained into advantages and limits of applying generic operators to BJSPT instances, and future research directions are highlighted.

Author Contributions: Conceptualization, J.L. and F.W.; Software, J.L.; Investigation, J.L. and F.W.; Writing–original draft preparation, J.L. and F.W.

Funding: This research received no external funding.

Acknowledgments: The authors would like to thank Felix Müller for his remarkable commitment in the preparation of the statistical figures.

Conflicts of Interest: The authors declare no conflict of interest.

References

1. Garey, M.R.; Johnson, D.S.; Sethi, R. The Complexity of Flowshop and Jobshop Scheduling. *Math. Oper. Res.* **1976**, *1*, 117–129. [CrossRef]
2. Van Laarhoven, P.J.; Aarts, E.H.; Lenstra, J.K. Job shop scheduling by simulated annealing. *Oper. Res.* **1992**, *40*, 113–125. [CrossRef]
3. Adams, J.; Balas, E.; Zawack, D. The shifting bottleneck procedure for job shop scheduling. *Manag. Sci.* **1988**, *34*, 391–401. [CrossRef]
4. Nowicki, E.; Smutnicki, C. A fast taboo search algorithm for the job shop problem. *Manag. Sci.* **1996**, *42*, 797–813. [CrossRef]
5. Bürgy, R.; Gröflin, H. The blocking job shop with rail-bound transportation. *J. Comb. Optim.* **2016**, *31*, 152–181. [CrossRef]
6. D'Ariano, A.; Pacciarelli, D.; Pranzo, M. A branch and bound algorithm for scheduling trains in a railway network. *Eur. J. Oper. Res.* **2007**, *183*, 643–657. [CrossRef]
7. Liu, S.Q.; Kozan, E. Scheduling trains as a blocking parallel-machine job shop scheduling problem. *Comput. Oper. Res.* **2009**, *36*, 2840–2852. [CrossRef]
8. Mati, Y.; Rezg, N.; Xie, X. A taboo search approach for deadlock-free scheduling of automated manufacturing systems. *J. Intell. Manuf.* **2001**, *12*, 535–552. [CrossRef]
9. Bürgy, R. A neighborhood for complex job shop scheduling problems with regular objectives. *J. Sched.* **2017**, *20*, 391–422. [CrossRef]
10. Heger, J.; Voss, T. Optimal Scheduling of AGVs in a Reentrant Blocking Job-shop. *Procedia CIRP* **2018**, *67*, 41–45. [CrossRef]
11. Lange, J.; Werner, F. Approaches to modeling train scheduling problems as job-shop problems with blocking constraints. *J. Sched.* **2018**, *21*, 191–207. [CrossRef]
12. Brizuela, C.A.; Zhao, Y.; Sannomiya, N. No-wait and blocking job-shops: Challenging problems for GA's. *Int. Conf. Syst. Man Cybern.* **2001**, *4*, 2349–2354.
13. Groeflin, H.; Klinkert, A. A new neighborhood and tabu search for the blocking job shop. *Discret. Appl. Math.* **2009**, *157*, 3643–3655. [CrossRef]
14. Mattfeld, D.C.; Bierwirth, C. An efficient genetic algorithm for job shop scheduling with tardiness objectives. *Eur. J. Oper. Res.* **2004**, *155*, 616–630. [CrossRef]
15. Lange, J. Solution Techniques for the Blocking Job Shop Scheduling Problem with Total Tardiness Minimization. Ph.D. Thesis, Otto-von-Guericke-Universität Magdeburg, Magdeburg, Germany, 2019. [CrossRef]
16. Lange, J.; Werner, F. A Permutation-Based Neighborhood for the Blocking Job-Shop Problem with Total Tardiness Minimization. In *Operations Research Proceedings 2017*; Springer International Publishing: Cham, Switzerland, 2018; pp. 581–586.
17. Mascis, A.; Pacciarelli, D. Job-shop scheduling with blocking and no-wait constraints. *Eur. J. Oper. Res.* **2002**, *143*, 498–517. [CrossRef]
18. Lange, J.; Bürgy, R. Mixed-Integer Programming Heuristics for the Blocking Job Shop Scheduling Problem. In Proceedings of the 14th Workshop on Models and Algorithms for Planning and Scheduling Problems, MAPSP 2019, Renesse, The Netherlands, 3–7 June 2019; pp. 58–60.
19. Nowicki, E.; Smutnicki, C. An advanced tabu search algorithm for the job shop problem. *J. Sched.* **2005**, *8*, 145–159. [CrossRef]
20. Balas, E.; Simonetti, N.; Vazacopoulos, A. Job shop scheduling with setup times, deadlines and precedence constraints. *J. Sched.* **2008**, *11*, 253–262. [CrossRef]
21. Bierwirth, C.; Kuhpfahl, J. Extended GRASP for the job shop scheduling problem with total weighted tardiness objective. *Eur. J. Oper. Res.* **2017**, *261*, 835–848. [CrossRef]
22. Pinedo, M.; Singer, M. A shifting bottleneck heuristic for minimizing the total weighted tardiness in a job shop. *Nav. Res. Logist. (NRL)* **1999**, *46*, 1–17. [CrossRef]
23. Wang, T.Y.; Wu, K.B. A revised simulated annealing algorithm for obtaining the minimum total tardiness in job shop scheduling problems. *Int. J. Syst. Sci.* **2000**, *31*, 537–542. [CrossRef]
24. De Bontridder, K.M.J. Minimizing total teighted tardiness in a generalized job shop. *J. Sched.* **2005**, *8*, 479–496. [CrossRef]

25. Essafi, I.; Mati, Y.; Dauzère-Pérès, S. A genetic local search algorithm for minimizing total weighted tardiness in the job-shop scheduling problem. *Comput. Oper. Res.* **2008**, *35*, 2599–2616. [CrossRef]

26. Bülbül, K. A hybrid shifting bottleneck-tabu search heuristic for the job shop total weighted tardiness problem. *Comput. Oper. Res.* **2011**, *38*, 967–983. [CrossRef]

27. Mati, Y.; Dauzère-Pérès, S.; Lahlou, C. A general approach for optimizing regular criteria in the job-shop scheduling problem. *Eur. J. Oper. Res.* **2011**, *212*, 33–42. [CrossRef]

28. Zhang, R.; Wu, C. A simulated annealing algorithm based on block properties for the job shop scheduling problem with total weighted tardiness objective. *Comput. Oper. Res.* **2011**, *38*, 854–867. [CrossRef]

29. González, M.Á.; González-Rodríguez, I.; Vela, C.R.; Varela, R. An efficient hybrid evolutionary algorithm for scheduling with setup times and weighted tardiness minimization. *Soft Comput.* **2012**, *16*, 2097–2113. [CrossRef]

30. Kuhpfahl, J.; Bierwirth, C. A study on local search neighborhoods for the job shop scheduling problem with total weighted tardiness objective. *Comput. Oper. Res.* **2016**, *66*, 44–57. [CrossRef]

31. Meloni, C.; Pacciarelli, D.; Pranzo, M. A rollout metaheuristic for job shop scheduling problems. *Ann. Oper. Res.* **2004**, *131*, 215–235.[CrossRef]

32. Oddi, A.; Rasconi, R.; Cesta, A.; Smith, S.F. Iterative Improvement Algorithms for the Blocking Job Shop. In Proceedings of the ICAPS, Atibaia, Brazil, 25–29 June 2012.

33. AitZai, A.; Boudhar, M. Parallel branch-and-bound and parallel PSO algorithms for job shop scheduling problem with blocking. *Int. J. Oper. Res.* **2013**, *16*, 14–37. [CrossRef]

34. Pranzo, M.; Pacciarelli, D. An iterated greedy metaheuristic for the blocking job shop scheduling problem. *J. Heuristics* **2016**, *22*, 587–611. [CrossRef]

35. Dabah, A.; Bendjoudi, A.; AitZai, A.; Taboudjemat, N.N. Efficient parallel tabu search for the blocking job shop scheduling problem. *Soft Comput.* **2019**. [CrossRef]

36. Gröflin, H.; Klinkert, A. Feasible insertions in job shop scheduling, short cycles and stable sets. *Eur. J. Oper. Res.* **2007**, *177*, 763–785. [CrossRef]

37. Graham, R.L.; Lawler, E.L.; Lenstra, J.K.; Rinnooy Kan, A. Optimization and approximation in deterministic sequencing and scheduling: A survey. In *Annals of Discrete Mathematics*; Elsevier: Amsterdam, The Netherlands, 1979; Volume 5, pp. 287–326.

38. Pinedo, M. *Scheduling: Theory, Algorithms, and Systems*; Springer: Berlin/Heidelberg, Germany, 2016.

39. Brucker, P.; Knust, S. *Complex Scheduling*; Springer: Berlin/Heidelberg, Germany, 2011.

40. Błażewicz, J.; Ecker, K.H.; Pesch, E.; Schmidt, G.; Weglarz, J. *Handbook on Scheduling: From Theory to Applications*; International Handbook on Information Systems; Springer: Berlin/Heidelberg, Germany, 2007. [CrossRef]

41. Lawrence, S. *Supplement to Resource Constrained Project Scheduling: An Experimental Investigation of Heuristic Scheduling Techniques*; GSIA, Carnegie Mellon University: Pittsburgh, PA, USA, 1984.

42. Mascis, A.; Pacciarelli, D. *Machine Scheduling via Alternative Graphs*; Technical Report; Universita degli Studi Roma Tre, DIA: Rome, Italy, 2000.

43. Bierwirth, C.; Mattfeld, D.C.; Watson, J.P. Landscape regularity and random walks for the job-shop scheduling problem. In *European Conference on Evolutionary Computation in Combinatorial Optimization*; Springer: Berlin/Heidelberg, Germany, 2004, pp. 21–30.

44. Schiavinotto, T.; Stützle, T. A review of metrics on permutations for search landscape analysis. *Comput. Oper. Res.* **2007**, *34*, 3143–3153. [CrossRef]

45. Werner, F. Some relations between neighbourhood graphs for a permutation problem. *Optimization* **1991**, *22*, 297–306. [CrossRef]

46. Anderson, E.J.; Glass, C.A.; Potts, C.N., Machine Scheduling. In *Local Search in Combinatorial Optimization*; Aarts, E.H.L.; Lenstra, J.K., Eds.; Wiley: Chichester, UK, 1997; Chapter 11, pp. 361–414.

47. Lange, J. A comparison of neighborhoods for the blocking job-shop problem with total tardiness minimization. In Proceedings of the 16th International Conference of Project Management and Scheduling 208, Rome, Italy, 17–20 April 2018; pp. 132–135.

Article

Modeling and Solving Scheduling Problem with m Uniform Parallel Machines Subject to Unavailability Constraints

Jihene Kaabi 

College of Information Technology, University of Bahrain, P.O. Box 32038 Manama, Bahrain; jkaapi@uob.edu.bh

Received: 7 November 2019; Accepted: 19 November 2019; Published: 21 November 2019

Abstract: The problem investigated in this paper is scheduling on uniform parallel machines, taking into account that machines can be periodically unavailable during the planning horizon. The objective is to determine planning for job processing so that the makespan is minimal. The problem is known to be NP-hard. A new quadratic model was developed. Because of the limitation of the aforementioned model in terms of problem sizes, a novel algorithm was developed to tackle big-sized instances. This consists of mainly two phases. The first phase generates schedules using a modified Largest Processing Time (*LPT*)-based procedure. Then, theses schedules are subject to further improvement during the second phase. This improvement is obtained by simultaneously applying pairwise job interchanges between machines. The proposed algorithm and the quadratic model were implemented and tested on variously sized problems. Computational results showed that the developed quadratic model could optimally solve small- to medium-sized problem instances. However, the proposed algorithm was able to optimally solve large-sized problems in a reasonable time.

Keywords: uniform parallel machines; unavailability constraints; makespan; quadratic programming; optimal algorithm

---

1. Introduction

In the industry field, machines are often supposed to be continuously available for processing assigned jobs. However, this assumption is not totally realistic in real-world cases. For instance, machines may be subject to unavailability periods due to many reasons, such as preventive maintenance [1], corrective maintenance [2], and tool-change activities [3]. There are two main concerns related to the temporary unavailability of a machine. The first is related to the increased costs caused by stopping the machine's activity, while the second is linked to the difficulty in taking decisions regarding the balance between resource unavailability and production. Therefore, a proper planning strategy in a manufacturing system is necessary for it to operate in the most cost-effective way.

Scheduling under machine-unavailability constraints has attracted the attention of many researchers, and many real applications can be found. In [4], the authors listed two applications in the aerospace industry where the machine must be stopped to change microdrilling tools after a fixed number of use times. Another application was mentioned by [5] related to electric-battery vehicles that require refuelling operations.

In this paper, we study a scheduling problem on m uniform parallel machines with multiple unavailability constraints with the objective to minimize the makespan, which is the completion time of the last assigned job. The reason behind the choice of such an objective is that minimizing the makespan can ensure a good load balance among the machines. We followed three-field $\alpha|\beta|\gamma$ classification, developed by [6], to represent the problem as $Qm, h_{ik}|a|\gamma$. In the first field, Q denotes uniform parallel machine setting, m represents the number of considered machines, and h_{ik} states that each machine is unavailable during k periods in the planning horizon. In the second field, β, a

indicates that the machines are subject to availability constraints. Lastly, the third field, γ, describes the objective to be minimized, that is, the completion time of the last processed job, denoted by C_{max}.

Many papers in the literature studied parallel machine-scheduling problems with availability constraints, but very few considered a uniform parallel machine setting. To the best of our knowledge, only two related papers exist so far, [7,8]. In [7], the authors studied the uniform parallel machine-scheduling problem where each machine could be unavailable during one period of time. The considered performance measures were total completion times and makespan. Two types of jobs were treated, namely, identical and nonidentical jobs. Linear programming models and optimal algorithms were developed to solve the problem where jobs are identical. For the case of nonidentical jobs, the authors proved that the problem is NP-hard, and proposed a quadratic program and a heuristic that were tested on large-sized problem instances. The online version of the problem was studied in [8]. The authors considered the case of two machines under the constraint of one periodically unavailable machine. The identical- and uniform-machine cases were investigated. The objective was to minimize the makespan. The solution approach consisted of optimal algorithms with competitive ratios.

Furthermore, most research papers studied the case of identical parallel machines. For example, [9–13] studied various identical parallel machine problems allowing various types of unavailable intervals for machines.

The shortage in research in this area, and the important applications of the investigated problem in reality motivated the author of this paper to explore this area more and contribute to the scientific research on it. Uniform parallel machine scheduling can be found in the manufacturing field where the same type of job can be processed on new and old machines that have different speeds. As an example, a printing task can take much more time on an old machine than on a new one.

In this paper, the main contributions are a quadratic programming-model (QM) formulation of a uniform parallel machine with multiple availability constraints and an algorithm that provides optimal solutions. To the best of our knowledge, the proposed QM is the first such formulation for scheduling on uniform parallel machine with availability constraints.

The content of this paper is organized as follows. Problem notations are laid in Section 2. In Section 3.1, a quadratic model for the problem with makespan as an objective is developed. Section 3.2 details an algorithm proposed for makespan-performance measurement. The proposed algorithm was tested on different problem instances, and results are displayed in Section 4. Finally, a general conclusion is formulated in Section 5.

2. Notations

For accuracy of description, by 'unavailability interval' we denote the time interval in which the machine is not available for processing any job, whereas the time interval between two consecutive unavailability intervals is called the 'availability interval' of the machine.

In this paper, we consider m uniform parallel machines that can process n jobs. Each job j, $j = 1, \ldots, n$ is characterized by processing time p_j and completion time C_j. We assumed that the jobs were ready at time 0 and could be processed once at any time, but could not be interrupted once started. Since we consider uniform parallel machines, each machine i, $i = 1, \ldots, m$, can process at most one job at a time at speed s_i. So, the processing time of any job j depends on the machine on which it is processed and is equal to $p_{ij} = p_j/s_i$, $i = 1, \ldots, m$; $j = 1, \ldots, n$. Without loss of generality, we assumed that jobs were indexed in LPT order, that is, $p_{i1} \geq p_{i2} \geq \ldots \geq p_{in}$. We assumed that the machine could process the next job once the previous one was finished. Thus, no setup time was considered. Let s_{ik} and e_{ik} be the starting and ending time of the kth unavailability period on machine i, respectively. Without loss of generality, we assumed that all machines were available at the beginning of the planning horizon. By L_{ik}, we denote the length of the kth availability interval on machine i.

The problem was to find a job assignment on machines that minimizes the makespan. As stated earlier, the problem of scheduling jobs on uniform parallel machines subject to unavailability constraints has not been studied before. Therefore, a mathematical formulation of the problem can be of great interest. Thus, in Section 3.1, we detail a mathematical model to describe the problem under consideration.

3. Proposed Solution Approach for $Qm, h_{in_i}|a|C_{max}$

In this section, we studied the scheduling problem on uniform parallel machine, where each machine i can be unavailable during n_i unavailability periods in its planning horizon. Thus, there are $n_i + 1$ availability intervals. The objective was to minimize the makespan.

It is easy to see that $Qm, h_{in_i}|a|C_{max}$ is NP-hard. To see this, let $s_i = 1$ for every machine i. Then the problem reduces to the identical parallel machine-scheduling problem under availability constraints that was proved to be NP-hard by [14].

3.1. Mathematical Model

Let

$$x_{ijk} = \begin{cases} 1 & \text{if job } j \text{ is executed on machine } i \text{ during } k^{th} \text{ availability interval} \\ 0 & \text{Otherwise.} \end{cases}$$

$$y_{ik} = \begin{cases} 1 & \text{if all jobs on machine } i \text{ are completed before the start of } k^{th} \text{ unavailability period} \\ 0 & \text{Otherwise.} \end{cases}$$

Using the above-listed decision variables, the problem can be modeled as a quadratic program as follows:

$$\text{Minimize } C_{max} = Max_j \, C_j \tag{1}$$

Subject to

$$\sum_{k=1}^{n_i+1} \sum_{j=1}^{n} p_{ij}x_{ijk} + \sum_{k=1}^{n_i-1} \left(\sum_{l=1}^{k}(e_{il} - s_{il}) \right) y_{ik} \leq \sum_{k=1}^{n_i} s_{ik}y_{ik} + d(1 - \sum_{k=1}^{n_i} y_{ik}) \ \ i = 1,\dots,m, \tag{2}$$

where d is a large positive number.

$$\sum_{k=1}^{n_i+1} \sum_{j=1}^{n} p_{ij}x_{ijk} + \sum_{k=1}^{n_i}(e_{ik} - s_{ik})(1 - \sum_{l=1}^{k} y_{il}) + \sum_{k=1}^{n_i} \left[(s_{ik} - e_{ik-1}) - \sum_{j=1}^{n} p_{ij}x_{ijk} \right] \left[1 - \sum_{l=1}^{k} y_{il} \right] \leq C_{max} \tag{3}$$
$$i = 1,\dots,m$$

$$\sum_{k=1}^{n_i} y_{ik} \leq 1 \ \ i = 1,\dots,m \tag{4}$$

$$\sum_{j=1}^{n} p_{ij}x_{ijk} \leq s_{ik} - e_{ik-1} \ \ i = 1,\dots,m \, ; k = 1,\dots,n_i \tag{5}$$

$$\sum_{i=1}^{m} \sum_{k=1}^{n_i+1} x_{ijk} = 1 \ \ j = 1,\dots,n \tag{6}$$

$$x_{ijk} \in \{0,1\} \ \ i = 1,\dots,m; \ j = 1,\dots,n \ k = 1,\dots,n_i+1 \tag{7}$$

$$y_{ik} \in \{0,1\} \ \ i = 1,\dots,m; \ k = 1,\dots,n_i \tag{8}$$

Equation (1) minimizes the makespan. Equation (2) guarantees that, when all jobs are completed before the start of the 1st unavailability period, the unavailability duration is not considered in the evaluation of the completion time of the last job assigned to machine i . There are m of these constraints. Equation (3) states that the completion time of the last job assigned to machine i is at most equal to the makespan. There are m of these constraints. Equation (4) guarantees that no more than one y_{ik} is equal one for a given machine i. There are m of these constraints. The total processing time of the jobs assigned to a given availability interval cannot exceed the length of that interval. This is shown by Equation (5). There are $m \sum n_i$ of these constraints. Equation (6) assures that, if a job is assigned to a machine, it can be processed on only one availability interval of that machine. There are n constraints of this type. Equations (7) and (8) define the non-negativity constraints about the decision variables used to develop the mathematical model.

The above quadratic model (QM) can be optimally solved by CPLEX for problem instances with up to 73 machines. Therefore, a good polynomial algorithm that can solve large and more complicated problems, and provide promising results is of great interest.

The Largest Processing Time algorithm (LPT) is a famous rule used to build heuristics for scheduling problems with a makespan criterion. For example, in [15] the authors proposed LPT-based heuristics to solve $Q2||Cmax$ and $Qm, ai||Cmax$ problems, respectively. The LPT rule sorts jobs into a nonincreasing order of their processing times and then assigns a job to the machine on which it can finish as early as possible.

3.2. Proposed-Solution Approach

The approach proposed to solve the problem of scheduling on parallel machines under unavailability constraints consists of two steps. The first step focuses on assigning jobs to different available machines using a newly proposed LPT-Based Heuristic, named $LPTBH$. The second step, named $LSHIP$, tries to improve solutions obtained by $LPTBH$. The Main Algorithm, named MA, is a combination of $LPTBH$ and $LSHIP$.

3.2.1. $LPTBH$ Heuristic Procedure

The main idea of $LPTBH$ is to divide the set of jobs N into two subsets. The first set includes jobs that can be assigned to one of the machines' availability intervals. The second set contains the remaining jobs. The $LPTBH$ consists of two phases. The first is the main phase, as it schedules the maximum of jobs. First, for every machine, a list of job candidates is formed on the basis of whether they could fit the machine's availability intervals except the last ones. This step is achieved by using the *Candidate_Search* procedure shown in Algorithm 1. Second, jobs in every constructed list are sorted in decreasing order of their processing times. Then, for every machine, starting from machine 1, select the first job in the candidate list of machine 1. If the selected job is only in that machine's list, assign it to the availability interval that can fit it. Otherwise, assign it to the machine on which it can finish as early as possible. The first phase ends when all the machines' job-candidate lists are empty. The remaining unscheduled jobs are input for the second phase. The pseudocode of the $LPTBH$ heuristic is shown in Algorithm 2. Table 1 lists notations used to develop Algorithms 1 and 2.

Table 1. Notations used in Algorithms 1 and 2.

Notation	Meaning
S	Set of all the jobs to be scheduled
$C_i,\ i=1,\ldots,m$	Completion time of last job assigned to machine i
$av_{ik},\ i=1,\ldots,m; k=1,\ldots,n_i$	Length of kth availability interval of machine i
$maxAv_i,\ i=1,\ldots,m$	Length of largest availability interval of machine i
$Lc_i,\ i=1,\ldots,m$	List of jobs that can be processed in any availability interval on machine i
LR	List of remaining unscheduled jobs

Algorithm 1 *Candidate_Search.*

1: **procedure** (Input $N = \{1, \ldots, n\}$, m, p_{ij}, $i = 1, \ldots, m$; $j = 1, \ldots, n$, $maxAv_i$, $i = 1, \ldots, m$)
2:
3:     **for** $i = 1$ to m **do**
4:
5:         **for** $j = 1$ to n **do**
6:
7:             **if** $(p_{ij} \leq maxAv_i)$ **then**
8:
9:                 $Lc_i = Lc_i \cup \{j\}$
10:
11:             **end if**
12:
13:         **end for**
14:
15:     **end for**
16:
17:     Sort the jobs in every Lc_i, $i = 1, \ldots, m$ in a nonincreasing order of their processing times.
18:
19: **end procedure**
20:

Algorithm 2 *LPTBH.*

1: **procedure** (Input $N = \{1, \ldots, n\}$, m, p_{ij}, $i = 1, \ldots, m$; $j = 1, \ldots, n$, N_i, $i = 1, \ldots, m$, S_{ik}, E_{ik}, $i = 1, \ldots, m$; $k = 1, \ldots, N_i$. Output $S = C_{max}$)
2:
3:     **for** $i = 1$ to m **do**
4:
5:         **for** $k = 1$ to N_i **do**
6:
7:             $av_{ik} = E_{ik} - S_{ik-1}$
8:
9:         **end for**
10:
11:         $maxAv_i = \max_k av_{ik}$
12:
13:         $C_i = 0$
14:
15:     **end for**
16:
17:     Call *Candidate_Search*
18:
19:     **while** $(S \neq \varnothing)$ **do**
20:
21:         Among the jobs of Lc_i, $i = 1, \ldots, m$, select the job with the highest processing time. Let l be that job and $i_l(s)$ the machine(s) to which it can be assigned.
22:
23:             **if** (l exists in more than one Lc_i) **then**
24:
25:                 Assign l to the machine on which it can finish as early as possible.
26:
27:             **else**Assign l to machine i_l
28:
29:                 Update C_{i_l}
30:
31:             **end if**
32:
33:         $S = S \setminus \{l\}$
34:
35:         Update av_{ik} of the machine to which job l was assigned.
36:
37:         Call *Candidate_Search*
38:
39:         **if** $(Lc_i = \varnothing, \forall\, i = 1, \ldots, m)$ **then**
40:
41:             **if** $(|S| \neq n)$ **then**
42:
43:                 $LR \leftarrow N \setminus S$
44:
45:                 Schedule the jobs of LR according to LPT rule.
46:
47:                 Calculate $C_i, \forall\, i = 1, \ldots, m$
48:
49:             **end if**
50:
51:             $S \leftarrow S \setminus LR$
52:
53:         **end if**
54:
55:     **end while**
56:
57:     $S = \max_i C_i$
58:
59:     **return** S.
60:
61: **end procedure**

3.2.2. Improvement Procedure *LSHIP*

The idea of the improvement procedure was inspired from a local-search heuristic proposed in [16], developed to solve the scheduling problem of parallel identical-batch processing machines. The aim of the improvement procedure was to try to balance the load of different machines so that the completion times of the last jobs in every machine are almost the same. This improvement can be achieved by interchanging pairs of jobs between the most loaded machine and other machines. The flowchart of the aforementioned heuristic is shown in Figure 1.

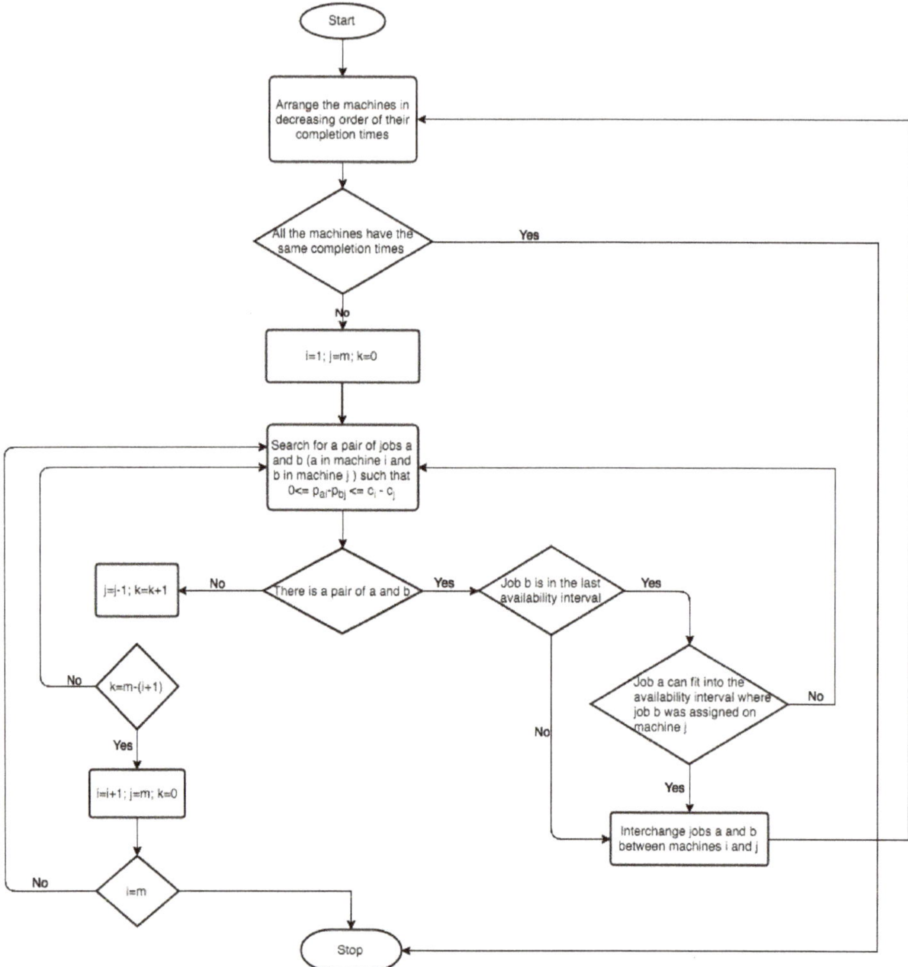

Figure 1. Flowchart of *LSHIP* procedure.

In order to illustrate the proposed heuristic, let us consider a problem instance with 2 machines and 10 jobs. Table 2 summarizes the input data, and Figures 2 and 3 show the Gantt charts of solutions obtained by *LPTBH* and *LSHIP*, respectively.

Table 2. Input data for 10 jobs and two machines.

Job j	1	2	3	4	5	6	7	8	9	10
p_{1j}	25	25	30	36	38	38	41	44	44	47
p_{2j}	10	10	12	14	15	15	16	17	17	19

Figure 2. Gantt chart of solution generated by $LPTBH$.

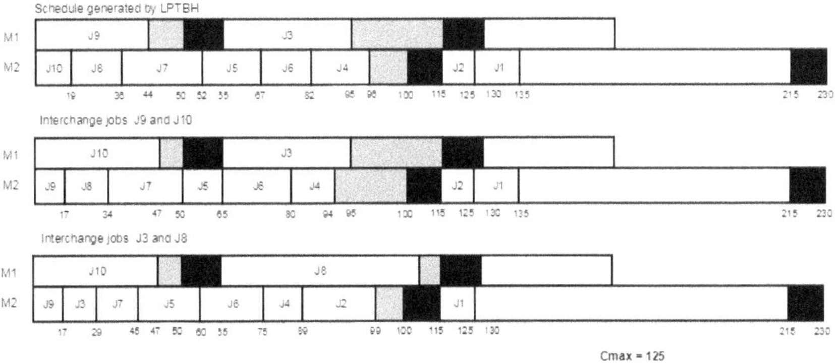

Figure 3. Gantt chart of the solution generated by $LSHIP$.

Note that after interchanging a pair of jobs between two machines, the $LSHIP$ procedure looks to shift jobs to the left whenever the idle time interval on the machine can fit them. In the above example, after interchanging jobs $J3$ and $J8$, $LSHIP$ shifted job $J2$ to the left since it could fit in the idle time interval.

4. Experiment Results

For the purpose of evaluating the performance of the proposed algorithm, many problem instances were tested. These were generated after examining the important factors that significantly impacted the performance of the proposed algorithm. The first factor was the number of jobs n to be processed that directly affects the machines' load. The second important factor is the number of machines m that has an impact on the assignment of jobs to machines. Job processing times may play a role in the efficiency of the proposed algorithm. Thus, we generated problem instances with different job processing times. The algorithm was coded in IntelliJ IDEA. In addition, the quadratic model was modelled in IBM ILOG CPLEX Optimization Studio 12.7. The proposed heuristic was implemented using programming language Java. We ran all test problems on an Intel Core i5 2.5 Gigahertz, 4 Gigabyte RAM Macintosh HD.

In order to avoid useless computational time, the program was stopped for two possible reasons. The first was when the CPLEX became unable to generate a solution within the time limit of 3600 s (1 h). The second reason was due to memory overflow. At this point, the best feasible solution found within the time limit was recorded.

4.1. Data Generation

A deep empirical study was conducted with the aim to generate datasets that would help to correctly analyze the efficiency of the proposed algorithm. By the end, two dataset series were considered, namely, $DS1$ and $DS2$. In fact, the way to generate dataset series $DS2$ was inspired from Graham's data-generation process [17] addressing $P||Cmax$ problems. The parameters used to generate $DS1$ and $DS2$ are summarized in Tables 3 and 4, respectively.

Table 3. $DS1$ parameters.

Number of machines (m)	$m \in \{2,3,5\}$
Number of jobs (n)	$n \in \{20,30,40,50,60,70,80\}$
Machine speed (s_i)	$s_i \in U(1,5)$
Job processing time (p_j)	$p_j \in U(5,50)$ and $p_j \in U(50,100)$
Number of unavailability periods (n_i)	$n_i = m \; \forall i = 1,\dots,m$
Duration of an unavailability period on machine i (t_i)	$t_i = 10$, if $p_j \in U(5,50)$ and $t_i = 15$, if $p_j \in U(50,100) \; \forall i = 1,\dots,m$
Length of time interval between two consecutive unavailability periods on machine i (T_i)	$Ti = 25i$, if $p_j \in U(5,50)$ and $T_i = 50i$, if $p_j \in U(50,100)$

Table 4. $DS2$ parameters.

Number of machines (m)	$m \in \{30,31,32,\dots,\dots,80\}$
Number of jobs (n)	$n = 2m + 1$
Machine speed (s_i)	$s_i \in U(1,5)$
Job processing time (p_j)	$p_j \in U(1,100)$
Number of unavailability periods (n_i)	$n_i = 2 \; \forall i = 1,\dots,m$
Duration of an unavailability period on machine i (t_i)	$t_i = 10 \; \forall i = 1,\dots,m$
Length of time interval between two consecutive unavailability periods on machine i (T_i)	$Ti = 20i$

The starting and ending times S_{ik} and E_{ik} of the unavailability periods were generated according to Equations (9) and (10), respectively.

$$s_{ik} = kT_i + (k-1)t_i \;\; i = 1,\dots,m; k = 1,\dots,n_i \tag{9}$$

$$e_{ik} = s_{ik} + t_i \;\; i = 1,\dots,m; k = 1,\dots,n_i \tag{10}$$

4.2. Experiments

In this section, we outline different experiments that were conducted to evaluate the performance of the QM and the proposed algorithm. In all experiments, Central Processing Unit time ($CPUt$) represents the time in seconds required to find the optimal or best feasible solution. Tables 5 and 6 show the results obtained by QP and MA for small and large job processing times, respectively.

Table 5 clearly shows that the proposed algorithm was generating optimal solutions with a CPU time of less than 1 second for all problem instances. Quadratic model QM was also able to provide optimal schedules in a reasonable time. By considering much longer processing times than in the previous data series, we still obtained optimal solutions in reasonable CPU time even though the quadratic model became slower than in the first batch of problem instances. The proposed algorithm outperformed the quadratic model in terms of computational time that was still less than 1 second. Table 6 confirms these observations.

Table 5. Comparison of QM and MA for datasets $DS1$ with $s_i \in U(1,5)$ and $p_j \in U(5,50)$.

m	n	$C_{max}(QM)$	$C_{max}(MA)$	$CPUt(QM)$	$CPUt(MA)$
2	20	66	66	1.16	0.01
2	30	66	66	1.16	0.01
2	40	229	229	0.46	0.02
2	50	463	463	0.86	0.05
2	60	343	343	1.26	0.03
2	70	401	401	0.71	0.04
2	80	780	780	1.25	0.03
3	20	87	87	1.72	0.01
3	30	218	218	3.58	0.04
3	40	244	244	3.91	0.02
3	50	165	165	2.81	0.02
3	60	282	282	4.73	0.02
3	70	246	246	4.01	0.04
3	80	298	298	6.46	0.04
5	20	28	28	1.52	0.02
5	30	60	60	3.25	0.03
5	40	85	85	7.24	0.03
5	50	196	196	114.24	0.06
5	60	93	93	14.19	0.04
5	70	120	120	27.02	0.06
5	80	174	174	121.58	0.05

Table 6. Comparison of QM and MA for datasets $DS1$ with $s_i \in U(1,5)$ and $p_j \in U(50,100)$.

m	n	$C_{max}(QM)$	$C_{max}(MA)$	$CPUt(QP)$	$CPUt(MA)$
2	20	169	169	1.76	0.02
2	30	409	409	1.25	0.03
2	40	319	319	1.03	0.01
2	50	622	622	0.73	0.02
2	60	567	567	0.6	0.03
2	70	659	659	0.79	0.04
2	80	689	689	1.10	0.02
3	20	130	130	1.83	0.01
3	30	239	239	3.24	0.04
3	40	359	359	2.09	0.02
3	50	365	365	8.45	0.03
3	60	407	407	17.98	0.04
3	70	561	561	23.65	0.06
3	80	702	702	3.53	0.03
5	20	94	94	3.00	0.01
5	30	143	143	5.87	0.02
5	40	277	277	9.55	0.06
5	50	272	272	473.35	0.06
5	60	208	208	14.41	0.06
5	70	382	382	796.08	0.06
5	80	339	339	223.71	0.06

In order to investigate the limitations of the proposed quadratic model, a second dataset series, namely, $DS2$ was considered. Table 7 reports the computational results for both QM and MA.

Table 7. Comparison of QM and MA for datasets $DS2$.

m	n	$C_{max}(QM)$	$C_{max}(MA)$	QP Optimal?	$CPUt(QM)$	$CPUt(MA)$
30	61	133	133	Yes	70.74	0.15
33	67	137	137	Yes	40.19	0.22
36	73	142	142	Yes	21.1	0.21
40	81	140	140	Yes	44.71	0.31
45	91	232	232	Yes	398.64	0.33
51	103	240	240	Yes	495.44	0.4
57	115	237	237	Yes	214.28	0.2
62	125	336	336	Yes	263.21	0.2
68	137	331	331	Yes	393.97	0.39
73	147	434	434	No	3603.54	0.42
76	153	439	439	No	3602.48	0.43
80	161	538	538	Yes	3602.9	0.27

The computational results displayed in Table 7 show that quadratic model QM was able to generate an optimal solution within a time limit for problems with up to 73 machines.

On the basis of the computational results shown in Table 7, the quadratic model was not able to generate optimal solutions in a reasonable time and for bigger problems. Therefore, proposed procedure MA was tested for large-sized problems and compared to an adapted form of $MLPT$, proposed earlier by the author of this paper in [7]. Table 8 reports the obtained results for problem instances with $m \in \{100, 200, 300, 400, 500, 600, , 700, 800, 1000\}$ and $n = 2m + 1$.

Table 8. Comparison of MA and $MLPT$.

m	n	$C_{max}(MLPT)$	$C_{max}(MA)$	$C_{max}(MLPT)/C_{max}(LSHIP)$	$CPUt(QP)$	$CPUt(MA)$
100	201	1120	880	1.27	0.47	0.49
200	401	1090	1050	1.03	2.16	2.5
300	601	1250	970	1.28	4.69	4.81
400	801	1110	960	1.15	9.12	9.55
500	1001	1120	760	1.47	16.24	15.48
600	1201	1260	1240	1.01	28.24	24.07
700	1401	1240	1160	1.06	41.92	62.24
800	1601	1260	1110	1.13	74.97	48.5
1000	2001	1110	1080	1.02	120.04	122.99

Table 8 shows that MA outperformed $MLPT$ for all problem instances with slightly higher CPU time than the time of $MLPT$ CPU for most instances. In addition, run time increased with problem size.

5. Conclusions and Future Work

In this paper, we studied the problem of parallel machine scheduling with multiple planned nonavailability periods. In the current literature, very few papers investigated this problem. The problem was formulated as a quadratic program and optimally solved using CPLEX for small- to moderately large-sized problems. In order to be able to solve large-sized problems, an algorithm consisting of two main phases was developed. The first phase searches for schedules on the basis of the LPT rule. The second aims to improve these schedules by considering simultaneous pairwise interchanges of jobs between machines. A deep computational study was conducted to test the efficiency of the proposed approach. Many datasets were carefully generated to help evaluate the algorithm. Computational results showed that the proposed algorithm generated optimal solutions for all considered problem sizes and outperformed an adapted form of a heuristic that was developed earlier by the author of this paper. Further investigation can be done to consider other criteria and more general versions of the problem, such as the dynamic case where jobs arrive one by one over the planning horizon.

Funding: This research received no external funding.

Conflicts of Interest: The authors declare no conflict of interest.

References

1. Azadeh, A.; Sheikhalishahi, M.; Firoozi, M.; Khalili, S. An integrated multi-criteria Taguchi computer simulation-DEA approach for optimum maintenance policy and planning by incorporating learning effects. *Int. J. Prod. Res.* **2013**, *51*, 5374–5385. [CrossRef]
2. Yazdani, M.; Khalili, S.M.; Jolai, F. A parallel machine scheduling problem with two-agent and tool change activities: An efficient hybrid metaheuristic algorithm. *Int. J. Comput. Integr. Manuf.* **2016**, *29*, 1075–1088. [CrossRef]
3. Azadeh, A.; Sheikhalishahi, M.; Khalili, S.M.; Firoozi, M. An integrated fuzzy simulation—Fuzzy data envelopment analysis approach for optimum maintenance planning. *Int. J. Comput. Integr. Manuf.* **2014**, *27*, 181–199. [CrossRef]
4. Low, C.; Ji, M.; Hsu, C.J.; Su, C.T. Minimizing the makespan in a single machine scheduling problems with flexible and periodic maintenance. *Appl. Math. Model.* **2010**, *34*, 334–342. [CrossRef]
5. Schneider, M.; Stenger, A.; Hof, J. An adaptive VNS algorithm for vehicle routing problems with intermediate stops. *Or Spectrum* **2015**, *37*, 353–387. [CrossRef]
6. Graham, R.L.; Lawler, E.L.; Lenstra, J.K.; Kan, A.R. Optimization and approximation in deterministic sequencing and scheduling: a survey. *Ann. Discrete Math.* **1979**, *5*, 287–326.
7. Kaabi, J.; Harrath, Y. Scheduling on uniform parallel machines with periodic unavailability constraints. *Int. J. Prod. Res.* **2019**, *57*, 216–227. [CrossRef]
8. Liu, M.; Zheng, F.; Chu, C.; Xu, Y. Optimal algorithms for online scheduling on parallel machines to minimize the makespan with a periodic availability constraint. *Theor. Comput. Sci.* **2011**, *412*, 5225–5231. [CrossRef]
9. Liao, L.W.; Sheen, G.J. Parallel machine scheduling with machine availability and eligibility constraints. *Eur. J. Oper. Res.* **2008**, *184*, 458–467. [CrossRef]
10. Mellouli, R.; Sadfi, C.; Chu, C.; Kacem, I. Identical parallel-machine scheduling under availability constraints to minimize the sum of completion times. *Eur. J. Oper. Res.* **2009**, *197*, 1150–1165. [CrossRef]
11. Fu, B.; Huo, Y.; Zhao, H. Approximation schemes for parallel machine scheduling with availability constraints. *Discret. Appl. Math.* **2011**, *159*, 1555–1565. [CrossRef]
12. Wang, X.; Cheng, T. A heuristic for scheduling jobs on two identical parallel machines with a machine availability constraint. *Int. J. Prod. Econ.* **2015**, *161*, 74–82. [CrossRef]
13. Gedik, R.; Rainwater, C.; Nachtmann, H.; Pohl, E.A. Analysis of a parallel machine scheduling problem with sequence dependent setup times and job availability intervals. *Eur. J. Oper. Res.* **2016**, *251*, 640–650. [CrossRef]
14. Lee, C.Y. Parallel machines scheduling with nonsimultaneous machine available time. *Discret. Appl. Math.* **1991**, *30*, 53–61. [CrossRef]
15. Mireault, P.; Orlin, J.B.; Vohra, R.V. A parametric worst case analysis of the LPT heuristic for two uniform machines. *Oper. Res.* **1997**, *45*, 116–125. [CrossRef]
16. Kashan, A.H.; Karimi, B.; Jenabi, M. A hybrid genetic heuristic for scheduling parallel batch processing machines with arbitrary job sizes. *Comput. Oper. Res.* **2008**, *35*, 1084–1098. [CrossRef]
17. Graham, R.L. Bounds on multiprocessing timing anomalies. *SIAM J. Appl. Math.* **1969**, *17*, 416–429. [CrossRef]

 *algorithms* 

Article

Some Results on Shop Scheduling with S-Precedence Constraints among Job Tasks

Alessandro Agnetis [1,*], **Fabrizio Rossi** [2] and **Stefano Smriglio** [2]

[1] Dipartimento di Ingegneria dell'Informazione e Scienze Matematiche, Università di Siena, SI 53100 Siena, Italy
[2] Dipartimento di Ingegneria e Scienze dell'Informazione e Matematica, Università di L'Aquila,
     AQ 67100 L'Aquila, Italy; fabrizio.rossi@univaq.it (F.R.); stefano.smriglio@univaq.it (S.S.)
* Correspondence: agnetis@diism.unisi.it

Received: 9 October 2019; Accepted: 18 November 2019; Published: 25 November 2019

Abstract: We address some special cases of job shop and flow shop scheduling problems with s-precedence constraints. Unlike the classical setting, in which precedence constraints among the tasks of a job are finish–start, here the task of a job cannot start before the task preceding it has started. We give polynomial exact algorithms for the following problems: a two-machine job shop with two jobs when recirculation is allowed (i.e., jobs can visit the same machine many times), a two-machine flow shop, and an m-machine flow shop with two jobs. We also point out some special cases whose complexity status is open.

Keywords: job shop; flow shop; s-precedence constraints; exact algorithms; complexity

---

1. Introduction

This paper addresses a variant of classical shop scheduling models. While, in classical job shop or flow shop (as well as in the large majority of scheduling problems with precedence constraints), the task of a job cannot start before the previous task of the same job has *finished*, we address a situation in which each task of a job cannot start before the previous task of the same job has *started*. These types of constraints are known in the literature as *s-precedence constraints*. Scheduling problems with s-precedence constraints have been introduced by Kim and Posner [1] in the case of parallel machines. They showed that makespan minimization is NP-hard, and developed a heuristic procedure deriving tight worst-case bounds on the relative error. Kim, Sung, and Lee [2] performed a similar analysis when the objective was the minimization of total completion time of the tasks, while Kim [3] extended the analysis to uniform machines. Tamir [4] analyzed a parallel-machine problem in which traditional finish–start precedence constraints coexisted with s-precedence constraints (that she renamed *selfish precedence constraints*, giving an enjoyable dramatized motivation of the model), and established various worst-case bounds for classical dispatching rules which refer to specific structures of precedence constraints. Indeed, s-precedence constraints also arise in project management, called *start–start* precedence constraints (Demeulemeester and Herroelen [5]), as a result of the elaboration of a work breakdown structure (WBS) and of the coordination among different operational units. To our knowledge, none has addressed job shop and flow shop problems with s-precedence constraints so far.

The problem can be formally introduced as follows. We are given a set of n h $J^1, J^2, \ldots, J^n$, to be processed on a shop with m machines denoted as $M_1, \ldots, M_m$. Each job J^k consists of a totally ordered set of *tasks*, $J^k = \{O_1^k, O_2^k, \ldots, O_{n_k}^k\}$, $k = 1, \ldots, n$. Task O_i^k requires processing time p_i^k on a given machine $M(O_i^k)$, $i = 1, \ldots, n_k$. Tasks cannot be preempted. Task O_i^k can only start after task O_{i-1}^k is started; i.e., there is an s-precedence constraint between tasks O_{i-1}^k and O_i^k, for all $k = 1, \ldots, n$, $i = 2, \ldots, n_i$.

A *schedule* is an assignment of starting times to all tasks so that at any time each machine processes at most one task and all s-precedence constraints are satisfied. The problem is to find a feasible schedule that minimizes makespan.

We characterize the complexity of special cases of the problem, considering a fixed number of jobs and machines. Shop problems with few jobs occur when modeling synchronization and conflicts among processes share common resources. Examples of this situation include scheduling robot moves in flexible robotic cells (Agnetis et al [6]), aircraft scheduling during taxiing at an airport so that no aircraft collides (Avella et al. [7]), or, in container terminals, the synchronization of crane gantry movements once transportation tasks have been assigned (Briskorn and Angeloudis [8]).

The structure of the paper is as follows. In Section 2 we consider the job shop scenario, and give a polynomial time algorithm for the problem in which $n = 2$, $m = 2$, and each job can visit the machines several times (that is, *recirculation* [9] is allowed). In Section 3 we focus on the flow shop scenario. We show that the two-machine flow shop can be solved in linear time and we give a polynomial time algorithm for the m-machine problem with two jobs. In Section 4 we briefly discuss cases with $n > 2$ and point out open problems.

2. The Job Shop with Two Jobs and Two Machines

In this section we describe a polynomial algorithm for the job shop problem with two jobs and two machines; i.e., $J2|n = 2, s - prec|C_{\max}$. For notation simplicity, in this section we denote the two jobs as A and B, consisting of the sequence of *tasks*, $A = \{A_1, A_2, \ldots, A_{n_A}\}$ and $B = \{B_1, B_2, \ldots, B_{n_B}\}$, respectively. Task A_i (B_h) requires processing time p_i^A (p_h^B) on machine $M(A_i)$ ($M(B_h)$).

Obviously, if two consecutive tasks of the same job, say, A_i and A_{i+1}, require the same machine, then A_{i+1} has to wait for the completion of A_i, but if the machines required by the two operations are different, i.e., $M(A_{i+1}) \neq M(A_i)$, then A_{i+1} can start even if A_i has not completed yet. So, unlike the classical job shop setting in which precedence relations are finish-start, in our model it may actually happen that A_{i+1} even *completes* before A_i (the same of course applies to job B).

Given a partial schedule, the first unscheduled tasks of the two jobs will be referred to as the *available tasks*. Suppose now that one of the two machines, say M', has just completed a task, while the other machine, say M'', is still busy. If both the available tasks require M'', we say that machine M' is *blocked* and this certainly results in idle time on M'.

We let $A[i]$ and $B[h]$ denote the first i tasks of A and the first h tasks of B; i.e., $A[i] = \{A_1, A_2, \ldots, A_i\}$ and $B[h] = \{B_1, B_2, \ldots, B_h\}$.

Given A and B, consider any two task subsequences $X \subseteq A$ and $Y \subseteq B$. We want to characterize the schedules of $X \cup Y$ such that each task starts right after the previous task on the same machine has completed. More formally, a schedule of $X \cup Y$ is a *no-idle subschedule* (NIS) if, across the span of such a subschedule, the only machine idle time occurs, on one machine, after all the tasks of $X \cup Y$ have *started*. When $X = A[i]$ and $Y = B[h]$, for some $1 \leq i \leq n_A$ and $1 \leq h \leq n_B$, then we say that the NIS is an *initial no-idle subschedule* (INIS).

Consider Figure 1 and the task set $A[2] \cup B[2]$. The subschedule in Figure 1a is *not* an INIS for $A[2] \cup B[2]$, since on M_1 there is idle time before B_2 starts. On the contrary, in the case depicted in Figure 1b, the subschedule of $A[2] \cup B[2]$ is an INIS. Note that if we restrict our attention to the task set $A[2] \cup B[1]$, then the subschedule in Figure 1a is an INIS.

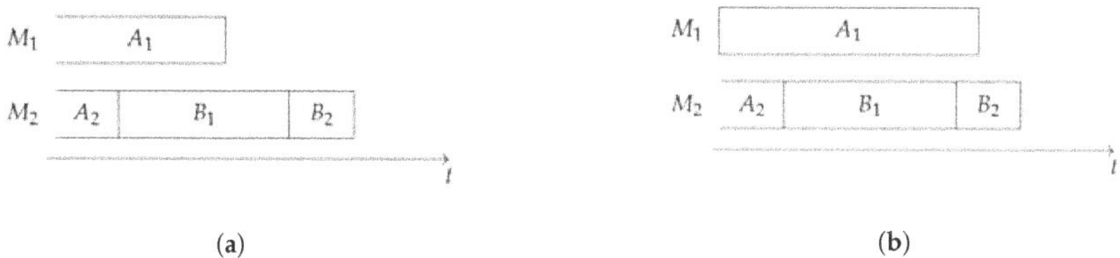

Figure 1. In instance (**a**), the set $A[2] \cup B[2]$ does not form an INIS (initial no-idle subschedule); in instance (**b**) it does.

2.1. Generating Initial No-Idle Subschedules

We denote by $A[i, M_j]$ and $B[h, M_j]$, the subset of tasks of $A[i]$ and $B[h]$ respectively, requiring machine M_j, $j = 1, 2$; i.e.,

$$A[i, M_j] = \{A_r : r \le i, M(A_r) = M_j\} \quad j = 1, 2,$$

$$B[h, M_j] = \{B_q : q \le h, M(B_q) = M_j\} \quad j = 1, 2.$$

We also let $P(A[i, M_j])$ and $P(B[h, M_j])$ indicate the total processing time of tasks in $A[i, M_j]$ and $B[h, M_j]$; i.e.,

$$P(A[i, M_j]) = \sum_{r \in A[i, M_j]} p_r^A,$$

$$P(B[h, M_j]) = \sum_{q \in B[h, M_j]} p_q^B.$$

If an INIS of tasks $A[i] \cup B[h]$ exists, its makespan is given by

$$\max\{P(A[i, M_1]) + P(B[h, M_1]), P(A[i, M_2]) + P(B[h, M_2])\}.$$

Proposition 1. *In any optimal schedule, there are indices i and h such that the subschedule involving tasks $A[i] \cup B[h]$ is an INIS.*

In fact, given an optimal schedule, consider the subschedule of the tasks scheduled on the two machines from time 0 to the end of the first idle interval of the schedule, assuming, e.g., that such an idle interval occurs on M_1. If the subschedule is not an INIS, we can iteratively remove the last task scheduled on M_2 in the subschedule, until the definition of INIS is met.

In view of Proposition 1, we are only interested in schedules in which the initial part is an INIS. However, not all initial no-idle subschedules are candidates to be the initial part of an optimal schedule.

We first address the following question. Can we determine all operation pairs (i, h) such that an INIS of $A[i] \cup B[h]$ exists? We show next that this question can be answered in polynomial time.

The idea is to build the no-idle partial schedules from the beginning of the schedule onward. To this aim, let us define an unweighted graph G, which we call *initial no-idle graph*. Nodes of G are denoted as (i, h), representing a NIS of $A[i] \cup B[h]$ (for shortness, we use (i, h) also to denote the corresponding INIS). If the schedule obtained appending B_{h+1} to schedule (i, h) is still an INIS, we insert node $(i, h + 1)$ and an arc from (i, h) to $(i, h + 1)$ in G. Symmetrically, if the schedule obtained appending A_{i+1} to (i, h) is an INIS, we insert $(i + 1, h)$ and an arc from (i, h) to $(i + 1, h)$ in G.

As illustrated later on (cases $(i) - (iv)$ below), while building the graph G, we can also determine whether or not a certain INIS can be the initial part of an optimal schedule. If it can, we call it a *target node*.

Consider any node (i, h) in G, and the machine completing soonest in the INIS. Ties can be broken arbitrarily, but to fix ideas, suppose that M_2 is still busy when M_1 completes. (Note that, since there is no idle time, M_1 completes at time $P(A[i, M_1]) + P(B[h, M_1])$.) If $i < n_A$ and $h < n_B$, the two available tasks are A_{i+1} and B_{h+1}, and four cases can occur.

(i) $M(A_{i+1}) = M(B_{h+1}) = M_2$. In this case, M_1 is necessarily idle until M_2 completes (Figure 2a). Hence, there is no way to continue an INIS, and therefore node (i, h) has no outgoing arcs. In this case, (i, h) is a target node.

(ii) $M(A_{i+1}) = M_1$ and $M(B_{h+1}) = M_2$. In this case, when A_i completes, the only way to continue an INIS is to start task A_{i+1} on M_1 (Figure 2b). Thus we generate node $(i + 1, h)$ and the arc from (i, h) to $(i + 1, h)$, which is the only outgoing arc of (i, h). In this case as well, (i, h) is a target node.

(iii) $M(A_{i+1}) = M_2$ and $M(B_{h+1}) = M_1$. A symmetrical discussion to the previous case holds; i.e., the only way to continue an INIS is to start task B_{h+1} on M_1 (Figure 2c), so we generate node $(i, h + 1)$ and the arc from (i, h) to $(i, h + 1)$, which is the only outgoing arc of (i, h). In this case also, (i, h) is a target node.

(iv) $M(A_{i+1}) = M(B_{h+1}) = M_1$. In this case, the INIS can be continued in two possible ways; i.e., scheduling either A_{i+1} or B_{h+1} on M_1 (Figure 2b,c respectively). Therefore, (i, h) has two outgoing arcs, pointing towards nodes $(i + 1, h)$ and $(i, h + 1)$, respectively. However, in this case (i, h) is *not* a target node, since there is no point in keeping M_1 idle until the completion of M_2.

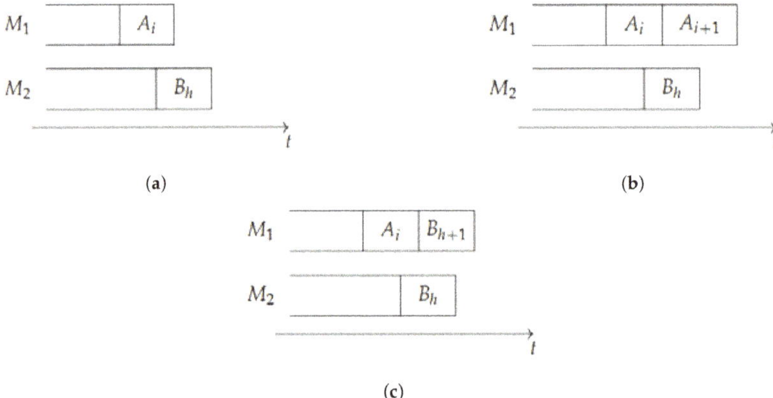

Figure 2. Possible scenarios when M_1 completes before M_2: (**a**) The INIS (i, h) cannot be continued, (**b**) it can only be continued scheduling A_{i+1}, and (**c**) it can only be continued scheduling B_{h+1}.

Clearly, if M_2 completes before M_1, in the four above cases the roles of M_1 and M_2 are exchanged. If either $i = n_A$ or $h = n_B$, the above cases simplify as follows, where we assume that $h = n_B$; i.e., job B is finished. (A symmetric discussion holds in $i = n_A$.)

(v) $M(A_{i+1}) = M_1$. In this case, we can continue an INIS starting task A_{i+1} on M_1. Thus we generate node $(i + 1, h)$ and the arc from (i, h) to $(i + 1, h)$, which is the only outgoing arc of (i, h). Node (i, h) is a target node.

(vi) $M(A_{i+1}) = M_2$. In this case, M_1 is necessarily idle until M_2 completes. Hence, there is no way to continue an INIS, and therefore node (i, h) has no outgoing arcs. In this case, (i, h) is a target node.

Again, the roles of the two machines are exchanged if M_2 frees up before M_1 in the partial schedule.

In conclusion, the question of whether a NIS exists for the task set $A[i] \cup B[h]$ is equivalent to asking whether node (i,h) can be reached from the dummy initial node $(0,0)$ on G.

A few words on complexity. Clearly, G has $O(n_A n_B)$ nodes, and each node has at most two outgoing arcs. The graph G can be built very efficiently. In fact, for each node (i,h), it can be checked in constant time, which condition holds among (i)–(iv) (or (v)–(vi) when one of the jobs is over), and hence whether or not it is a target node.

2.2. Minimizing the Makespan

Now we can address the main question. How to schedule the tasks on the two machines so that the overall makespan is minimized. The key idea here is that any active schedule can be seen as the juxtaposition of no-idle subschedules. In fact, suppose that after processing a certain task A_i, one machine stays idle until the other machine completes task B_h. It is important to observe that this may happen for one of *two* reasons:

- When a machine completes, it is blocked because both available tasks require the other machine;
- When a machine completes, there *is* one task the machine can process, but it might be profitable to wait for the other machine to free up another task.

Note that in both of these two cases (i,h) is a target node of G. On the contrary, if a machine completes a task while the other machine is still busy, and both available tasks require that machine (i.e., (i,h) is not a target node of G), with no loss of generality we can assume that the machine *will* immediately start one of them, since otherwise the schedule might turn out non-active (there is no point in waiting for the other machine to complete its task).

If t denotes the makespan of an INIS, the schedule after t is completely independent from the schedule before t. In other words, the optimal solution from t onward is the optimal solution of a problem in which t is indeed time 0, and the two jobs are $A \setminus A[i]$ and $B \setminus B[h]$. Hence, to address the overall problem, the idea is to build another, higher-level graph in which the arcs specify portions of the overall schedule.

Suppose that (i,h) is a target node of graph G, and consider the task sets $A \setminus A[i]$ and $B \setminus B[h]$. We can build a new *no-idle graph* on these sets, and call it $G(i,h)$. (Correspondingly, the graph previously denoted as G can be renamed $G(0,0)$.) Suppose that (r,q) is a target node in graph $G(i,h)$. This means that the tasks of the set $\{A_{i+1}, A_{i+2}, \ldots, A_r\} \cup \{B_{h+1}, B_{h+2}, \ldots, B_q\}$ form a NIS, that we denote by $[(i+1, h+1) \to (r,q)]$. It is convenient to extend the previous notation, letting $A[i+1, r, M_j]$ denote the set of tasks of $A[i+1, r]$ that require machine M_j, and analogously we let $B[h+1, q, M_j]$ be the set of tasks of $B[h+1, q]$ that require M_j. Their total processing times are denoted as $P(A[i+1, r, M_j])$ and $P(B[h+1, q, M_j])$. (The set previously denoted as $A[i, M_j]$ should now be written $A[0, i, M_j]$.)

We next introduce the (weighted) graph $\mathcal{G}$ as follows. As in G, nodes denote task pairs (i,h). There is an arc $[(i,h),(r,q)]$ if (r,q) is a target node in the graph $G(i,h)$; i.e., if the NIS $[(i+1, h+1) \to (r,q)]$ exists. Such an arc $[(i,h),(r,q)]$ is weighted by the length of the corresponding NIS; i.e.,

$$\max\{P(A[i+1,r,M_1]) + P(B[h+1,q,M_1]), P(A[i+1,r,M_2]) + P(B[h+1,q,M_2])\}.$$

Moreover, $\mathcal{G}$ contains a (dummy) initial node $(0,0)$ while the final node is (n_A, n_B). At this point the reader should have no difficulty in figuring out that the following theorem holds.

Theorem 1. *Given an instance of $J2|n = 2, s - prec|C_{\max}$, the optimal schedule corresponds to the shortest path from $(0,0)$ to (n_A, n_B) on $\mathcal{G}$, and its weight gives the minimum makespan.*

Now, let us discuss complexity issues. The graph $\mathcal{G}$ can indeed be generated starting from $(0,0)$, and moving schedules forward. From each node (i,h) of $\mathcal{G}$, we can generate the corresponding no-idle graph $G(i,h)$, and add to $\mathcal{G}$ all target nodes of $G(i,h)$. We then connect node (i,h) in $\mathcal{G}$ to each of these nodes, weighing the arc with the corresponding length of the NIS. If a target node was already present in $\mathcal{G}$, we only add the corresponding new arc. Complexity analysis is, therefore, quite simple. There are $O(n_A n_B)$ nodes in $\mathcal{G}$. Each of these nodes has a number of outgoing arcs, whose weight can be computed in $O(n_A n_B)$. Clearly, finding the shortest path on $\mathcal{G}$ is not the bottleneck step, and therefore, the following result holds.

Theorem 2. $J2|n = 2, s - prec|C_{\max}$ *can be solved in* $O(n_A^2 n_B^2)$.

Example 1. *Consider the following instance, in which job A has four tasks and job B two tasks.*

job	1	2	3	4
ine A	$5,M_1$	$1,M_1$	$4,M_2$	$6,M_1$
B	$4,M_2$	$7,M_2$	–	–

Figure 3a depicts the graph $G(0,0)$, in which all nodes are target nodes. Figure 3b shows the INIS $[(0,0) \to (2,2)]$. At the end of this INIS, machine M_1 is blocked, since the next task of A requires M_2 and job B is already finished. Notice that $[(0,0) \to (2,2)]$ is the longest INIS which can be built, but the optimal solution does not contain it. Figure 4a shows the best schedule which can be attained when the INIS $[(0,0) \to (2,2)]$, having makespan 17. Figure 4b shows the optimal schedule, having makespan 16. The optimal schedule consists of two no-idle subschedules; namely, the INIS $[(0,0) \to (1,1)]$ (containing tasks A_1 and B_1 and corresponding to arc $[(0,0),(1,1)]$ on $\mathcal{G}$), and the NIS $[(2,2) \to (4,2)]$ (containing tasks A_2, A_3, A_4 and B_2 and corresponding to arc $[(1,1),(4,2)]$ on $\mathcal{G}$). For illustrative purposes, Figure 5 shows the graph $G(1,1)$. Notice that in such a graph, $(2,1)$ is not a target node.

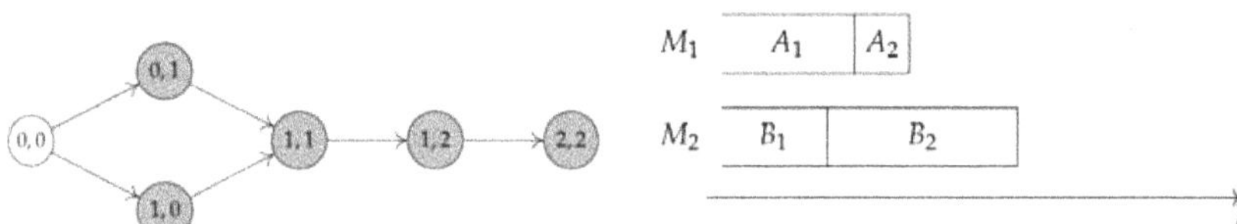

Figure 3. (**a**) The graph $G(0,0)$ in the example. (**b**) The INIS $[(0,0) \to (2,2)]$.

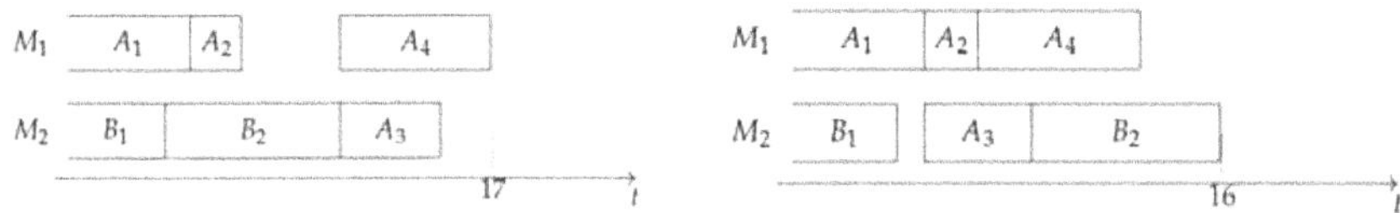

Figure 4. (**a**) The best schedule starting with the INIS $[(0,0) \to (2,2)]$, and (**b**) the optimal schedule in the example.

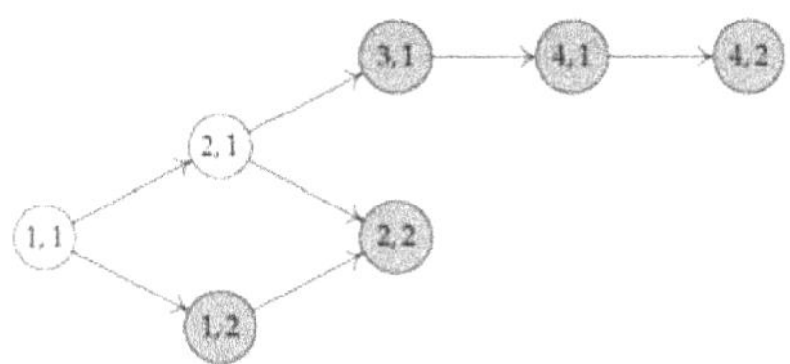

Figure 5. Graph $G(1,1)$ in the example.

3. Flow Shop

In this section we consider the flow shop problem, i.e., $F|s-prec|C_{\max}$, in which the job set J contains n jobs, and job J^k requires processing time p_j^k on machine M_j (here we use index j for both tasks and machines, as there is exactly one task per machine). While in the classical problem $F||C_{\max}$ a job cannot start on machine M_j before it is completed on M_{j-1}, in $Fm|s-prec|C_{\max}$, a job can start on machine M_j as soon as it is started on M_{j-1}.

3.1. Two-Machine Flow Shop ($F2|s-prec|C_{\max}$)

We next consider the two-machine flow shop problem, so p_1^k and p_2^k denote the processing times of job J_k on M_1 and M_2 respectively, $k = 1, \ldots, n$. Note that, as in the classical $F2||C_{\max}$, with no loss of generality we can assume that in any feasible schedule the machine M_1 processes all the jobs consecutively with no idle time between them. We next show that problem $F2|s-prec|C_{\max}$ can be solved in linear time.

Proposition 2. *Given an instance of $F2|s-prec|C_{\max}$, there always exists a schedule σ^* having makespan* $\max\{\sum_{k=1}^n p_1^k, \sum_{k=1}^n p_2^k\}$, *which is therefore optimal.*

Proof. Given an instance of $F2|s-prec|C_{\max}$, partition the jobs into two sets, J' and J'', such that $J' = \{J^k|p_1^k \le p_2^k\}$ and $J'' = J \setminus J'$. Then, build σ^* by scheduling, on both machines, first all jobs of J' in arbitrary order, and then all jobs of J'', also in arbitrary order. If we let $C(1)$ and $C(2)$ denote the completion time of the last job of J' on M_1 and M_2 respectively, one has $C(1) < C(2)$. From the definition of J', one gets that up to $C(2)$, no idle time occurs on M_2. From then on, all jobs of J'' are scheduled, and two cases may occur. (i) No idle time ever occurs on M_2, in which case the makespan equals $\max\{\sum_{k=1}^n p_1^k, \sum_{k=1}^n p_2^k\}$. ($ii$) Some idle time occurs on M_2. Consider the first time that M_2 is idle and M_1 is still processing a job J_k. Upon completion of J_k, the two machines will simultaneously start the next job, say, $J_{\bar{k}}$, but since $J_{\bar{k}} \in J''$, M_1 will still be processing it while M_2 returns idle. Since all remaining jobs belong to J'', this will happen for each job until the end of the schedule. In particular, when the last job is scheduled, again, M_2 completes first, so in conclusion, the makespan of σ^* is $\sum_{k=1}^n p_1^k$. $\square$

The above proof contains the solution algorithm. For each job J^k, put it into J' if $p_1^k \le p_2^k$ and in J'' otherwise. Then, schedule all jobs of J' followed by all jobs of J'' (in any order). Since establishing whether a job belongs to J' or J'' can be done in constant time, and since jobs can be sequenced in arbitrary order within each set, we can conclude with the following result.

Theorem 3. *$F2|s-prec|C_{\max}$ can be solved in $O(n)$.*

While $F2|s-prec|C_{\max}$ appears even simpler than the classical $F2||C_{\max}$, one may wonder whether other simplifications occur for $m > 2$. While the complexity status of $Fm|s-prec|C_{\max}$ is open, we point out a difference between $Fm||C_{\max}$ and $Fm|s-prec|C_{\max}$, which may suggest that the problem with s-precedence constraints is not necessarily easier than the classical counterpart.

It is well known [10] that in $Fm||C_{\max}$ there always exists an optimal schedule in which the job sequences on M_1 and M_2 are identical, and the same holds for machines M_{m-1} and M_m. (As a consequence, for $m \leq 3$ the optimal schedule is a permutation schedule.) This is no more true in $Fm|s-prec|C_{\max}$, even with only two jobs.

Example 2. *Consider an instance with three machines and two jobs, A and B:*

j	1	2	3
ine A	$4,M_1$	$6,M_2$	$1,M_3$
B	$10,M_1$	$4,M_2$	$9,M_3$
ine			

Scheduling the jobs in the order AB on all three machines, one gets $C_{max} = 15$, and the makespan is attained on machine M_3 (see Figure 6a). If they are scheduled in the order BA on all three machines, $C_{max} = 16$, and in this case the value of the makespan is attained on M_2 (Figure 6b). If jobs are scheduled in the order AB on M_1 and BA on M_2 and M_3, then $C_{max} = 14$ (all three machines complete at the same time, Figure 6c), and this is the optimal schedule.

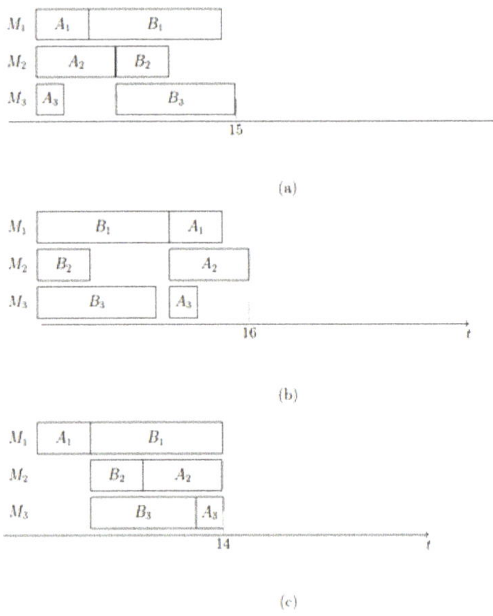

Figure 6. Three schedules for Example 2.

3.2. Flow Shop with Two Jobs and m Machines ($Fm|n = 2, s-prec|C_{\max}$)

In what follows we give an Algorithm 1 that solves $Fm|n = 2, s-prec|C_{max}$. Again we denote the two jobs with A and B, and by p_j^A and p_j^B the processing time of jobs A and B on machine $M_j, j = 1, \ldots, m$, respectively. Notice that a schedule is fully specified by the order of the two jobs on each machine, either AB or BA. In what follows, for a given schedule and a given machine, we call *leader* the job scheduled first and *follower* the other job. So if, on a given machine, the jobs are sequenced in the order AB, then, on that machine, A is the leader and B is the follower.

Algorithm 1 For finding a schedule with $C_{max} \leq K$ if it exists.

1: Initialize $F_A(0) = F_B(0) = 0$;
2: **for** $u = 1, \ldots, m$ **do**
3:     **for** $v = 1, \ldots, m$ **do**
4:        Compute $L_A(u,v)$, $L_B(u,v)$, $S_A(u,v)$ and $S_B(u,v)$ via (1), (2), (3) and (4) respectively;
5:     **end for**
6: **end for**
7: **for** $v = 1, \ldots, m$ **do**
8:     Compute $F_A(v)$ and $F_B(v)$ via (5) and (6);
9: **end for**
10: **if** $F_A(m) < +\infty$ or $F_B(m) < +\infty$ **then**
11:     $C_{max} \leq K$;
12: **else**
13:     $C_{max} > K$.
14: **end if**

Given any feasible schedule, we can associate with it a decomposition of the m machines into *blocks*, each consisting of a maximal set of consecutive machines in which the two jobs are scheduled in the same order. We denote the block consisting of machines $M_u, M_{u+1}, \ldots, M_v$ as $<M_u, M_v>$ (see Figure 7). In a block, due to the s-precedence constraints, all the tasks of the leader job start at the same time. Given a block $<M_u, M_v>$, we can compute a number of quantities. (Assume for the moment that $v < m$.) If, in $<M_u, M_v>$, A is the leader, then we call *leader span* the length of the longest A-task in the block, and denote it with $L_A(u,v)$:

$$L_A(u,v) = \max_{u \leq j \leq v} \{p_j^A\}, \tag{1}$$

and similarly, if B is the leader, the leader span is given by:

$$L_B(u,v) = \max_{u \leq j \leq v} \{p_j^B\}. \tag{2}$$

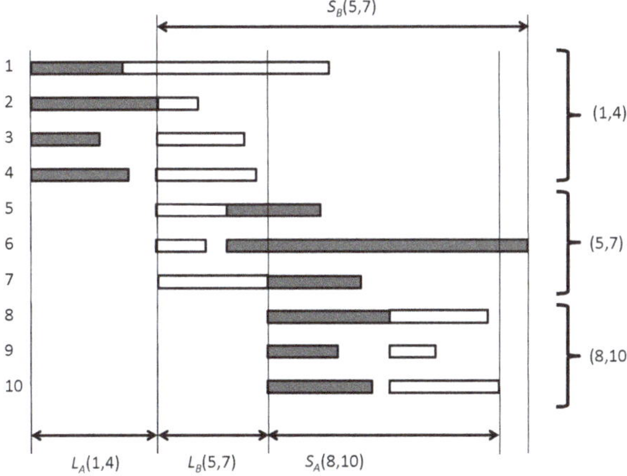

Figure 7. A sample schedule for $Fm|n = 2, s - prec|C_{\max}$. The tasks of job A are in grey.

Notice that, due to the definition of block, in the block that follows $<M_u, M_v>$, the roles of leader and follower are exchanged. Hence, the time at which the leader completes its longest task in $<M_u, M_v>$ is also the start time of the other job's tasks in the next block.

Given a block $<M_u, M_v>$, suppose again that A is the leader. We let $S_A(u,v)$ indicate the *span* of block $<M_u, M_v>$; i.e., the difference between the maximum completion time of a B-task and the start time of all A-tasks in $<M_u, M_v>$. This is given by:

$$S_A(u,v) = \max_{u \leq j \leq v} \{ \max_{u \leq h \leq j} \{p_h^A\} + p_j^B \}, \tag{3}$$

and exchanging the roles of leader and follower in $<M_u, M_v>$, we get

$$S_B(u,v) = \max_{u \leq j \leq v} \{ \max_{u \leq h \leq j} \{p_h^B\} + p_j^A \}. \tag{4}$$

Notice that trivial lower and upper bounds for the minimum makespan are given by

$$LB = \max\{ \max_{1 \leq j \leq m} \{p_j^A\}, \max_{1 \leq j \leq m} \{p_j^B\} \}$$

and

$$UB = \max_{1 \leq j \leq m} \{p_j^A\} + \max_{1 \leq j \leq m} \{p_j^B\}$$

respectively. In what follows we address the problem of determining a schedule having a makespan not larger than K, or prove that it does not exist. Assuming that all processing time values are integers, a binary search over the interval $[LB, UB]$ allows one to establish the value of the minimum makespan.

As we already observed, a relevant difference between $Fm||C_{max}$ and $Fm|s-prec|C_{max}$ is that, in a feasible schedule for $Fm|s-prec|C_{max}$, the value of C_{max} may not be attained on the last machine, but rather on any machine. This fact requires carefully handling by the algorithm.

Let $F_A(v)$ be the minimum sum of leader spans of all blocks from M_1 to M_v, when A is the leader of the last block (i.e., the block including M_v). Similarly, $F_B(v)$ is the same when B is the leader of the last block. In order to write a functional equation for $F_A(v)$ and $F_B(v)$, we introduce the notation $\delta(x) = 0$ if $x \leq 0$ and $\delta(x) = +\infty$ if $x > 0$.

Hence, we write

$$F_A(v) = \min_{0 \leq u \leq v} \{F_B(u) + L_A(u+1,v) + \delta(F_B(u) + S_A(u+1,v) - K)\}. \tag{5}$$

The first terms accounts for the fact that in the previous block the leader is B, while the rightmost term ($\delta(\cdot)$) rules out solutions in which the sum of the start time of the last block and the span of the block itself exceeds K. Symmetrically, one has:

$$F_B(v) = \min_{0 \leq u \leq v} \{F_A(u) + L_B(u+1,v) + \delta(F_A(u) + S_B(u+1,v) - K)\}. \tag{6}$$

Expressions (5) and (6) are computed for $v = 1, \ldots, m$. If at least one of the values $F_A(m)$ and $F_B(m)$ has a finite value, a schedule of makespan not exceeding K exists. The values of machine index for which each minimum in (5) and (6) is attained define the blocks of the schedule, which can, therefore, be easily backtracked.

Equations (5) and (6) must be initialized, simply letting $F_A(0) = F_B(0) = 0$.

Notice that in general one cannot infer the value of the minimum makespan schedule directly from this procedure. If the minimum in the computation of $F_A(m)$ has been attained for, say, machine M_u, it does not imply that $F_B(u) + S_A(u+1,m)$ is indeed the minimum makespan. This is because the overall makespan may be due to a previous machine, and the algorithm has no control on this aspect. For instance, in the sample schedule of Figure 7 the makespan is attained on machine M_6. However, its actual value has no relevance, so long as it does not exceed K, since it does not affect the values $F_A(v)$ and $F_B(v)$ subsequently computed.

Concerning complexity, each computation of (5) and (6) requires $O(m)$ comparisons. Since the whole procedure is repeated at each step of a binary search over $[LB, UB]$, the following result holds.

Theorem 4. *Problem $Fm|n = 2, s - prec|C_{max}$ can be solved in $O(m^2 log(UB - LB))$.*

4. Further Research

In this paper we established some preliminary complexity results for perhaps the most basic cases of shop problems with s-precedence constraints. Here, we briefly elaborate on possible research directions.

- *Job shop problem with three jobs.* The job shop problem with more than two jobs is NP-hard. This is a direct consequence of the fact that $J|s-prec|C_{max}$ can be viewed as a generalization of $J||C_{max}$, which is NP-hard with three jobs [11].

Theorem 5. *$J|n = 3, s - prec|C_{max}$ is NP-hard.*

Proof. Consider an instance I of $J|n = 3|C_{max}$, in which O_i^k denotes the i-th task of job J^k in I, having processing time p_i^k on machine $M(O_i^k)$.

We can define an instance I' of $J|n = 3, s - prec|C_{max}$ with the same number of machines. The three jobs of I' are obtained replacing each task O_i^k of I with a sequence of two tasks $O_{i'}^k$ and $O_{i''}^k$, in which $O_{i'}^k$ precedes $O_{i''}^k$, $O_{i'}^k$ has length p_i^k and requires machine $M(O_i^k)$, while $O_{i''}^k$ has sufficiently small length $\epsilon > 0$ and also requires machine $M(O_i^k)$. As a consequence, in $J|s-prec|C_{max}$, the task O_{i+1}^k cannot start before $O_{i'}^k$ is started, but since $M(O_{i'}^k) = M(O_{i''}^k) = M(O_i^k)$, this can only occur after $O_{i'}^k$ is finished. So, for sufficiently small ϵ, a feasible schedule for I' having makespan $\leq K + m\epsilon$ exists if and only if a feasible schedule for I exists having makespan $\leq K$. $\square$

Notice that the above reduction cannot be applied to $F|n = 3, s - prec|C_{max}$, since in the flow shop each job visits all machines exactly once. In fact, the complexity of $Fm|s-prec|C_{max}$ is open, even for fixed $m \geq 3$ or fixed $n \geq 3$.

- *Open problems with two jobs.* The approach in Section 2 for $J2|n = 2, s - prec|C_{max}$ cannot be trivially extended to more than two machines. The complexity of this case is open. Additionally, an open issue is whether a more efficient algorithm can be devised for $J2|n = 2, s - prec|C_{max}$, and a strongly polynomial algorithm for $Fm|n = 2, s - prec|C_{max}$.

Author Contributions: Conceptualization, A.A. F.R. and S.S.; methodology, A.A. F.R. and S.S.; validation, A.A. F.R. and S.S.; formal analysis, A.A. F.R. and S.S.; investigation, A.A. F.R. and S.S.; resources, A.A. F.R. and S.S.; writing–original draft preparation, A.A. F.R. and S.S.; writing–review and editing, A.A. F.R. and S.S.; visualization, A.A. F.R. and S.S.; supervision, A.A. F.R. and S.S.

Funding: This research received no external funding.

Conflicts of Interest: The authors declare no conflict of interest.

References

1. Kim, E.-S.; Posner, M.E. Parallel machine scheduling with s-precedence constraints. *IIE Trans.* **2010**, *42*, 525–537. [CrossRef]
2. Kim, E.-S.; Sung, C.-S.; Lee, I.-S. Scheduling of parallel machines to minimize total completion time subject to s-precedence constraints. *Comput. Oper. Res.* **2009**, *36*, 698–710. [CrossRef]

3. Kim, E.-S. Scheduling of uniform parallel machines with s-precedence constraints. *Math. Comput. Model.* **2011**, *54*, 576–583. [CrossRef]

4. Tamir, T. Scheduling with Bully Selfish Jobs. *Theory Comput. Syst.* **2012**, *50*, 124–146. [CrossRef]

5. Demeulemeester, E.L.; Herroelen, W.S. *Project Scheduling, a Research Handbook*; Springer Science & Business Media: Berlin, Germany, 2006.

6. Agnetis, A.; Lucertini, M.; Nicolò, F. Flow Management in Flexible Manufacturing Cells with Pipeline Operations. *Manag. Sci.* **1993**, *39*, 294–306. [CrossRef]

7. Avella, P.; Boccia, M.; Mannino, C.; Vasilev, I. Time-indexed formulations for the runway scheduling problem. *Transp. Sci.* **2017**, *51*, 1031–1386. [CrossRef]

8. Briskorn, D.; Angeloudis, P. Scheduling co-operating stacking cranes with predetermined container sequences. *Discret. Appl. Math.* **2016**, *201*, 70–85. [CrossRef]

9. Bertel, S.; Billaut, J.-C. A genetic algorithm for an industrial multiprocessor flow shop scheduling problem with recirculation, *Eur. J. Oper. Res.* **2004**, *159*, 651–662. [CrossRef]

10. Pinedo, M. *Scheduling, Theory, Algorithms and Systems*; Springer: Cham, Switzerland, 2018.

11. Brucker, P.; Sotskov, Y.N.; Werner, F. Complexity of shop scheduling problems with fixed number of jobs: A survey. *Math. Methods Oper. Res.* **2007**, *65*, 461–481. [CrossRef]

Article

Linking Scheduling Criteria to Shop Floor Performance in Permutation Flowshops

Jose M. Framinan [1,*] and **Rainer Leisten** [2]

1   Industrial Management, School of Engineering, University of Seville, 41004 Sevilla, Spain
2   Industrial Engineering, Faculty of Engineering, University of Duisburg-Essen, 47057 Duisburg, Germany; leisten@uni-due.de
*   Correspondence: framinan@us.es

Received: 30 October 2019; Accepted: 4 December 2019; Published: 7 December 2019

Abstract: The goal of manufacturing scheduling is to allocate a set of jobs to the machines in the shop so these jobs are processed according to a given criterion (or set of criteria). Such criteria are based on properties of the jobs to be scheduled (e.g., their completion times, due dates); so it is not clear how these (short-term) criteria impact on (long-term) shop floor performance measures. In this paper, we analyse the connection between the usual scheduling criteria employed as objectives in flowshop scheduling (e.g., makespan or idle time), and customary shop floor performance measures (e.g., work-in-process and throughput). Two of these linkages can be theoretically predicted (i.e., makespan and throughput as well as completion time and average cycle time), and the other such relationships should be discovered on a numerical/empirical basis. In order to do so, we set up an experimental analysis consisting in finding optimal (or good) schedules under several scheduling criteria, and then computing how these schedules perform in terms of the different shop floor performance measures for several instance sizes and for different structures of processing times. Results indicate that makespan only performs well with respect to throughput, and that one formulation of idle times obtains nearly as good results as makespan, while outperforming it in terms of average cycle time and work in process. Similarly, minimisation of completion time seems to be quite balanced in terms of shop floor performance, although it does not aim exactly at work-in-process minimisation, as some literature suggests. Finally, the experiments show that some of the existing scheduling criteria are poorly related to the shop floor performance measures under consideration. These results may help to better understand the impact of scheduling on flowshop performance, so scheduling research may be more geared towards shop floor performance, which is sometimes suggested as a cause for the lack of applicability of some scheduling models in manufacturing.

Keywords: scheduling; shop floor performance; flowshop; manufacturing

---

1. Introduction

To handle the complexity of manufacturing decisions, these have been traditionally addressed in a hierarchical manner, in which the overall problem is decomposed into a number of sub-problems or decision levels [1]. Given a decision level, pertinent decisions are taken according to specific local criteria. It is clear that, for this scheme to work efficiently, the decisions among levels should be aligned to contribute to the performance of the whole system. Among the different decisions involved in manufacturing, here we focus on scheduling decisions. Scheduling (some authors use the term "detailed scheduling") is addressed usually after medium-term production planning decisions have been considered, since production planning decision models do not usually make distinction between products within a family, and do not take into account sequence-dependent costs, or detailed machine capacity [2]. A short-term detailed scheduling model usually assumes that there are several jobs—each one with its own characteristics—that have to be scheduled so one or more *scheduling criteria* are minimised. The schedule is then released to the shop floor, so the events in the shop floor are executed according to the sequence and timing suggested by the schedule [3]. Therefore, there is a clear impact of the chosen scheduling criteria on (medium/long term) shop floor performance, which is eventually reflected on shop floor performance measures such as the throughput of the system (number of jobs dispatched by time unit), cycle time (average time that the jobs spend in the manufacturing system), or work in process. As these performance measures can be linked to key aspects of the competitiveness of the company (e.g., throughput is related to capacity and resource utilisation, while cycle time and work in process are related to lead times and inventory holding costs), the chosen scheduling criterion may have an important impact in the performance of the company, so it is important to assess the impact of different scheduling criteria on shop floor performance measures. However, perhaps for historical reasons, the connection between shop floor performance measures and scheduling criteria has been neglected by the literature since, to the best of our knowledge, there are not contributions addressing this topic. In general, the lack of understanding and quantification of these connections has led to a number of interrelated issues:

- Some widely employed scheduling criteria have been subject of criticism due to their apparent lack of applicability to real-world situations (see, e.g., the early comments in [4] on Johnson's famous paper, or [5] and [1] on the lack of real-life application of makespan minimisation algorithms), which suggest a poor alignment of these criteria with the companies' goals.
- Some justifications for using specific scheduling criteria are given without a formal proof. For instance, it is usual in the scheduling literature to mention that minimising the completion time in a flowshop leads to minimising work-in-process, whereas this statement—as we discuss in Section 2.2—is not correct from a theoretical point of view.
- Some scheduling criteria employed in manufacturing have been borrowed from other areas. For instance, the minimisation of the completion time variance is taken from the computer scheduling context; therefore their potential advantages on manufacturing have to be tested.
- There are different formulations for some scheduling criteria intuitively linked to shop floor performance: While machine idle time minimisation can be seen, at least approximately, as related to increasing the utilisation of the system, there are alternative, non-equivalent, manners to formulate idle time. Therefore, it remains an open question to know which formulation is actually better in terms of effectively increasing the utilisation of the system.
- Finally, since it is customary that different, conflicting goals have to be balanced in the shop floor (such as balancing work in process, and throughput), it would be interesting to know the contribution of the different scheduling criteria to shop floor performance in order to properly balance them.

Note that, in two cases, the linkages between scheduling criteria and shop floor performance measures can be theoretically established. More specifically, it can be formally proved that makespan minimisation implies maximising the throughput, and that completion time minimisation implies the minimising the average cycle time. However, for the rest of the cases such relationships cannot be theoretically proved, so they have to be tested via experimentation. To do so, in this paper we carry out an extensive computational study under a different variety of scheduling criteria, shop floor performance measures, and instance parameters.

Since the mathematical expression of the scheduling criteria is layout-dependent, we have to focus on a particular production environment. More specifically, in this paper we assume a flow shop layout where individual jobs are not committed to a specific due date. The main reason for the choice is that flow line environments are probably the most common setting in repetitive manufacturing. Regarding not considering individual due dates for jobs, it should be mentioned that both scheduling criteria and shop floor performance measures differ greatly from due date related settings to non due date related ones, and therefore this aspect must be subject of a separate analysis. Finally, we also assume that all jobs to be scheduled are known in advance.

The results of the experiments carried out in this paper show that

1. There are several scheduling criteria (most notably the completion time variance and one definition of idle time) which are poorly related with any of the indicators considered for shop floor performance.
2. Makespan minimisation is heavily oriented towards increasing throughput, but it yields poor results in terms of average completion time and work-in-process. This confines its suitability to manufacturing scenarios with very high utilisation costs as compared to those associated with cycle time and inventory.
3. Minimisation of one definition of idle times results in sequences with only a marginal worsening in terms of throughput, but a substantial improvement in terms of cycle time and inventory. Therefore, this criterion emerges as an interesting one when the alignment with shop floor performance is sought.
4. Minimisation of completion times also provides quite balanced schedules in terms of shop floor performance measures; note that it does not lead to the minimisation of WIP, as recurrently stated in the literature.

The rest of the paper is organised as follows: In the next section, the scheduling criteria and shop floor performance measures to be employed in the experimentation are discussed, as well as the theoretically provable linkages among them. The methodology adopted in the computational experience is presented in Section 3.2. The results are discussed in Section 4. Finally, Section 5 is devoted to outline the main conclusions and to highlight areas for future research.

2. Background and Related Work

In this section, we first present the usual scheduling criteria employed in the literature, while in Section 2.2 we discuss the usual shop floor performance measures, together with the relationship with the scheduling criteria that can be formally proved. For the sake of brevity, we keep the detailed explanations on both criteria and performance measures at minimum, so the interested reader is referred to the references given for formal definitions.

2.1. Scheduling Criteria

Undoubtedly, the most widely employed scheduling criterion is the makespan minimisation (usually denoted as C_{max}) or maximum flow time (see, e.g., [6] for a recent review on research in flowshop sequencing with makespan objective). Another important measure is the (total or average) total completion time or $\sum C_j$. Although less employed in scheduling research than makespan, total completion time has also received a lot of attention, particularly during the last years. Just to mention a few recent papers, we note the contributions in [7,8].

An objective also considered in the literature is the minimisation of machine idle time, which can be defined in (at least) three different ways [9]:

*   The idle time, as well as the head and tail, of every single machine, i.e., the time before the first job is started on a machine and the time after the last job is finished on a machine, but the whole schedule has started on the first machine and has not been finished yet on the last machine, can be included into the idle time or not. In a static environment, including all heads and tails means that idle time minimisation is equivalent to minimisation of makespan (see, e.g., in [4]). This case would not have to be considered further.
*   Excluding heads and tails would give an idle time within the schedule, implicitly assuming that the machines could be used for other tasks/jobs outside the current problem before and after the current schedule passes/has passed the machine. This definition of idle time is also known as "core idle time" (see, e.g., in [10–12]) and it has been used by [13] and by [14] in the context of a multicriteria problem. We denote this definition of idle time as $\sum IT_j$.

- Including machine heads in the idle time computation whereas the tails are not included means that the machines are reserved for the schedule before the first job of the schedule arrives but are released for other jobs outside the schedule as soon as the last job has left the current machine. In the following, we denote this definition as $\sum ITH_j$. This definition is first encountered in [15] and in [16] and it has been used recently as a secondary criterion for the development of tie-breaking rules for makespan minimisation algorithms (see, e.g., [17,18]).

Figure 1 illustrates these differences in idle time computation for an example of two jobs on three machines. The light grey time-periods (IT and Head) are included in our idle time definition whereas the Tail is not. In the literature, an equivalent expression for heads and tails are Front Delay and Back Delay, respectively, see in [19] or [9].

M1	p(1,1)		p(2,1)			‚Tail‘
M2		p(1,2)	IT		p(2,2)	
M3	‚Head‘			p(1,3)	IT	p(2,3)

Figure 1. Different components of machine idle time.

Finally, the last criterion under consideration is the Completion Time Variance (CTV). CTV was originally introduced by [20] in the computer scheduling context, where it is desirable to organise the data files in on-line computing systems so that the file access times are as uniform as possible. It has been subsequently applied in the manufacturing scheduling context as it is stated to be an appropriate objective for just-in-time production systems, or any other situation where a uniform treatment of the jobs is desirable (see, e.g., in [21–24]). In the flow shop/job shop scheduling context, it has been employed by [25–32].

2.2. Shop Floor Performance Measures

Shop floor performance is usually measured using different indicators. Among classical texts, Goldratt [33] mentions throughput, inventory, and operating expenses as key manufacturing performance measures. Nahmias [34] mentions the following manufacturing objectives: meet due dates, minimise WIP, minimise cycle time, and achieve a high resource utilisation. Wiendahl [35] identifies four main objectives in the production process: short lead times, low schedule deviation, low inventories, and high utilisation. Hopp and Spearman [1] list the following manufacturing objectives: high throughput, low inventory, high utilisation, short cycle times, and high product variety. Li et al. [36] cites utilisation and work-in-process as the two main managerial concerns in manufacturing systems. Finally, throughput and lateness are identified by several authors (e.g., [37,38]) as the main performance indicators in manufacturing.

Although these objectives have remained the same during decades [39], their relative importance has changed across time [40], and also depends on the specific manufacturing sector (for instance, in the semiconductor industry, average cycle time is regarded as the most important objective, see, e.g., [41] or [42]). According to the references reviewed above, we consider three performance measures: Throughput (TH), Work-In-Process (WIP), and Average Cycle Time (ACT) as shop floor performance indicators. With respect to other indicators mentioned in the reviewed references, note that one of them is not relevant in the deterministic environment to which this analysis is constrained (low schedule deviation), while other is not specifically related to shop floor operation (high product variety). Furthermore, as our study does not assume individual due dates for jobs, we exclude due date related measures, although we wish to note that, quite often short cycle times are employed as an indicator of due date adherence [38,43]. Finally, we prove below that utilisation and throughput are directly related, so utilisation does not need to be considered in addition to throughput.

Regarding the relationship of the shop floor performance measures with the scheduling criteria, it is easy to check that TH the throughput may be defined in terms of $C_{max}(S)$ the makespan of a sequence S of n jobs, i.e.,

$$TH(S) = \frac{n}{C_{max}(S)} \tag{1}$$

As a result, throughput is inversely proportional to makespan. Note that the utilisation $U(S)$ can be defined as (see, e.g., [36]):

$$U(S) = \frac{\sum_i \sum_j p_{ij}}{C_{max}(S)} \tag{2}$$

therefore, it is clear that $U(S) = \frac{\sum_i \sum_j p_{ij}}{n} \cdot TH(S)$, and, as $\frac{\sum_i \sum_j p_{ij}}{n}$ is constant for a given instance, then it can bee seen that the two indicators are fully related.

Accordingly, ACT average cycle time can be expressed in terms of the completion time, see, e.g., [44]:

$$ACT(S) = \frac{\sum C_j(S)}{n} \tag{3}$$

It follows that the total completion time is proportional to ACT. Since TH, ACT and WIP are linked through Little's law, the following equation holds.

$$WIP(S) = TH(S) \cdot ACT(S) = \frac{\sum C_j(S)}{C_{max}(S)} \tag{4}$$

From Equation (4), it may be seen that total completion time and WIP minimisation are not exactly equivalent, although it is a common statement in the scheduling flowshop literature: It is easy to show that the two criteria are equivalent for the single-machine case, but this does not necessarily hold for the flowshop case.

As, apart from the two theoretical equivalences above discussed, there are no straightforward relationship between the scheduling criteria and the shop floor performance measures, such relationships should be empirically discovered over a high number of problem instances. This computational experience must take into account that the results might be possibly influenced by the instance sizes and the processing times employed. The methodology to carry out the experimentation is described in the next section.

3. Computational Experience

The following approach is adopted to asses how the minimisation of a certain scheduling criterion impacts on the different shop floor indicators:

1. Build a number of scheduling instances of different sizes and with different mechanisms for generating the processing times. The procedure to build these test-beds is described in Section 3.1.
2. For each one of these instances, find the sequences optimising each one of the scheduling criteria under consideration. For small-sized instances, the optimal solutions can be found, while for the biggest instances, a good solution found by a heuristic approach is employed. The procedure for this step is described in Section 3.2.
3. For each one of these five optimal (or good) sequences, compute their corresponding values of TH, WIP, and ACT. This can be done in a straightforward manner according to Equations (1)–(4).
4. Analyse the so-obtained results. This is carried out in Section 4.

3.1. Testbed Setting

Although, in principle, a possible option to obtain flowshop instances to perform our research may be to extract these data from real-life settings, this option poses a number of difficulties. First, obtaining such data is a representative number is complicated. There are only few references publishing real data in the literature (see [45,46]). It may be thus required to obtain such data from primary sources, which may be a research project itself. Second, processing time data are highly industry-dependent, and it is likely that a sector-by-sector analysis would be required, which in turn makes the analysis even more complicate and increases the need of obtaining additional data. Finally, extracting these data from industry would make processing times to be external (independent) variables in the analysis.

Therefore, we generate these data according to test-bed generation methods available in the literature. For the flowshop layout in our research, this means establishing the problem size (number of jobs and machines) and processing times of each job on each machine.

With respect to the values of the number of jobs n and m machines, we have chosen the following: $n \in \{20, 50, 100, 200\}$, and $m \in \{10, 20, 50\}$. For each problem size, 30 instances have been generated. This number has been chosen so that the results have a relatively high statistical significance.

Regarding the generation of the processing times, methods for generating processing times can be classified in random and correlated. In random methods, processing times are assumed to be independent from the jobs and the machines, i.e., they are generated by sampling them from a random interval using a uniform distribution [a, b]. The most usual values for this interval are [1,99] (see, e.g., in [47,48]), while in some other cases even wider intervals are employed (e.g., [49] uses [1,200]). Random methods intend to produce difficult problem instances, as it is known that, at least with respect to certain scheduling criteria, this generation method yields the most difficult problems [50,51]. As foreseeable, random processing times are not found in practice [52]. Instead of random processing times, in real-life manufacturing environments it is encountered a mixture of job-correlation and machine-correlation for the processing times, as some surveys suggest (e.g., [53]). To model this correlation, several methods have been proposed, such as those of [54–56], or [57]. Among these, the latest method synthesises the others. This method allows obtaining problem instances with mixed correlation between jobs and machines. The amplitude of the interval from which the distribution means of the processing times are uniformly sampled depends on a parameter $\alpha \in [0,1]$. For low values of α, differences among the processing times in the machines are small, while the opposite occurs for large values of α. For a detailed description of the implementation, the reader is referred to [57].

Finally, it is to note that several works claim the Erlang distribution to better capture the distribution of processing times (e.g., [4,19], or [58]), yet these do not specify whether this has been confirmed in real-life settings. Therefore, we discard this approach.

In [57], the processing times for each job on each machine p_{ij} are generated according to the following steps.

1. Set the upper and lower bounds of processing times, Dur_{LB} and Dur_{UB}, respectively, and a factor α controlling the correlation of the processing times.
2. Obtain the value $Interval_{st}$ by drawing a uniform sample from the interval $[Dur_{LB}, Dur_{UB} + Width_{eff}]$, where $Width_{eff} = rint(\alpha \cdot (Dur_{UB} - Dur_{LB}))$.
3. For each machine j, obtain $D_j = [d_j^{lb}, d_j^{ub}] = [\mu_j - d_j^{hw}, \mu_j + d_j^{hw}]$, where μ_j is sampled from the interval $[Interval_{st}, Interval_{st} + Width_{eff}]$ and d_j^{hw} is uniformly sampled from the interval $[1, 5]$.
4. For each job i, a real value $rank_i$ is uniformly sampled from the interval $[0, 1]$. Then, the processing times p_{ij} are obtained in the following manner: $p_{ij} = rint(rank_i \cdot (d_j^{ub} - d_j^{lb})) + d_j^{lb} + \eta$, where η is a 'noise factor' obtained by uniformly sampling from the interval $[-2, 2]$.
5. p_{ij} are ensured to be within the upper and lower bounds, i.e. if $p_{ij} < Dur_{LB}$, then $p_{ij} = Dur_{LB}$. Analogously, if $p_{ij} > Dur_{UB}$, then $p_{ij} = Dur_{UB}$.

The parameter α controls the degree of correlation, so for the case $\alpha = 0.0$, there is no correlation among jobs and machines. In our research, we consider four different ways to generate processing times:

- LC (Medium Correlation): Processing times are drawn according to the procedure described above and $\alpha = 0.1$.
- MC (Medium Correlation): Processing times are drawn according to the procedure described above and $\alpha = 0.5$.
- HC (High Correlation): Processing times are drawn according to the procedure described above and $\alpha = 0.9$.
- NC (No Correlation): Processing times are drawn from a uniform distribution [1,99]. This represents the "classical" noncorrelated assumption in many scheduling papers.

3.2. Optimisation of Scheduling Criteria

For each one of the problem instances, the sequences minimising each one of the considered scheduling criteria are obtained. For small problem sizes (i.e., $n \in \{5, 10\}$), this has been done by exhaustive search. As for bigger problem sizes, using exhaustive search or any other exact method is not feasible in view of the NP-hardness of these decision problems, we have found the best sequence (with respect to each of the scheduling criteria considered) by using an efficient metaheuristic, which is allowed a long CPU time interval. More specifically, we have built a tabu search algorithm (see, e.g., [59]). The basic outline of the algorithm is as follows.

- The neighbourhood definition includes the sum of the general pairwise interchange and insertion neighbourhoods. Both neighbourhood definitions are widely used in the literature.
- The size of the tabu list L has been set to the maximum value between the number of jobs and the number of machines, i.e., $L = \max n, m$. As the size of the list is used to avoid getting trapped into local optima, the idea is keeping a list size related to the size of the neighbourhood.
- As stopping criterion, the algorithm terminates after a number of iterations without improvement. This number has been set as the minimum of $10 \cdot n$. This ensures a large minimum number of iterations, while increasing this number of iterations with the problem size.

4. Computational Results

4.1. Dominance Relationships among Scheduling Criteria

A first goal of the experiments is to establish which scheduling criterion is more related to the different shop floor performance measures. To check the statistical significance of the results, we test a number of hypotheses using a one-sided test for the differences of means of paired samples (see, e.g., [60]) for every combination of m and n. More specifically, for each pair of scheduling criteria (A, B) and a shop floor performance measure ζ, we would like to know whether the sequence resulting from the minimisation of scheduling criteria A yields a better value for ζ, denoted as $\zeta(A)$, than the sequence resulting from the minimisation of scheduling criteria B. More specifically, we want to establish the significance of the null hypothesis $H_0 : \zeta(A)$ *better than* $\zeta(B)$ to determine whether criterion A is more aligned with SF indicator ζ than criterion B, or vice versa. Note that *better than* may express different ordinal relations depending on the performance measure, i.e., it is better to have a higher TH, but it is better to have lower ACT and WIP, therefore we specifically test the following three hypotheses for every combination of scheduling criteria A and B:

$$H_0 : TH(A) > TH(B)$$

$$H_1 : TH(A) \leq TH(B)$$

with respect to throughput, and

$$H_0 : WIP(A) \leq WIP(B)$$

$$H_1 : WIP(A) > WIP(B)$$

and

$$H_0 : ACT(A) \leq ACT(B)$$

$$H_1 : ACT(A) > ACT(B)$$

with respect to average completion time and work in process, respectively.

The results for each pair of scheduling criteria (A, B) are shown in Tables 1–10 for the different testbeds, where p-values are given as the maximum level of significance to reject H_0 (p represents the limit value to reject hypothesis H_0 resulting from a t-test, i.e., for every level of significance $\alpha \leq p$, H_0 would have to be rejected, whereas for every $\alpha > p$, H_0 would not be rejected. A high p indicates that H_0 can be rejected with high level of significance, and therefore H_1 can be accepted.) To express it in an informal way: a value close to zero in the column corresponding to the performance measure ζ in the table comparing the pair of scheduling criteria (A, B) indicates that minimizing criterion A leads to better values of ζ than minimizing criterion B, whereas a high value indicates the opposite.

To make an example of the interpretation of the procedure adopted, let us take the column TH for any of the testbeds in Table 1 (all zeros). This column shows the p-values obtained by testing the null hypothesis that makespan minimisation produces solutions with higher throughput than those produced by using flowtime minimisation as a scheduling criterion. Since these p-values are zero for all problem sizes, then the null hypothesis cannot be rejected. As a consequence, we can be quite confident (statistically speaking) that makespan minimisation is more aligned with throughput increase than completion time minimisation.

In view of the results of the tables, the following comments can be done.

- Regarding Table 1, it is clear that makespan outperforms the total completion time regarding throughput, and that the total completion time outperforms the makespan regarding average cycle time. These results are known from theory and, although they could have been omitted, we include them for symmetry. The table also shows that completion time outperforms makespan with respect to work in process, a result that cannot be theoretically predicted. This results is obtained for all instance sizes and different methods to generate the processing times. As a consequence, if shop floor performance is measured using primarily one indicator, C_{max} would be the most aligned objective with respect to throughput, whereas $\sum C_j$ would be the most aligned with respect to cycle time and work in process.
- From Table 2, it can be seen that makespan outperforms $\sum ITH_j$ with respect to throughput, and, in general, with respect to ACT (with the exception of small problem instances for certain processing times' generation). Finally, regarding work in process, in general, makespan outperforms $\sum ITH_j$ if $n > m$, whereas the opposite occurs if $m \geq n$.
- Tables 3 and 4 show an interesting result: despite the problem size and/or the distribution of the processing times, makespan outperforms both $\sum IT_j$ and CTV for all three shop floor performance measures considered. This result reveals that the minimisation of CTV or $\sum IT_j$ are poorly linked to shop floor performance, as least compared to makespan minimisation.
- Table 5 show that, regardless the generation of processing times and/or the problem size, completion time performs worse than $\sum ITH_j$ for makespan, whereas it outperforms it in terms of average cycle time and work in process.
- Table 6 show that, with few exception cases, the completion time outperforms $\sum IT_j$ for all three SF indicators.
- In Table 7, a peculiar pattern can be observed: while it can be that $\sum C_j$ dominates CTV with respect to the three SF indicators, this is not the case for the random processing times, as in this case the makespan values obtained by CTV are higher than those observed for the total completion time.
- In Tables 8 and 9 it can be seen that $\sum ITH_j$ outperforms both $\sum IT_j$ and CTV for all instance sizes and all generation of the processing times. Regarding considering the heads or not in the idle time

function, this result makes clear that idle time minimisation including the heads is better with respect to all shop floor performance measures considered.

- Finally, in Table 10 it can be seen that the relative performance of $\sum IT_j$ and CTV with respect to the indicators depends on the type of testbed and on the problem instance size. However, in view of the scarce alignment of both scheduling criteria with any SF already detected in Tables 3, 4, 8 and 9, these results do not seem relevant for the purpose of our analysis.
- If a trade-off between two shop floor performance measures is sought, for each pair of indicators it is possible to represent the set of *efficient* scheduling criteria in a multi-objective manner, i.e., criteria for which no other criterion in the set obtains better results with respect to both two indicators considered. This set is represented in Table 11, and it can be seen that completion time minimisation is the only efficient criterion to minimise both WIP and ACT. In contrast, if TH is involved in the trade-off, a better value for TH (and worse for ACT and WIP) can be obtained by minimising $\sum ITH_j$, and a further better value for TH (at the expenses of worsening ACT and WIP) would be obtained by minimising C_{max}.

Table 1. Maximum level of p-values regarding the pair $(C^*_{max}, \sum C^*_j)$ for different testbeds.

		LC			MC			HC			NC		
n	m	TH	ACT	WIP	TH	ACT	WIP	TH	ACT	WIP	TH	ACT	WIP
5	5	0.0	100.0	100.0	0.0	100.0	100.0	0.0	100.0	100.0	0.0	100.0	100.0
5	10	0.0	100.0	100.0	0.0	100.0	100.0	0.0	100.0	100.0	0.0	100.0	100.0
10	5	0.0	100.0	100.0	0.0	100.0	100.0	0.0	100.0	100.0	0.0	100.0	100.0
10	10	0.0	100.0	100.0	0.0	100.0	100.0	0.0	100.0	100.0	0.0	100.0	100.0
20	10	0.0	100.0	100.0	0.0	100.0	100.0	0.0	100.0	100.0	0.0	100.0	100.0
20	20	0.0	100.0	100.0	0.0	100.0	100.0	0.0	100.0	100.0	0.0	100.0	100.0
20	50	0.0	100.0	100.0	0.0	100.0	100.0	0.0	100.0	100.0	0.0	100.0	100.0
50	10	0.0	100.0	100.0	0.0	100.0	100.0	0.0	100.0	100.0	0.0	100.0	100.0
50	20	0.0	100.0	100.0	0.0	100.0	100.0	0.0	100.0	100.0	0.0	100.0	100.0
50	50	0.0	100.0	100.0	0.0	100.0	100.0	0.0	100.0	100.0	0.0	100.0	100.0
		0.0	100.0	100.0	0.0	100.0	100.0	0.0	100.0	100.0	0.0	100.0	100.0

Table 2. p-values for rejecting the hypotheses regarding the pair $(C^*_{max}, \sum ITH^*_j)$ for different testbeds.

		LC			MC			HC			NC		
n	m	TH	ACT	WIP	TH	ACT	WIP	TH	ACT	WIP	TH	ACT	WIP
5	5	0.0	0.0	0.0	0.0	100	97.1	0.0	0.0	0.0	0.0	0.0	0.0
5	10	0.0	0.0	85.9	0.0	100	100.0	0.0	0.0	100.0	0.0	0.0	100.0
10	5	0.0	0.0	0.0	0.0	100	0.1	0.0	0.0	0.0	0.0	0.0	100.0
10	10	0.0	0.0	74.9	0.0	100	100.0	0.0	0.0	0.2	0.0	0.0	100.0
20	10	0.0	0.0	0.0	0.0	0.16	100.0	0.0	0.0	0.1	0.0	0.0	100.0
20	20	0.0	0.0	100.0	0.0	0	97.9	0.0	0.0	7.4	0.0	0.0	100.0
20	50	0.0	0.0	100.0	0.0	0	100.0	0.0	0.0	100.0	0.0	0.0	100.0
50	10	0.0	0.0	0.0	0.0	0	18.1	0.0	0.0	0.0	0.0	0.0	0.0
50	20	0.0	0.0	0.0	0.0	0	0.0	0.0	0.0	100.0	0.0	0.0	0.0
50	50	0.0	0.0	100.0	0.0	0	100.0	0.0	0.0	100.0	0.0	0.0	100.0
		0.0	0.0	46.1	0.0	40.0	71.3	0.0	0.0	40.8	0.0	0.0	70.0

Table 3. p-values for rejecting the hypotheses H_0 regarding the pair $(C^*_{max}, \sum IT^*_j)$ for different testbeds.

		LC			MC			HC			NC		
n	m	TH	ACT	WIP	TH	ACT	WIP	TH	ACT	WIP	TH	ACT	WIP
5	5	0.0	0.0	0.0	0.0	0.0	0.0	0.0	0.0	0.0	0.0	0.0	0.0
5	10	0.0	0.0	0.0	0.0	0.0	0.0	0.0	0.0	0.0	0.0	0.0	0.0
10	5	0.0	0.0	0.0	0.0	0.0	0.0	0.0	0.0	0.0	0.0	0.0	0.0
10	10	0.0	0.0	0.0	0.0	0.0	0.0	0.0	0.0	0.0	0.0	0.0	0.0
20	10	0.0	0.0	0.0	0.0	0.0	0.0	0.0	0.0	0.0	0.0	0.0	0.0

Table 3. *Cont.*

n	m	LC			MC			HC			NC		
		TH	ACT	WIP	TH	ACT	WIP	TH	ACT	WIP	TH	ACT	WIP
20	20	0.0	0.0	0.0	0.0	0.0	0.0	0.0	0.0	0.0	0.0	0.0	0.0
20	50	0.0	0.0	0.0	0.0	0.0	0.0	0.0	0.0	0.0	0.0	0.0	0.0
50	10	0.0	0.0	0.0	0.0	0.0	0.0	0.0	0.0	0.0	0.0	0.0	0.0
50	20	0.0	0.0	0.0	0.0	0.0	0.0	0.0	0.0	0.0	0.0	0.0	0.0
50	50	0.0	0.0	0.0	0.0	0.0	0.0	0.0	0.0	0.0	0.0	0.0	0.0
		0.0	0.0	0.0	0.0	0.0	0.0	0.0	0.0	0.0	0.0	0.0	0.0

Table 4. *p*-values for rejecting the hypotheses H_0 regarding the pair (C^*_{max}, CTV^*) for different testbeds.

n	m	LC			MC			HC			NC		
		TH	ACT	WIP	TH	ACT	WIP	TH	ACT	WIP	TH	ACT	WIP
5	5	0.0	0.0	0.0	0.0	0.0	0.0	0.0	0.0	0.0	0.0	0.0	0.0
5	10	0.0	0.0	0.0	0.0	0.0	0.0	0.0	0.0	0.0	0.0	0.0	0.0
10	5	0.0	0.0	0.0	0.0	0.0	0.0	0.0	0.0	0.0	0.0	0.0	0.0
10	10	0.0	0.0	0.0	0.0	0.0	0.0	0.0	0.0	0.0	0.0	0.0	0.0
20	10	0.0	0.0	0.0	0.0	0.0	0.0	0.0	0.0	0.0	0.0	0.0	0.0
20	20	0.0	0.0	0.0	0.0	0.0	0.0	0.0	0.0	0.0	0.0	0.0	0.0
20	50	0.0	0.0	0.0	0.0	0.0	0.0	0.0	0.0	0.0	0.0	0.0	0.0
50	10	0.0	0.0	0.0	0.0	0.0	0.0	0.0	0.0	0.0	0.0	0.0	0.0
50	20	0.0	0.0	0.0	0.0	0.0	0.0	0.0	0.0	0.0	0.0	0.0	0.0
50	50	0.0	0.0	0.0	0.0	0.0	0.0	0.0	0.0	0.0	0.0	0.0	0.0
		0.0	0.0	0.0	0.0	0.0	0.0	0.0	0.0	0.0	0.0	0.0	0.0

Table 5. *p*-values for rejecting the hypotheses H_0 regarding the pair $(\sum C^*_j, ITH^*)$ for different testbeds.

n	m	LC			MC			HC			NC		
		TH	ACT	WIP	TH	ACT	WIP	TH	ACT	WIP	TH	ACT	WIP
5	5	100.0	0.0	0.0	100.0	0.0	0.0	100.0	0.0	0.0	100.0	0.0	0.0
5	10	100.0	0.0	0.0	100.0	0.0	0.0	100.0	0.0	0.0	100.0	0.0	0.0
10	5	100.0	0.0	0.0	100.0	0.0	0.0	100.0	0.0	0.0	100.0	0.0	0.0
10	10	100.0	0.0	0.0	100.0	0.0	0.0	100.0	0.0	0.0	100.0	0.0	0.0
20	10	100.0	0.0	0.0	100.0	0.0	0.0	100.0	0.0	0.0	100.0	0.0	0.0
20	20	100.0	0.0	0.0	100.0	0.0	0.0	100.0	0.0	0.0	100.0	0.0	0.0
20	50	100.0	0.0	0.0	100.0	0.0	0.0	100.0	0.0	0.0	100.0	0.0	0.0
50	10	100.0	0.0	0.0	100.0	0.0	0.0	100.0	0.0	0.0	100.0	0.0	0.0
50	20	100.0	0.0	0.0	100.0	0.0	0.0	100.0	0.0	0.0	100.0	0.0	0.0
50	50	100.0	0.0	0.0	100.0	0.0	0.0	100.0	0.0	0.0	100.0	0.0	0.0
		100.0	0.0	0.0	100.0	0.0	0.0	100.0	0.0	0.0	100.0	0.0	0.0

Table 6. *p*-values for rejecting the hypotheses H_0 regarding the pair $(\sum C^*_j, IT^*)$ for different testbeds.

n	m	LC			MC			HC			NC		
		TH	ACT	WIP	TH	ACT	WIP	TH	ACT	WIP	TH	ACT	WIP
5	5	0.0	0.0	0.0	0.0	0.0	0.0	0.0	0.0	0.0	0.0	0.0	0.0
5	10	0.0	0.0	0.0	0.0	0.0	0.0	0.0	0.0	0.0	0.0	0.0	0.0
10	5	100.0	0.0	0.0	0.0	0.0	0.0	0.0	0.0	0.0	99.8	0.0	0.0
10	10	0.0	0.0	0.0	0.0	0.0	0.0	0.0	0.0	0.0	0.0	0.0	0.0
20	10	0.0	0.0	0.0	0.0	0.0	0.0	0.0	0.0	0.0	100.0	0.0	0.0
20	20	0.0	0.0	0.0	0.0	0.0	0.0	0.0	0.0	0.0	0.0	0.0	0.0
20	50	95.4	0.0	0.0	0.0	0.0	0.0	0.0	0.0	0.0	96.9	0.0	0.0
50	10	0.0	0.0	0.0	0.0	0.0	0.0	0.0	0.0	0.0	94.6	0.0	0.0
50	20	0.0	0.0	0.0	0.0	0.0	0.0	0.0	0.0	0.0	100.0	0.0	0.0
50	50	98.9	0.0	0.0	0.0	0.0	0.0	0.0	0.0	0.0	0.0	0.0	0.0
		29.4	0.0	0.0	0.0	0.0	0.0	0.0	0.0	0.0	49.1	0.0	0.0

Table 7. Maximum level of significance for rejecting the hypotheses H_0 regarding the pair $(\sum C_j^*, CTV^*)$ for different testbeds.

		LC			MC			HC			NC		
n	m	TH	ACT	WIP	TH	ACT	WIP	TH	ACT	WIP	TH	ACT	WIP
5	5	0.1	0.0	0.0	0.0	0.0	0.0	0.0	0.0	0.0	97.4	0.0	0.0
5	10	99.9	0.0	0.0	27.5	0.0	0.0	93.6	0.0	0.0	100.0	0.0	0.0
10	5	100.0	0.0	0.0	0.0	0.0	0.0	0.0	0.0	0.0	99.4	0.0	0.0
10	10	0.0	0.0	0.0	0.0	0.0	0.0	0.0	0.0	0.0	100.0	0.0	0.0
20	10	0.0	0.0	0.0	0.0	0.0	0.0	0.0	0.0	0.0	100.0	0.0	0.0
20	20	0.0	0.0	0.0	0.0	0.0	0.0	0.0	0.0	0.0	100.0	0.0	0.0
20	50	100.0	0.0	0.0	0.0	0.0	0.0	0.0	0.0	0.0	100.0	0.0	0.0
50	10	0.0	0.0	0.0	0.0	0.0	0.0	0.0	0.0	0.0	0.0	0.0	0.0
50	20	0.0	0.0	0.0	0.0	0.0	0.0	0.0	0.0	0.0	0.0	0.0	0.0
50	50	0.0	0.0	0.0	0.0	0.0	0.0	0.0	0.0	0.0	100.0	0.0	0.0
		30.0	0.0	0.0	2.8	0.0	0.0	9.4	0.0	0.0	79.7	0.0	0.0

Table 8. p-values for rejecting the hypotheses H_0 regarding the pair (ITH^*, IT^*) for different testbeds.

		LC			MC			HC			NC		
n	m	TH	ACT	WIP	TH	ACT	WIP	TH	ACT	WIP	TH	ACT	WIP
5	5	0.0	0.0	0.0	0.0	0.0	0.0	0.0	0.0	0.0	0.0	0.0	0.0
5	10	0.0	0.0	0.0	0.0	0.0	0.0	0.0	0.0	0.0	0.0	0.0	0.0
10	5	100.0	0.0	0.0	0.0	0.0	0.0	0.0	0.0	0.0	99.8	0.0	0.0
10	10	0.0	0.0	0.0	0.0	0.0	0.0	0.0	0.0	0.0	0.0	0.0	0.0
20	10	0.0	0.0	0.0	0.0	0.0	0.0	0.0	0.0	0.0	100.0	0.0	0.0
20	20	0.0	0.0	0.0	0.0	0.0	0.0	0.0	0.0	0.0	0.0	0.0	0.0
20	50	95.4	0.0	0.0	0.0	0.0	0.0	0.0	0.0	0.0	96.9	0.0	0.0
50	10	0.0	0.0	0.0	0.0	0.0	0.0	0.0	0.0	0.0	94.6	0.0	0.0
50	20	0.0	0.0	0.0	0.0	0.0	0.0	0.0	0.0	0.0	100.0	0.0	0.0
50	50	98.9	0.0	0.0	0.0	0.0	0.0	0.0	0.0	0.0	0.0	0.0	0.0
		29.4	0.0	0.0	0.0	0.0	0.0	0.0	0.0	0.0	49.1	0.0	0.0

Table 9. p-values for rejecting the hypotheses H_0 regarding the pair (ITH_j^*, CTV^*) for different testbeds.

		LC			MC			HC			NC		
n	m	TH	ACT	WIP	TH	ACT	WIP	TH	ACT	WIP	TH	ACT	WIP
5	5	0.0	0.0	0.0	0.0	0.0	0.0	0.0	0.0	0.0	0.0	0.0	0.0
5	10	0.0	0.0	0.0	0.0	0.0	0.0	0.0	0.0	0.0	0.0	0.0	0.0
10	5	100.0	0.0	0.0	0.0	0.0	0.0	0.0	0.0	0.0	99.8	0.0	0.0
10	10	0.0	0.0	0.0	0.0	0.0	0.0	0.0	0.0	0.0	0.0	0.0	0.0
20	10	0.0	0.0	0.0	0.0	0.0	0.0	0.0	0.0	0.0	100.0	0.0	0.0
20	20	0.0	0.0	0.0	0.0	0.0	0.0	0.0	0.0	0.0	0.0	0.0	0.0
20	50	95.4	0.0	0.0	0.0	0.0	0.0	0.0	0.0	0.0	96.9	0.0	0.0
50	10	0.0	0.0	0.0	0.0	0.0	0.0	0.0	0.0	0.0	94.6	0.0	0.0
50	20	0.0	0.0	0.0	0.0	0.0	0.0	0.0	0.0	0.0	100.0	0.0	0.0
50	50	98.9	0.0	0.0	0.0	0.0	0.0	0.0	0.0	0.0	0.0	0.0	0.0
		29.4	0.0	0.0	0.0	0.0	0.0	0.0	0.0	0.0	49.1	0.0	0.0

Table 10. p-values for rejecting the hypotheses H_0 regarding the pair (IT^*, CTV^*) for different testbeds.

		LC			MC			HC			NC		
n	m	TH	ACT	WIP	TH	ACT	WIP	TH	ACT	WIP	TH	ACT	WIP
5	5	100.0	100.0	100.0	100.00	100.0	67.97	7.29	0.03	0.00	100.0	100.0	100.0
5	10	100.0	100.0	0.0	100.0	100.00	0.00	100.0	100.0	0.0	100.0	100.0	0.0
10	5	0.0	99.96	100.0	6.96	100.0	100.0	0.0	100.0	100.0	7.94	100.0	100.0
10	10	100.0	100.0	100.0	100.0	100.00	98.09	100.00	100.0	100.0	100.0	100.0	100.0
20	10	100.0	100.0	100.0	100.0	100.0	100.0	100.0	100.0	100.0	0.0	100.0	100.0
20	20	100.0	100.0	100.0	100.0	100.0	100.0	100.0	100.0	100.0	100.0	100.0	100.0
20	50	100.0	100.0	64.3	100.0	100.0	100.0	100.0	100.0	100.0	100.0	100.0	0.0
50	10	100.0	100.0	100.0	100.0	100.0	100.0	8.4	100.0	100.0	0.0	100.0	100.0
50	20	1.1	100.0	100.0	100.0	100.0	100.0	100.0	100.0	100.0	0.0	100.0	100.0
50	50	0.0	1.2	100.0	100.0	100.0	100.0	0.0	100.0	100.0	100.0	100.0	100.0
		70.1	89.0	86.4	100.0	100.0	100.0	63.6	100.0	88.9	66.7	100.0	80.0

Table 11. Efficient criteria for each pair of SF indicators.

SF Indicators	Efficient Scheduling Criteria
(WIP, ACT)	$\sum C_j$
(WIP, TH)	$\sum C_j, \sum ITH_j, C_{max}$
(ACT, TH)	$\sum C_j, \sum ITH_j, C_{max}$

4.2. Ranking of Scheduling Criteria

In this section, we further try to explore the trade-off among the different criteria by answering the following question: Once we choose certain scheduling criterion according to the aforementioned ranking, how are the gains (or losses) that we can expect in the different shop floor performance measures when we switch from one scheduling criterion to another. More formally, we intend to quantify the difference between picking one scheduling criterion or another for a given shop floor performance measure. To address this issue, we define the RD_{PM} or Relative Deviation with respect to a given PM (performance measure) in the following manner.

$$RD(A)_{PM} = \frac{PM(S_A) - PM(S_{A^+})}{PM(S_{A^+})} \cdot 100 \qquad (5)$$

where $PM(S_A)$ is the value of PM obtained for the sequence S_A which minimises scheduling criterion A. Analogously, S_{A^+} is the sequence obtained by minimising scheduling criterion A^+, being A^+ the scheduling criterion ranking immediately behind A for the performance measurement PM. When A is the scheduling criterion ranking last for PM, then RD is set to zero.

Note that this definition of RD allows us to obtain more information than the mere rank of scheduling criteria. For instance, let us consider the scheduling criteria A, B, and C, which rank (ascending order) with respect to the performance measure PM in the following manner: B, C, A. This information (already obtained in Section 4.1) simply states that B (C) is more aligned that C (A) with respect to performance measure PM, but does not convey information on whether there are substantial differences between the three criteria for PM, or not. This information can be obtained by measuring the corresponding RD: If $RD(B)$ is zero or close to zero, it implies that B and C yield similar values for PM, and therefore there is not so much difference (with respect to PM) between minimizing B, or C. In contrast, a high value of $RD(C)$ indicates a great benefit (with respect to PM) when switching from minimizing A to minimizing C.

Since RD is defined for a specific instance, we use the Average Relative Deviation (ARD) for comparison across the testbed, consisting in averaging the RDs. The results of the experiments for the different testbeds with respect to ARD are shown in Tables 12–15, together with the rank of each criterion for each problem size. In addition, the cumulative ARD of the scheduling criteria for each shop floor performance measures are shown in Figure 2 for the different testbed. In view of the results, we give the following comments.

Table 12. Average Relative Deviation (ARD) and ranks (in parentheses) of the scheduling criteria for the random test-bed.

n	m	TH					ACT					WIP				
		C_{max}	$\sum C_j$	$\sum ITH_j$	$\sum IT_j$	CTV	C_{max}	$\sum C_j$	$\sum ITH_j$	$\sum IT_j$	CTV	C_{max}	$\sum C_j$	$\sum ITH_j$	$\sum IT_j$	CTV
5	5	2.10	3.68	6.03	0.00	0.42	0.55	13.44	4.39	0.00	0.96	2.52	6.34	9.24	0.00	4.74
		(1)	(4)	(2)	(5)	(3)	(2)	(1)	(3)	(5)	(4)	(2)	(1)	(3)	(5)	(4)
5	10	2.79	1.89	1.48	0.00	1.50	2.33	9.88	1.24	0.64	0.00	1.51	5.85	5.38	0.00	2.66
		(1)	(4)	(2)	(5)	(3)	(3)	(1)	(2)	(4)	(5)	(2)	(1)	(3)	(5)	(4)
10	5	1.82	0.00	8.51	0.13	0.37	4.98	18.48	0.54	0.00	1.43	1.24	10.03	12.90	0.00	1.00
		(1)	(5)	(2)	(3)	(4)	(3)	(1)	(2)	(5)	(4)	(2)	(1)	(3)	(5)	(4)
10	10	5.13	1.30	2.66	0.00	2.36	2.17	13.59	2.71	0.00	0.88	2.26	7.20	6.87	0.00	4.37
		(1)	(4)	(2)	(5)	(3)	(3)	(1)	(2)	(5)	(4)	(2)	(1)	(3)	(5)	(4)
20	10	3.77	0.00	5.21	0.45	2.17	2.51	19.88	0.97	0.00	3.47	2.68	11.21	8.72	0.00	2.21
		(1)	(5)	(2)	(3)	(4)	(3)	(1)	(2)	(5)	(4)	(2)	(1)	(3)	(5)	(4)
20	20	5.71	0.67	1.93	0.00	1.92	1.53	14.62	1.01	0.00	1.36	4.47	7.21	4.14	0.00	3.83
		(1)	(4)	(2)	(5)	(3)	(3)	(1)	(2)	(5)	(4)	(2)	(1)	(3)	(5)	(4)
20	50	4.79	0.00	3.27	0.13	0.02	0.28	10.66	1.83	0.30	0.00	2.84	5.20	2.36	0.00	2.90
		(1)	(5)	(3)	(4)	(2)	(3)	(1)	(2)	(4)	(5)	(2)	(1)	(3)	(5)	(4)
50	10	1.70	1.87	5.61	0.11	0.00	1.95	23.84	2.92	0.00	4.02	3.75	18.67	9.11	0.00	2.03
		(1)	(4)	(2)	(3)	(5)	(2)	(1)	(3)	(5)	(4)	(2)	(1)	(3)	(5)	(4)
50	20	5.11	0.45	3.49	0.66	0.00	0.53	17.96	3.41	0.00	0.18	5.56	10.31	4.26	0.00	2.37
		(1)	(4)	(2)	(3)	(5)	(2)	(1)	(4)	(5)	(3)	(2)	(1)	(3)	(5)	(4)
50	50	6.04	0.57	0.53	0.00	1.27	1.12	11.93	1.21	0.00	0.03	4.50	6.12	1.75	0.00	2.96
		(1)	(4)	(2)	(5)	(3)	(4)	(1)	(2)	(5)	(3)	(2)	(1)	(3)	(5)	(4)
		3.90	1.04	3.87	0.15	1.00	1.80	15.43	2.02	0.09	1.23	3.13	8.81	6.47	0.00	2.91

Table 13. Average Relative Deviation (ARD) and ranks (in parentheses) of the scheduling criteria for the LC test-bed.

n	m	TH					ACT					WIP				
		C_{max}	$\sum C_j$	$\sum ITH_j$	$\sum IT_j$	CTV	C_{max}	$\sum C_j$	$\sum ITH_j$	$\sum IT_j$	CTV	C_{max}	$\sum C_j$	$\sum ITH_j$	$\sum IT_j$	CTV
5	5	0.08	0.06	0.33	0.00	0.15	0.05	0.83	0.32	0.00	0.05	0.14	0.42	0.71	0.00	0.20
		(1)	(3)	(2)	(5)	(4)	(2)	(1)	(3)	(5)	(4)	(2)	(1)	(3)	(5)	(4)
5	10	0.08	0.09	0.13	0.00	0.03	0.07	0.45	0.00	0.02	0.00	0.08	0.22	0.22	0.00	0.09
		(1)	(4)	(2)	(5)	(3)	(3)	(1)	(2)	(4)	(5)	(2)	(1)	(3)	(5)	(4)
10	5	0.03	0.00	0.31	0.05	0.04	0.03	1.08	0.39	0.00	0.10	0.06	0.66	0.74	0.00	0.05
		(1)	(5)	(2)	(4)	(3)	(2)	(1)	(3)	(5)	(4)	(2)	(1)	(3)	(5)	(4)
10	10	0.15	0.15	0.32	0.00	0.16	0.31	1.00	0.00	0.00	0.27	0.15	0.55	0.77	0.00	0.42
		(1)	(3)	(2)	(5)	(4)	(3)	(1)	(2)	(5)	(4)	(2)	(1)	(3)	(5)	(4)
20	10	0.09	0.14	0.29	0.00	0.07	0.19	1.33	0.44	0.00	0.21	0.27	0.96	0.87	0.00	0.28
		(1)	(3)	(2)	(5)	(4)	(2)	(1)	(3)	(5)	(4)	(2)	(1)	(3)	(5)	(4)
20	20	0.34	0.26	0.83	0.00	0.14	0.25	2.54	0.05	0.00	0.63	0.29	1.46	1.39	0.00	0.74
		(1)	(3)	(2)	(5)	(4)	(3)	(1)	(2)	(5)	(4)	(2)	(1)	(3)	(5)	(4)
20	50	0.29	0.00	0.39	0.02	0.02	0.05	0.89	0.19	0.00	0.00	0.10	0.38	0.63	0.00	0.03
		(1)	(5)	(2)	(4)	(3)	(3)	(1)	(2)	(5)	(4)	(2)	(1)	(3)	(5)	(4)
50	10	0.24	0.30	0.88	0.00	0.24	0.26	10.83	0.98	0.00	1.26	0.49	9.72	2.06	0.00	1.64
		(1)	(3)	(2)	(5)	(4)	(2)	(1)	(3)	(5)	(4)	(2)	(1)	(3)	(5)	(4)
50	20	0.07	0.08	0.13	0.01	0.00	0.07	0.94	0.20	0.00	0.14	0.13	0.74	0.42	0.00	0.14
		(1)	(3)	(2)	(4)	(5)	(2)	(1)	(3)	(5)	(4)	(2)	(1)	(3)	(5)	(4)
50	50	0.52	0.42	1.20	0.06	0.00	0.33	3.37	0.31	0.00	0.42	0.21	1.97	2.24	0.05	0.00
		(1)	(4)	(2)	(3)	(5)	(3)	(1)	(2)	(5)	(4)	(2)	(1)	(3)	(4)	(5)
		0.19	0.15	0.48	0.01	0.09	0.16	2.33	0.29	0.00	0.31	0.19	1.71	1.01	0.01	0.36

Table 14. Average Relative Deviation (ARD) and ranks (in parentheses) of the scheduling criteria for the MC test-bed.

n	m	TH C_{max}	$\sum C_j$	$\sum ITH_j$	$\sum IT_j$	CTV	ACT C_{max}	$\sum C_j$	$\sum ITH_j$	$\sum IT_j$	CTV	WIP C_{max}	$\sum C_j$	$\sum ITH_j$	$\sum IT_j$	CTV
5	5	0.06	0.07	0.19	0.00	0.11	0.20	0.52	0.01	0.00	0.00	0.05	0.28	0.48	0.00	0.10
		(1)	(3)	(2)	(5)	(4)	(3)	(1)	(2)	(5)	(4)	(2)	(1)	(3)	(5)	(4)
5	10	0.16	0.01	0.39	0.00	0.24	0.08	0.79	0.10	0.11	0.00	0.05	0.35	0.74	0.00	0.12
		(1)	(3)	(2)	(5)	(4)	(3)	(1)	(2)	(4)	(5)	(2)	(1)	(3)	(5)	(4)
10	5	0.03	0.05	0.21	0.01	0.00	0.02	0.86	0.17	0.00	0.19	0.04	0.60	0.44	0.00	0.18
		(1)	(3)	(2)	(4)	(5)	(3)	(1)	(2)	(5)	(4)	(2)	(1)	(3)	(5)	(4)
10	10	0.17	0.15	0.27	0.00	0.17	0.39	0.80	0.09	0.00	0.02	0.08	0.46	0.91	0.00	0.18
		(1)	(3)	(2)	(5)	(4)	(3)	(1)	(2)	(5)	(4)	(2)	(1)	(3)	(5)	(4)
20	10	0.07	0.07	0.16	0.00	0.06	0.23	0.99	0.04	0.00	0.14	0.02	0.81	0.51	0.00	0.20
		(1)	(3)	(2)	(5)	(4)	(3)	(1)	(2)	(5)	(4)	(2)	(1)	(3)	(5)	(4)
20	20	0.09	0.08	0.22	0.00	0.07	0.08	0.77	0.01	0.00	0.20	0.08	0.47	0.39	0.00	0.26
		(1)	(3)	(2)	(5)	(4)	(3)	(1)	(2)	(5)	(4)	(2)	(1)	(3)	(5)	(4)
20	50	0.18	0.06	0.44	0.00	0.06	0.07	0.87	0.12	0.00	0.04	0.05	0.38	0.69	0.00	0.09
		(1)	(3)	(2)	(5)	(4)	(3)	(1)	(2)	(5)	(4)	(2)	(1)	(3)	(5)	(4)
50	10	0.05	0.10	0.13	0.00	0.04	0.01	2.05	0.31	0.00	0.30	0.06	1.88	0.53	0.00	0.36
		(1)	(3)	(2)	(5)	(4)	(3)	(1)	(2)	(5)	(4)	(2)	(1)	(3)	(5)	(4)
50	20	0.09	0.11	0.10	0.00	0.05	0.02	1.12	0.21	0.00	0.25	0.11	0.94	0.43	0.00	0.29
		(1)	(3)	(2)	(5)	(4)	(3)	(1)	(2)	(5)	(4)	(2)	(1)	(3)	(5)	(4)
50	50	0.09	0.07	0.16	0.00	0.04	0.09	0.58	0.05	0.00	0.09	0.04	0.39	0.35	0.00	0.13
		(1)	(3)	(2)	(5)	(4)	(3)	(1)	(2)	(5)	(4)	(2)	(1)	(3)	(5)	(4)
		0.10	0.08	0.23	0.00	0.08	0.12	0.94	0.11	0.01	0.12	0.06	0.66	0.55	0.00	0.19

Table 15. Average Relative Deviation (ARD) and ranks (in parentheses) of the scheduling criteria for the HC test-bed.

n	m	TH					ACT					WIP				
		C_{max}	$\sum C_j$	$\sum ITH_j$	$\sum IT_j$	CTV	C_{max}	$\sum C_j$	$\sum ITH_j$	$\sum IT_j$	CTV	C_{max}	$\sum C_j$	$\sum ITH_j$	$\sum IT_j$	CTV
5	5	0.05	0.08	0.26	0.01	0.00	0.07	0.59	0.16	0.05	0.00	0.11	0.29	0.48	0.04	0.00
		(1)	(3)	(2)	(4)	(5)	(2)	(1)	(3)	(4)	(5)	(2)	(1)	(3)	(4)	(5)
5	10	0.16	0.24	0.20	0.00	0.02	0.12	0.52	0.02	0.03	0.00	0.12	0.18	0.38	0.00	0.16
		(1)	(4)	(2)	(5)	(3)	(3)	(1)	(2)	(4)	(5)	(2)	(1)	(3)	(5)	(4)
10	5	0.01	0.06	0.35	0.05	0.00	0.03	0.99	0.33	0.00	0.14	0.03	0.67	0.66	0.00	0.12
		(1)	(3)	(2)	(4)	(5)	(2)	(1)	(3)	(5)	(4)	(2)	(1)	(3)	(5)	(4)
10	10	0.08	0.14	0.20	0.00	0.09	0.01	0.61	0.22	0.00	0.13	0.09	0.31	0.56	0.00	0.22
		(1)	(3)	(2)	(5)	(4)	(2)	(1)	(3)	(5)	(4)	(2)	(1)	(3)	(5)	(4)
20	10	0.07	0.08	0.17	0.00	0.08	0.02	0.97	0.26	0.00	0.24	0.07	0.72	0.50	0.00	0.28
		(1)	(3)	(2)	(5)	(4)	(2)	(1)	(3)	(5)	(4)	(2)	(1)	(3)	(5)	(4)
20	20	0.11	0.07	0.18	0.00	0.12	0.00	0.70	0.11	0.00	0.19	0.11	0.42	0.35	0.00	0.30
		(1)	(3)	(2)	(5)	(4)	(2)	(1)	(3)	(5)	(4)	(2)	(1)	(3)	(5)	(4)
20	50	0.11	0.11	0.20	0.00	0.05	0.01	0.43	0.07	0.00	0.07	0.03	0.20	0.37	0.00	0.12
		(1)	(3)	(2)	(5)	(4)	(3)	(1)	(2)	(4)	(5)	(2)	(1)	(3)	(5)	(4)
50	10	0.02	0.03	0.12	0.00	0.00	0.03	1.27	0.21	0.00	0.13	0.06	1.14	0.37	0.00	0.14
		(1)	(3)	(2)	(4)	(5)	(2)	(1)	(3)	(5)	(4)	(2)	(1)	(3)	(5)	(4)
50	20	0.05	0.06	0.11	0.00	0.04	0.14	0.98	0.01	0.00	0.16	0.04	0.83	0.32	0.00	0.19
		(1)	(3)	(2)	(5)	(4)	(3)	(1)	(2)	(5)	(4)	(2)	(1)	(3)	(5)	(4)
50	50	0.07	0.06	0.19	0.03	0.00	0.05	0.67	0.04	0.00	0.13	0.02	0.45	0.37	0.00	0.09
		(1)	(3)	(2)	(4)	(5)	(3)	(1)	(2)	(5)	(4)	(2)	(1)	(3)	(5)	(4)
		0.07	0.09	0.20	0.01	0.04	0.05	0.77	0.14	0.01	0.12	0.07	0.52	0.44	0.00	0.16

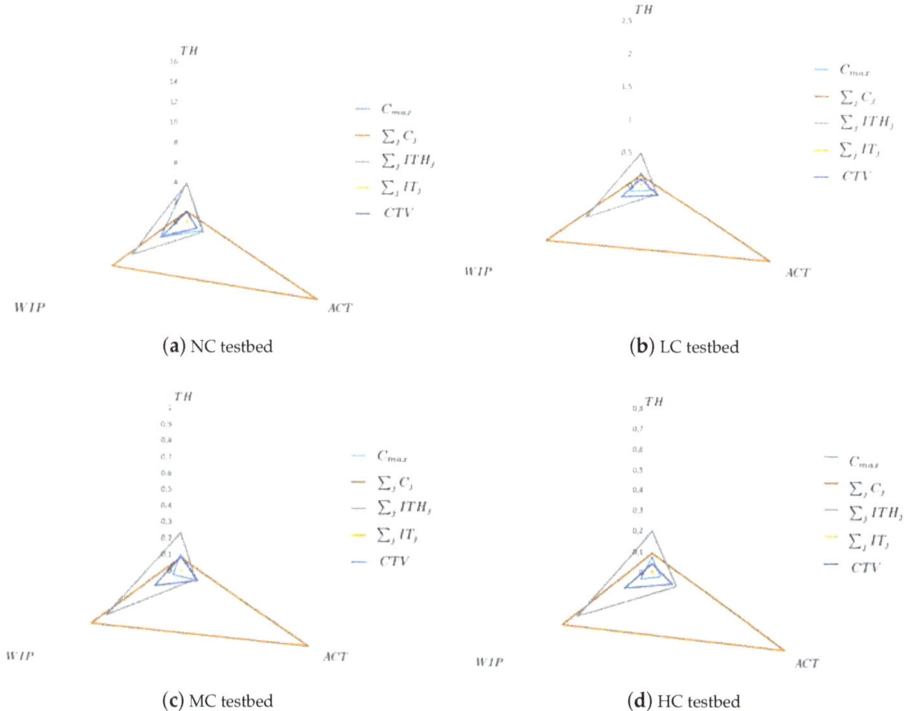

(a) NC testbed

(b) LC testbed

(c) MC testbed

(d) HC testbed

Figure 2. Relative performance of the criteria for the different testbeds.

- $\sum ITH_j$ emerges as an interesting criterion as its performance is only marginally worse than C_{max} with respect to TH—particularly in the NC testbed, see Figure 2a, but it obtains better values regarding ACT and WIP. Similarly, although it performs worse than $\sum C_j$ for ACT and WIP, it performs better in terms of throughput.
- The differences in ARD for throughput are, in general, smaller than those for ACT and WIP. For the correlated test-beds (LC to HC), the differences never reach 1%. This speaks for the little difference between minimising any of the scheduling measures if throughput maximisation is sought. The highest differences are encountered for the random test-bed (~6%).
- The differences in all measures for structured instances are smaller than for random test-bed. For instance, whereas makespan ranks first for TH (theoretically predictable), the maximum ARD for a given problem size in the random test-bed is 6.04%, whereas this is reduced to 0.52% for LC, and to 0.16% for HC. Analogously, the maximum differences between the completion time (ranking first for ACT) and the next criterion raise up to 23.84% for the random test-bed while dropping to 1.27% for HC. This means that the structured problems are easier than random problems because the distribution of the processing times flattens the objective functions, at least with respect to the considered shop floor performance measures.

5. Conclusions and Further Research

An extensive computational study has been carried out in order to analyse the links between several scheduling criteria in a flowshop and well-known shop floor performance measures. These results give some insights into the nature of these links, which can be summarised as follows.

- Roughly speaking, we could divide the considered scheduling criteria into two big categories: those tightly related to any (some) shop floor performance measure, and those poorly related to

SF performance. Among the later, we may classify CTV and $\sum IT_j$. Nevertheless, this is not meant to say that these criteria are not useful. However, from a shop floor performance perspective, it may be interesting to investigate whether these scheduling criteria relate to other performance measures. Perhaps extending the analysis to a due date scenario might yield some positive answer.

- Makespan matches (as theoretical predicted) throughput maximisation better than any other considered criteria. However, it turns out that differences between its minimisation and the minimisation of other criteria with respect to throughput are very small. Additionally, given the relatively poor performance of makespan with respect to ACT, one might ask whether makespan minimisation pays off for many manufacturing scenarios in terms of shop floor performance as compared, e.g., to completion time or $\sum ITH_j$ minimisation. A positive answer seems to be confined to these scenarios where costs associated to cycle time are almost irrelevant as compared to costs related to machine utilisation. The fact that this situation is not common in many manufacturing scenarios may lead to the lack of practical descriptions on the application of this criteria already discussed by [4].

- Completion time minimisation matches extremely well both work in process and average cycle time minimisation (the latter being theoretical predictable), better than any other criteria. In addition, the rest of the scheduling criteria perform much worse. Therefore, completion time minimisation emerges as a major criterion when it comes to increase shop floor performance. This empirical reasoning indicates the interest of the research on completion time minimisation rather than on other criteria, at least within the flowshop scheduling context.

- The minimisation of idle time (including the heads) performs better than completion time with respect to throughput. However, its performance is substantially worse than completion time regarding ACT and WIP. Hence, it seems an interesting criterion when throughput maximisation is the most important performance measure but work-in-process costs are not completely irrelevant.

- With respect to the influence of the test-bed design on the results, there are noticeable differences between the overall results obtained in the correlated test-beds (LC-HC), and those obtained from the random test-bed. In general, the introduction of structured processing times seems to reduce the differences between the scheduling criteria. At a first glance, this means that random processing times make it difficult to achieve a good shop floor performance by the application of a specific scheduling criterion. It is widely know that random problems produce difficult instances in the sense that there were high differences between bad and good schedules (with respect to a given scheduling criterion), at least for the makespan criterion. In view of the results of the experiments, we can also assert that these also translate into shop floor performance measures.

From these results, some aspects warrant future research:

- $\sum ITH_j$ emerges as an interesting scheduling criterion, with virtues in between makespan and completion time. For most of the problem settings, it compares to makespan in terms of cycle time, and it outperforms total completion time in terms of throughput. In view of these results, perhaps it is interesting devoting more efforts to flowshop minimisation with this criterion, which so far has been used only as a secondary tie-breaking rule. Interestingly, the results in this paper might suggest that its excellent performance in terms of tie-breaking rule is motivated by its alignment with shop floor performance.

- While it is possible to perfectly match the shop floor objectives of throughput and average cycle time with scheduling criteria (makespan and completion time, respectively), WIP cannot be linked to a scheduling criterion in a straightforward manner. Although the minimisation of completion time achieves the best results for WIP minimisation among the tested criteria, "true" work-in-process optimization is not the same as completion time minimisation. Here, the quotient between total completion time and makespan emerges as a "combined" scheduling criteria which may be worth of research as it matches an important shop floor performance measure such as work-in-process minimisation.

- The results of the present study are limited by the shop layout (i.e., the permutation flowshop) and the scheduling criteria (i.e., not due date-related criteria) considered. Therefore, an obvious extension of this study is to analyse other environments and scheduling measures. Particularly, the inclusion of due date related criteria could provide some additional insights on the linkage between these and the shop floor performance measures, as well as between the due date and non-due date scheduling criteria.

Author Contributions: Methodology, J.M.F. and R.L.; Writing—original draft, J.M.F.; Writing —review & editing, J.M.F. and R.L.

Funding: This research received no external funding.

Conflicts of Interest: The authors declare no conflict of interest.

References

1. Hopp, W.; Spearman, M. *Factory Physics. Foundations of Manufacturing Management*, 3rd ed.; Irwin: New York, NY, USA, 2008.
2. Framinan, J.; Leisten, R.; Ruiz, R. *Manufacturing Scheduling Systems: An Integrated View on Models, Methods and Tools*; Springer: Berlin/Heidelberg, Germany, 2014; pp. 1–400.
3. Aytug, H.; Lawley, M.A.; McKay, K.; Mohan, S.; Uzsoy, R. Executing production schedules in the face of uncertainties: A review and some future directions. *Eur. J. Oper. Res.* **2005**, *161*, 86–110. [CrossRef]
4. Conway, R.; Maxwell, W.L.; Miller, L.W. *Theory of Scheduling*; Dover: Mineola, NY, USA, 1967.
5. Dudek, R.A.; Panwalkar, S.S.; Smith, M.L. The Lessons of Flowshop Scheduling Research. *Oper. Res.* **1992**, *40*, 7–13, [CrossRef]
6. Fernandez-Viagas, V.; Ruiz, R.; Framinan, J. A new vision of approximate methods for the permutation flowshop to minimise makespan: State-of-the-art and computational evaluation. *Eur. J. Oper. Res.* **2017**, *257*, 707–721. [CrossRef]
7. Fernandez-Viagas, V.; Framinan, J. A beam-search-based constructive heuristic for the PFSP to minimise total flowtime. *Comput. Oper. Res.* **2017**, *81*, 167–177. [CrossRef]
8. Fernandez-Viagas, V.; Framinan, J. A new set of high-performing heuristics to minimise flowtime in permutation flowshops. *Comput. Oper. Res.* **2015**, *53*, 68–80. [CrossRef]
9. Framinan, J.; Leisten, R.; Rajendran, C. Different initial sequences for the heuristic of Nawaz, Enscore and Ham to minimize makespan, idletime or flowtime in the static permutation flowshop sequencing problem. *Int. J. Prod. Res.* **2003**, *41*, 121–148. [CrossRef]
10. Benkel, K.; Jørnsten, K.; Leisten, R. Variability aspects in flowshop scheduling systems. In Proceedings of the 2015 International Conference on Industrial Engineering and Systems Management (IESM), Seville, Spain, 21–23 October 2015; pp. 118–127.
11. Maassen, K.; Perez-Gonzalez, P.; Framinan, J.M. Relationship between common objective functions, idle time and waiting time in permutation flowshop scheduling. In Proceedings of the 29th European Conference on Operational Research (EURO 2018), Valencia, Spain, 8–11 July 2018.
12. Maassen, K.; Perez-Gonzalez, P. Diversity of processing times in permutation flow shop scheduling problems. In Proceedings of the 66th Operations Research Conference, Dresden, Germany, 3–6 September 2019.
13. Liao, C.J.; Tseng, C.T.; Luarn, P. A discrete version of particle swarm optimization for flowshop scheduling problems. *Comput. Oper. Res.* **2007**, *34*, 3099–3111. [CrossRef]
14. Liu, W.; Jin, Y.; Price, M. A new Nawaz-Enscore-Ham-based heuristic for permutation flow-shop problems with bicriteria of makespan and machine idle time. *Eng. Optim.* **2016**, *48*, 1808–1822. [CrossRef]
15. Sridhar, J.; Rajendran, C. Scheduling in flowshop and cellular manufacturing systems with multiple objectives-a genetic algorithmic approach. *Prod. Plan. Control* **1996**, *7*, 374–382. [CrossRef]
16. Ho, J.; Chang, Y.L. A new heuristic for the n-job, M-machine flow-shop problem. *Eur. J. Oper. Res.* **1991**, *52*, 194–202. [CrossRef]
17. Fernandez-Viagas, V.; Framinan, J. On insertion tie-breaking rules in heuristics for the permutation flowshop scheduling problem. *Comput. Oper. Res.* **2014**, *45*, 60–67. [CrossRef]
18. Fernandez-Viagas, V.; Framinan, J. A best-of-breed iterated greedy for the permutation flowshop scheduling problem with makespan objective . *Comput. Oper. Res.* **2019**, *112*, 104767. [CrossRef]

19. King, J.; Spachis, A. Heuristics for flow-shop scheduling. *Int. J. Prod. Res.* **1980**, *18*, 345–357. [CrossRef]
20. Merten, A.; Muller, M. Variance minimization in single machine sequencing problems. *Manag. Sci.* **1972**, *18*, 518–528. [CrossRef]
21. Kanet, J.J. Minimizing variation of flow time in single machine systems. *Manag. Sci.* **1981**, *27*, 1453–1464. [CrossRef]
22. Baker, K.R.; Scudder, G.D. Sequencing with earliness and tardiness penalties. A review. *Oper. Res.* **1990**, *38*, 22–36. [CrossRef]
23. Gupta, M.; Gupta, Y.; Bector, C. Minimizing the flow-time variance in single-machine systems. *J. Oper. Res. Soc.* **1990**, *41*, 767–779. [CrossRef]
24. Cai, X.; Cheng, T. Multi-machine scheduling with variance minimization. *Discret. Appl. Math.* **1998**, *84*, 55–70. [CrossRef]
25. Cai, X. V-shape property for job sequences that minimize the expected completion time variance. *Eur. J. Oper. Res.* **1996**, *91*, 118–123. [CrossRef]
26. Marangos, C.; Govande, V.; Srinivasan, G.; Zimmers, E., Jr. Algorithms to minimize completion time variance in a two machine flowshop. *Comput. Ind. Eng.* **1998**, *35*, 101–104. [CrossRef]
27. Gowrishankar, K.; Rajendran, C.; Srinivasan, G. Flow shop scheduling algorithms for minimizing the completion time variance and the sum of squares of completion time deviations from a common due date. *Eur. J. Oper. Res.* **2001**, *132*, 643–665. [CrossRef]
28. Leisten, R.; Rajendran, C. Variability of completion time differences in permutation flow shop scheduling. *Comput. Oper. Res.* **2015**, *54*, 155–167. [CrossRef]
29. Ganesan, V.; Sivakumar, A.; Srinivasan, G. Hierarchical minimization of completion time variance and makespan in jobshops. *Comput. Oper. Res.* **2006**, *33*, 1345–1367. [CrossRef]
30. Gajpal, Y.; Rajendran, C. An ant-colony optimization algorithm for minimizing the completion-time variance of jobs in flowshops. *Int. J. Prod. Econ.* **2006**, *101*, 259–272. [CrossRef]
31. Krishnaraj, J.; Pugazhendhi, S.; Rajendran, C.; Thiagarajan, S. A modified ant-colony optimisation algorithm to minimise the completion time variance of jobs in flowshops. *Int. J. Prod. Res.* **2012**, *50*, 5698–5706. [CrossRef]
32. Krishnaraj, J.; Pugazhendhi, S.; Rajendran, C.; Thiagarajan, S. Simulated annealing algorithms to minimise the completion time variance of jobs in permutation flowshops. *Int. J. Ind. Syst. Eng.* **2019**, *31*, 425–451. [CrossRef]
33. Goldratt, E. *The Haystack Syndrome: Shifting Information out of the Data Ocean*; North River Press: Croton-on-Hudson, NY, USA, 1996.
34. Nahmias, S. *Production and Operations Analysis*; Irwin: Homewood, IL, USA, 1993.
35. Wiendahl, H.P. *Load-Oriented Manufacturing Control*; Springer: Berlin/Heidelberg, Germany, 1995.
36. Li, W.; Dai, H.; Zhang, D. The Relationship between Maximum Completion Time and Total Completion Time in Flowshop Production. *Procedia Manuf.* **2015**, *1*, 146–156. [CrossRef]
37. Land, M. Parameters and sensitivity in workload control. *Int. J. Prod. Econ.* **2006**, *104*, 625–638. [CrossRef]
38. Thürer, M.; Stevenson, M.; Land, M.; Fredendall, L. On the combined effect of due date setting, order release, and output control: An assessment by simulation. *Int. J. Prod. Res.* **2019**, *57*, 1741–1755. [CrossRef]
39. Land, M. Workload in Job Shop, Grasping the Tap. Ph.D. Thesis, University of Groningen, Groningen, The Netherlands, 2004.
40. Wiendahl, H.P.; Glassner, J.; Petermann, D. Application of load-oriented manufacturing control in industry. *Prod. Plan. Control* **1992**, *3*, 118–129. [CrossRef]
41. Grewal, N.S.; Bruska, A.C.; Wulf, T.M.; Robinson, J.K. Integrating targeted cycle-time reduction into the capital planning process. In Proceedings of the 1998 Winter Simulation Conference, Washington, DC, USA, 13–16 December 1998; Volume 2, pp. 1005–1010.
42. Leachman, R.; Kang, J.; Lin, V. SLIM: Short cycle time and low inventory in manufacturing at Samsung electronics. *Interfaces* **2002**, *32*, 61–77. [CrossRef]
43. Sandell, R.; Srinivasan, K. Evaluation of lot release policies for semiconductor manufacturing systems. In Proceedings of the 1996 Winter Simulation Conference, Coronado, CA, USA, 8–11 December 1996; pp. 1014–1022.
44. Abedini, A.; Li, W.; Badurdeen, F.; Jawahir, I. Sustainable production through balancing trade-offs among three metrics in flow shop scheduling. *Procedia CIRP* **2019**, *80*, 209–214. [CrossRef]

45. Bestwick, P.F.; Hastings, N. New bound for machine scheduling. *Oper. Res. Q.* **1976**, *27*, 479–487. [CrossRef]
46. Lahiri, S.; Rajendran, C.; Narendran, T. Evaluation of heuristics for scheduling in a flowshop: A case study. *Prod. Plan. Control* **1993**, *4*, 153–158. [CrossRef]
47. Taillard, E. Benchmarks for Basic Scheduling Problems. *Eur. J. Oper. Res.* **1993**, *64*, 278–285. [CrossRef]
48. Vallada, E.; Ruiz, R.; Framinan, J. New hard benchmark for flowshop scheduling problems minimising makespan. *Eur. J. Oper. Res.* **2015**. *240*, 666–677. [CrossRef]
49. Demirkol, E.; Mehta, S.; Uzsoy, R. Benchmarks for shop scheduling problems. *Eur. J. Oper. Res.* **1998**, *109*, 137–141. [CrossRef]
50. Campbell, H.G.; Dudek, R.A.; Smith, M.L. A Heuristic Algorithm for the *n* Job, *m* Machine Sequencing Problem. *Manag. Sci.* **1970**, *16*, B-630–B-637. [CrossRef]
51. Dannenbring, D.G. An evaluation of flowshop sequencing heuristics. *Manag. Sci.* **1977**, *23*, 1174–1182. [CrossRef]
52. Amar, A.D.; Gupta, J. Simulated versus real life data in testing the efficiency of scheduling algorithms. *IIE Trans.* **1986**, *18*, 16–25. [CrossRef]
53. Panwalkar, S.S.; Dudek, R.; Smith, M.L. Sequencing research and the industrial scheduling problem. In *Symposium on the Theory of Scheduling and Its Applications*; Springer: Berlin/Heidelberg, Germany, 1973; pp. 29–38.
54. Rinnooy Kan, A. *Machine Scheduling Problems*; Martinus Nijhoff: The Hague, The Netherlands, 1976.
55. Lageweg, B.; Lenstra, J.; Rinnooy Kan, A. A general bounding scheme for the permutation flow-shop problem. *Oper. Res.* **1978**, *26*, 53–67. [CrossRef]
56. Reeves, C. A genetic algorithm for flowshop sequencing. *Comput. Oper. Res.* **1995**, *22*, 5–13. [CrossRef]
57. Watson, J.P.; Barbulescu, L.; Whitley, L.; Howe, A. Contrasting structured and random permutation flow-shop scheduling problems: Search-space topology and algorithm perfomance. *INFORMS J. Comput.* **2002**, *14*, 98–123. [CrossRef]
58. Park, Y.; Pegden, C.; Enscore, E. A survey and evaluation of static flowshop scheduling heuristics. *Int. J. Prod. Res.* **1984**, *22*, 127–141. [CrossRef]
59. Hoos, H.H.; Stützle, T. *Stochastic Local Search: Foundations and Applications*; Elsevier: Amsterdam, The Netherlands, 2005.
60. Montgomery, D.C. *Design and Analysis of Experiments*; John Wiley & Sons: Hoboken, NJ, USA, 2006.

Article

Two-Machine Job-Shop Scheduling Problem to Minimize the Makespan with Uncertain Job Durations

Yuri N. Sotskov [1,*], **Natalja M. Matsveichuk** [2] and **Vadzim D. Hatsura** [3]

[1] United Institute of Informatics Problems, National Academy of Sciences of Belarus, Surganova Street 6, 220012 Minsk, Belarus
[2] Department of Automated Production Management Systems, Belarusian State Agrarian Technical University, Nezavisimosti Avenue 99, 220023 Minsk, Belarus; matsveichuk@tut.by
[3] Department of Electronic Computing Machines, Belarusian State University of Informatics and Radioelectronics, P. Brovki Street 6, 220013 Minsk, Belarus; vadimgatsura@gmail.com
* Correspondence: sotskov48@mail.ru; Tel.: +375-17-284-2120

Received: 30 October 2019; Accepted: 16 December 2019; Published: 20 December 2019

Abstract: We study two-machine shop-scheduling problems provided that lower and upper bounds on durations of n jobs are given before scheduling. An exact value of the job duration remains unknown until completing the job. The objective is to minimize the makespan (schedule length). We address the issue of how to best execute a schedule if the job duration may take any real value from the given segment. Scheduling decisions may consist of two phases: an off-line phase and an on-line phase. Using information on the lower and upper bounds for each job duration available at the off-line phase, a scheduler can determine a minimal dominant set of schedules (DS) based on sufficient conditions for schedule domination. The DS optimally covers all possible realizations (scenarios) of the uncertain job durations in the sense that, for each possible scenario, there exists at least one schedule in the DS which is optimal. The DS enables a scheduler to quickly make an on-line scheduling decision whenever additional information on completing jobs is available. A scheduler can choose a schedule which is optimal for the most possible scenarios. We developed algorithms for testing a set of conditions for a schedule dominance. These algorithms are polynomial in the number of jobs. Their time complexity does not exceed $O(n^2)$. Computational experiments have shown the effectiveness of the developed algorithms. If there were no more than 600 jobs, then all 1000 instances in each tested series were solved in one second at most. An instance with 10,000 jobs was solved in 0.4 s on average. The most instances from nine tested classes were optimally solved. If the maximum relative error of the job duration was not greater than 20%, then more than 80% of the tested instances were optimally solved. If the maximum relative error was equal to 50%, then 45% of the tested instances from the nine classes were optimally solved.

Keywords: scheduling; uncertain duration; flow-shop; job-shop; makespan criterion

---

1. Introduction

A lot of real-life scheduling problems involve different forms of uncertainties. For dealing with uncertain scheduling problems, several approaches have been developed in the literature. A stochastic approach assumes that durations of the jobs are random variables with specific probability distributions known before scheduling. There are two types of stochastic scheduling problems [1], where one is on stochastic jobs and another is on stochastic machines. In the stochastic job problem, each job duration is assumed to be a random variable following a certain probability distribution. With an objective of minimizing the expected makespan, the flow-shop problem was considered in References [2–4]. In the stochastic machine problem, each job duration is a constant, while each completion time of the job is a random variable due to the machine breakdown or nonavailability. In References [5–7], flow-shop problems to stochastically minimize the makespan or total completion time have been considered.

If there is no information to determine a probability distribution for each random duration of the job, other approaches have to be used [8–10]. In the approach of seeking a robust schedule [8,11–13], a decision maker prefers a schedule that hedges against the worst-case scenario. A fuzzy approach [14–16] allows a scheduler to find best schedules with respect to fuzzy durations of the jobs. A stability approach [17–20] is based on the stability analysis of optimal schedules to possible variations of the durations. In this paper, we apply the stability approach to the two-machine job-shop scheduling problem with given segments of job durations. We have to emphasize that uncertainties of the job durations considered in this paper are due to external forces in contrast to scheduling problems with controllable durations [21–23], where the objective is to determine optimal durations (which are under the control of a decision maker) and to find an optimal schedule for the jobs with optimal durations.

2. Contributions and New Results

We study the two-machine job-shop scheduling problem with uncertain job durations and address the issue of how to best execute a schedule if each duration may take any value from the given segment. The main aim is to determine a minimal dominant set of schedules (DS) that would contain at least one optimal schedule for each feasible scenario of the distribution of durations of the jobs.

It is shown how an uncertain two-machine job-shop problem may be decomposed into two uncertain two-machine flow-shop problems. We prove several sufficient conditions for the existence of a small dominant set of schedules. In particular, the sufficient and necessary conditions are proven for the existence of a single pair of job permutations, which is optimal for the two-machine job-shop problem with any possible scenario. We investigated properties of the optimal pairs of job permutations for the uncertain two-machine job-shop problem.

In the stability approach, scheduling decisions may consist of two phases: an off-line phase and an on-line phase. Using information on the lower and upper bounds on each job duration available at the off-line phase, a scheduler can determine a small (or minimal) dominant set of schedules based on sufficient conditions for schedule dominance. The DS optimally covers all scenarios in the sense that, for each possible scenario, there exists at least one schedule in the DS that is optimal. The DS enables a scheduler to quickly make an on-line scheduling decision whenever additional information on completing some jobs becomes available. The stability approach enables a scheduler to choose a schedule, which is optimal for the most possible scenarios.

In this paper, we develop algorithms for testing a set of conditions for a schedule dominance. The developed algorithms are polynomial in the number of jobs. Their asymptotic complexities do not exceed $O(n^2)$, where n is a number of the jobs. Computational experiments have shown effectiveness of the developed algorithms: if there were no more than 600 jobs, then all 1000 instances in each tested series were solved in no more than one second. For the tested series of instances with 10,000 jobs, all 1000 instances of a series were solved in 344 seconds at most (on average, 0.4 s per one instance).

The paper is organized as follows. In Section 3, we present settings of the uncertain scheduling problems. The related literature and closed results are discussed in Section 4. In Section 4.2, we describe in detail the results published for the uncertain two-machine flow-shop problem. These results are used in Section 5, where we investigate properties of the optimal job permutations used for processing a set of the given jobs. Some proofs of the claims are given in Appendix A. In Section 6, we develop algorithms for constructing optimal schedules if the proven dominance conditions hold. In Section 7, we report on the wide computational experiments for solving a lot of randomly generated instances. Tables with the obtained computational results are presented in Appendix B. The paper is concluded in Section 8, where several directions for further researches are outlined.

3. Problem Settings and Notations

Using the notation $\alpha|\beta|\gamma$ [24], the two-machine job-shop scheduling problem with minimizing the makespan is denoted as $J2|n_i \leq 2|C_{max}$, where $\alpha = J2$ denotes a job-shop system with two available machines, n_i is the number of stages for processing a job, and $\gamma = C_{max}$ denotes the criterion of minimizing the makespan. In the problem $J2|n_i \leq 2|C_{max}$, the set $\mathcal{J} = \{J_1, J_2, ..., J_n\}$ of the given jobs have to be processed on machines from the set $\mathcal{M} = \{M_1, M_2\}$. All jobs are available for processing from the initial time $t = 0$. Let O_{ij} denote an operation of the job $J_i \in \mathcal{J}$ processed on machine $M_j \in \mathcal{M}$. Each machine can process a job $J_i \in \mathcal{J}$ no more than once provided that preemption of each operation O_{ij} is not allowed. Each job $J_i \in \mathcal{J}$ has its own processing order (machine route) on the machines in $\mathcal{M}$.

Let $\mathcal{J}_{1,2}$ denote a subset of the set $\mathcal{J}$ of the jobs with the same machine route (M_1, M_2), i.e., each job $J_i \in \mathcal{J}_{1,2}$ has to be processed first on machine M_1 and then on machine M_2. Let $\mathcal{J}_{2,1} \subseteq \mathcal{J}$ denote a subset of the jobs with the opposite machine route (M_2, M_1). Let $\mathcal{J}_k \subseteq \mathcal{J}$ denote a set of the jobs, which has to be processed only on machine $M_k \in \mathcal{M}$. The partition $\mathcal{J} = \mathcal{J}_1 \cup \mathcal{J}_2 \cup \mathcal{J}_{1,2} \cup \mathcal{J}_{2,1}$ holds. We denote $m_h = |\mathcal{J}_h|$, where $h \in \{1; 2; 1, 2; 2, 1\}$.

We first assume that the duration p_{ij} of each operation O_{ij} is fixed before scheduling. The considered criterion C_{max} is the minimization of the makespan (schedule length) as follows:

$$C_{max} := \min_{s \in S} C_{max}(s) = \min_{s \in S}\{\max\{C_i(s) : J_i \in \mathcal{J}\}\},$$

where $C_i(s)$ denotes a completion time of the job $J_i \in \mathcal{J}$ in the schedule s and S denotes a set of semi-active schedules existing for the problem $J2|n_i \leq 2|C_{max}$. A schedule is called semi-active if no job (operation) can be processed earlier without changing the processing order or violating some given constraints [1,25,26].

Jackson [27] proved that the problem $J2|n_i \leq 2|C_{max}$ is polynomially solvable and that the optimal schedule for this problem may be determined as a pair (π', π'') of the job permutations (calling it a *Jackson's pair of permutations*) such that $\pi' = (\pi_{1,2}, \pi_1, \pi_{2,1})$ is a sequence of all jobs from the set $\mathcal{J}_1 \cup \mathcal{J}_{1,2} \cup \mathcal{J}_{2,1}$ processed on machine M_1 and $\pi'' = (\pi_{2,1}, \pi_2, \pi_{1,2})$ is a sequence of all jobs from the set $\mathcal{J}_2 \cup \mathcal{J}_{1,2} \cup \mathcal{J}_{2,1}$ processed on machine M_2. Job J_j belongs to the permutation π_h if $J_j \in \mathcal{J}_h$.

In a Jackson's pair (π', π'') of the job permutations, the order for processing jobs from set $\mathcal{J}_1$ (from set $\mathcal{J}_2$, respectively) may be arbitrary, while for the permutation $\pi_{1,2}$, the following inequality holds for all indexes k and m, $1 \leq k < m \leq m_{1,2}$:

$$\min\{p_{i_k 1}, p_{i_m 2}\} \leq \min\{p_{i_m 1}, p_{i_k 2}\} \tag{1}$$

(for the permutation $\pi_{2,1}$, the following inequality holds for all indexes k and m, $1 \leq k < m \leq m_{2,1}$) [28]:

$$\min\{p_{j_k 2}, p_{j_m 1}\} \leq \min\{p_{j_m 2} p_{j_k 1}\} \tag{2}$$

The aim of this paper is to investigate the uncertain two-machine job-shop scheduling problem. Therefore, we next assume that duration p_{ij} of each operation O_{ij} is unknown before scheduling; namely, in the realization of a schedule, a value of p_{ij} may be equal to any real number no less than the given lower bound l_{ij} and no greater than the given upper bound u_{ij}. Furthermore, it is assumed that probability distributions of random durations of the jobs are unknown before scheduling. Such a job-shop scheduling problem is denoted as $J2|l_{ij} \leq p_{ij} \leq u_{ij}, n_i \leq 2|C_{max}$. The problem $J2|l_{ij} \leq p_{ij} \leq u_{ij}, n_i \leq 2|C_{max}$ is called an uncertain scheduling problem in contrast to the deterministic scheduling problem $J2|n_i \leq 2|C_{max}$. Let a set of all possible vectors $p = (p_{1,1}, p_{1,2}, \ldots, p_{n1}, p_{n2})$ of the job durations be determined as follows: $T = \{p : l_{ij} \leq p_{ij} \leq u_{ij}, J_i \in \mathcal{J}, M_j \in \mathcal{M}\}$. Such a vector $p = (p_{1,1}, p_{1,2}, \ldots, p_{n1}, p_{n2}) \in T$ of the possible durations of the jobs is called a scenario.

It should be noted that the problem $J2|l_{ij} \leq p_{ij} \leq u_{ij}, n_i \leq 2|C_{max}$ is mathematically incorrect. Indeed, in most cases, a single pair of job permutations which is optimal for all possible scenarios

$p \in T$ for the uncertain problem $J2|l_{ij} \leq p_{ij} \leq u_{ij}, n_i \leq 2|C_{max}$ does not exist. Therefore, in the general case, one cannot find an optimal solution for this uncertain scheduling problem.

For a fixed scenario $p \in T$, the uncertain problem $J2|l_{ij} \leq p_{ij} \leq u_{ij}, n_i \leq 2|C_{max}$ turns into the deterministic problem $J2|n_i \leq 2|C_{max}$ associated with scenario p. The latter deterministic problem is an individual one and we denote it as $J2|p, n_i \leq 2|C_{max}$. For any fixed scenario $p \in T$, there exists a Jackson's pair of the job permutations that is optimal for the individual deterministic problem $J2|p, n_i \leq 2|C_{max}$ associated with scenario p.

Let $S_{1,2}$ denote a set of all permutations of $m_{1,2}$ jobs from the set $\mathcal{J}_{1,2}$, where $|S_{1,2}| = m_{1,2}!$. Let $S_{2,1}$ denote a set of all permutations of $m_{2,1}$ jobs from the set $\mathcal{J}_{2,1}$, where $|S_{2,1}| = m_{2,1}!$. Let $S = <S_{1,2}, S_{2,1}>$ be a subset of the Cartesian product $(S_{1,2}, \pi_1, S_{2,1}) \times (S_{2,1}, \pi_2, S_{1,2})$ such that each element of the set S is a pair of job permutations $(\pi', \pi'') \in S$, where $\pi' = (\pi^i_{1,2}, \pi_1, \pi^j_{2,1})$ and $\pi'' = (\pi^j_{2,1}, \pi_2, \pi^i_{1,2})$, $1 \leq i \leq m_{1,2}!, 1 \leq j \leq m_{2,1}!$. The set S determines all semi-active schedules and vice versa.

Remark 1. *As an order for processing jobs from set $\mathcal{J}_1$ (from set $\mathcal{J}_2$) may be arbitrary in the Jackson's pair of job permutations (π', π''), in what follows, we fix both permutations π_1 and π_2 in the increasing order of the indexes of their jobs. Thus, both permutations π_1 and π_2 are now fixed, and so their upper indexes are omitted in each permutation from the pair $(\pi', \pi'') = ((\pi^i_{1,2}, \pi_1, \pi^j_{2,1}), (\pi^j_{2,1}, \pi_2, \pi^i_{1,2}))$.*

Due to Remark 1, the equality $|S| = m_{1,2}! \cdot m_{2,1}!$ holds. The following definition is used for a J-solution for the uncertain problem $J2|l_{ij} \leq p_{ij} \leq u_{ij}, n_i \leq 2|C_{max}$.

Definition 1. *A minimal (with respect to the inclusion) set of pairs of job permutations $S(T) \subseteq S$ is called a J-solution for the problem $J2|l_{ij} \leq p_{ij} \leq u_{ij}, n_i \leq 2|C_{max}$ with set J of the given jobs if, for each scenario $p \in T$, the set $S(T)$ contains at least one pair $(\pi', \pi'') \in S$ of the job permutations, which is optimal for the individual deterministic problem $J2|p, n_i \leq 2|C_{max}$ associated with scenario p.*

From Definition 1, it follows that, for any proper subset S' of the set $S(T)$ $S' \subset S(T)$, there exists at least one scenario $p' \in T$ such that set S' does not contain an optimal pair of job permutations for the individual deterministic problem $J2|p', n_i \leq 2|C_{max}$ associated with scenario p', i.e., set $S(T)$ is a minimal (with respect to the inclusion) set possessing the property indicated in Definition 1.

The uncertain job-shop problem $J2|l_{ij} \leq p_{ij} \leq u_{ij}, n_i \leq 2|C_{max}$ is a generalization of the uncertain flow-shop problem $F2|l_{ij} \leq p_{ij} \leq u_{ij}|C_{max}$, where all jobs from the set $\mathcal{J}$ have the same machine route. Two flow-shop problems are associated with the individual job-shop problem $J2|l_{ij} \leq p_{ij} \leq u_{ij}, n_i \leq 2|C_{max}$. In one of these flow-shop problems, an optimal schedule for processing jobs $\mathcal{J}_{1,2}$ has to be determined, i.e., $\mathcal{J}_{2,1} = \mathcal{J}_1 = \mathcal{J}_2 = \emptyset$. In another flow-shop problem, an optimal schedule for processing jobs $\mathcal{J}_{2,1}$ has to be determined, i.e., $\mathcal{J}_{1,2} = \mathcal{J}_1 = \mathcal{J}_2 = \emptyset$. Thus, a solution of the problem $J2|l_{ij} \leq p_{ij} \leq u_{ij}, n_i \leq 2|C_{max}$ may be based on solutions of the two associated problems $F2|l_{ij} \leq p_{ij} \leq u_{ij}|C_{max}$ with job set $\mathcal{J}_{1,2}$ and with job set $\mathcal{J}_{2,1}$.

The permutation $\pi_{1,2}$ of all jobs from set $\mathcal{J}_{1,2}$ (the permutation $\pi_{2,1}$ of all jobs from set $\mathcal{J}_{2,1}$, respectively) is called a *Johnson's permutation*, if the inequality in Equation (1) holds for the permutation $\pi_{1,2}$ (the inequality in Equation (2) holds for the permutation $\pi_{2,1}$, respectively). As it is proven in Reference [28], a Johnson's permutation is optimal for the deterministic problem $F2||C_{max}$.

4. A Literature Review and Closed Results

In this section, we address uncertain shop-scheduling problems if it is impossible to obtain probability distributions for random durations of the given jobs. In particular, we consider the uncertain two-machine flow-shop problem with the objective of minimizing the makespan. This problem is well studied and there are a lot of results published in the literature, unlike the uncertain job-shop problem.

4.1. Uncertain Shop-Scheduling Problems

The stability approach was proposed in Reference [17] and developed in Reference [18,29–31] for the C_{max} criterion, and in References [19,32–35] for the total completion time criterion $\sum C_i := \min_{s \in S} \sum_{J_i \in \mathcal{J}} C_i(s)$. The stability approach combines a stability analysis of the optimal schedules, a multi-stage decision framework, and the solution concept of a minimal dominant set $S(T)$ of schedules, which optimally covers all possible scenarios. The main aim of the stability approach is to construct a schedule which remains optimal for most scenarios of the set T. The minimality of the dominant set $S(T)$ is useful for the two-phase scheduling described in Reference [36].

At the off-line phase, one can construct set $S(T)$, which enables a scheduler to make a quick scheduling decision at the on-line phase whenever additional local information becomes available. The knowledge of the minimal dominant set $S(T)$ enables a scheduler to execute best a schedule and may end up executing a schedule optimally in many cases of the problem $F2|l_{ij} \leq p_{ij} \leq u_{ij}|C_{max}$ [36]. In Reference [17], a formula for calculating the *stability radius* of an optimal schedule is proven, i.e., the largest value of independent variations of the job durations in a schedule such that this schedule remains optimal. In Reference [19], a stability analysis of a schedule minimizing the total completion time was exploited in the branch-and-bound method for solving the job-shop problem $Jm|l_{ij} \leq p_{ij} \leq u_{ij}| \sum C_i$ with m machines. In Reference [29], for the two-machine flow-shop problem $F2|l_{ij} \leq p_{ij} \leq u_{ij}|C_{max}$, sufficient conditions have been identified when the transposition of two jobs minimizes the makespan.

Reference [37] addresses the total completion time objective in the flow-shop problem with uncertain durations of the jobs. A geometrical algorithm has been developed for solving the flow-shop problem $Fm|l_{ij} \leq p_{ij} \leq u_{ij}, n = 2| \sum C_i$ with m machines and two jobs. For this problem with two or three machines, sufficient conditions are determined such that the transposition of two jobs minimizes $\sum C_i$. Reference [38] is devoted to the case of separate setup times with the criterion of minimizing the makespan or total completion time. The job durations are fixed while each setup time is relaxed to be a distribution-free random variable within the given lower and upper bounds. Local and global dominance relations have been determined for the flow-shop problem with two machines.

Since, for the problem $F2|l_{ij} \leq p_{ij} \leq u_{ij}|C_{max}$ there often does not exist a single permutation of n jobs $\mathcal{J} = \mathcal{J}_{1,2}$ which remains optimal for all possible scenarios, an additional criterion may be introduced for dealing with uncertain scheduling problems. In Reference [39], a robust solution minimizing the worst-case deviation from optimality was proposed to hedge against uncertainties. While the deterministic problem $F2||C_{max}$ is polynomially solvable (the optimal Johnson's permutation may be constructed for the problem $F2||C_{max}$ in $O(n \log n)$ time), finding a job permutation minimizing the worst-case regret for the uncertain counterpart with a finite set of possible scenarios is NP hard.

In Reference [40], a binary NP hardness has been proven for finding a pair $(\pi_k, \pi_k) \in S$ of identical job permutations that minimizes the worst-case absolute regret for the uncertain two-machine flow-shop problem with the criterion C_{max} even for two possible scenarios. Minimizing the worst-case regret implies a time-consuming search over the set of $n!$ job permutations. In order to overcome this computational complexity in some cases, it is useful to consider a minimal dominant set of schedules $S(T)$ instead of the whole set S. To solve the flow-shop problem $F2|l_{ij} \leq p_{ij} \leq u_{ij}|C_{max}$ with job set $\mathcal{J}$, one can restrict a search within the set $S(T)$.

We next describe in detail the results published for the flow-shop problem $F2|l_{ij} \leq p_{ij} \leq u_{ij}|C_{max}$ since we use them for solving the job-shop problem $J2|l_{ij} \leq p_{ij} \leq u_{ij}, n_i \leq 2|C_{max}$ in Sections 5–7.

4.2. Closed Results

Since each permutation π' uniquely determines a set of the earliest completion times $C_i(\pi')$ of the jobs $J_i \in \mathcal{J}$ for the problem $F2||C_{max}$, one can identify the permutation π', $((\pi', \pi') \in S)$, with the semi-active schedule [1,25,26] determined by the permutation π'. Thus, the set S becomes a set of $n!$ pairs (π', π') of identical permutations of $n = m_{1,2}$ jobs from the set $\mathcal{J} = \mathcal{J}_{1,2}$ since the order for processing jobs $\mathcal{J}_{1,2}$ on both machines may be the same in the optimal schedule [28]. Therefore, the above Definition 1 is supplemented by the following remark.

Remark 2. *For the problem $F2|l_{ij} \leq p_{ij} \leq u_{ij}|C_{max}$ considered in this section, it is assumed that a J-solution $S(T)$ is a minimal dominant set of Johnson's permutations of all jobs from the set $\mathcal{J}_{1,2}$, i.e., for each scenario $p \in T$, the set $S(T)$ contains at least one optimal pair (π_k, π_k) of identical Johnson's permutations π_k such that the inequality in Equation (1) holds.*

In Reference [36], it is shown how to delete redundant pairs of (identical) permutations from the set S for constructing a J-solution for the problem $F2|l_{ij} \leq p_{ij} \leq u_{ij}|C_{max}$ with job set $\mathcal{J} = \mathcal{J}_{1,2}$. The order of jobs $J_v \in \mathcal{J}_{1,2}$ and $J_w \in \mathcal{J}_{1,2}$ is fixed in the J-solution if there exists at least one Johnson's permutation of the form $\pi_k = (s_1, J_v, s_2, J_w, s_3)$ for any scenario $p \in T$. In Reference [29], the sufficient conditions are proven for fixing the order of two jobs from set $\mathcal{J} = \mathcal{J}_{1,2}$. If one of the following conditions holds, then for each scenario $p \in T$, there exists a permutation $\pi_k = (s_1, J_v, s_2, J_w, s_3)$ that is a Johnson's one for the problem $F2|p|C_{max}$ associated with scenario p:

$$u_{v1} \leq l_{v2} \text{ and } u_{w2} \leq l_{w1}, \tag{3}$$

$$u_{v1} \leq l_{v2} \text{ and } u_{v1} \leq l_{w1}, \tag{4}$$

$$u_{w2} \leq l_{w1} \text{ and } u_{w2} \leq l_{v2}. \tag{5}$$

If at least one condition in Inequalities (3)–(5) holds, then there exists a J-solution $S(T)$ for the problem $F2|l_{ij} \leq p_{ij} \leq u_{ij}|C_{max}$ with fixed order $J_v \to J_w$ of jobs, i.e., job J_v has to be located before job J_w in any permutation π_i, $(\pi_i, \pi_i) \in S(T)$. If both conditions in Inequalities (4) and (5) do not hold, then there is no J-solution $S(T)$ with fixed order $J_v \to J_w$ in all permutations π_i, $(\pi_i, \pi_i) \in S(T)$. If no analogous condition holds for the opposite order $J_w \to J_v$, then at least one permutation with job J_v located before job J_w or that with job J_w located before job J_v have to be included in any J-solution $S(T)$ for the problem $F2|l_{ij} \leq p_{ij} \leq u_{ij}|C_{max}$. Theorem 1 is proven in Reference [41].

Theorem 1. *There exists a J-solution $S(T)$ for the problem $F2|l_{ij} \leq p_{ij} \leq u_{ij}|C_{max}$ with fixed order $J_v \to J_w$ of the jobs J_v and J_w in all permutations π_k, $(\pi_k, \pi_k) \in S(T)$ if and only if at least one condition of Inequalities (4) or (5) holds.*

In Reference [41], the necessary and sufficient conditions have been proven for the case when a single-element J-solution $S(T) = \{(\pi_k, \pi_k)\}$ exists for the problem $F2|l_{jm} \leq p_{jm} \leq u_{jm}|C_{max}$. The partition $\mathcal{J} = \mathcal{J}^0 \cup \mathcal{J}^1 \cup \mathcal{J}^2 \cup \mathcal{J}^*$ of the set $\mathcal{J} = \mathcal{J}_{1,2}$ is considered, where

$\mathcal{J}^0 = \{J_i \in \mathcal{J} : u_{i1} \leq l_{i2}, u_{i2} \leq l_{i1}\}$,
$\mathcal{J}^1 = \{J_i \in \mathcal{J} : u_{i1} \leq l_{i2}, u_{i2} > l_{i1}\} = \{J_i \in \mathcal{J} \setminus \mathcal{J}^0 : u_{i1} \leq l_{i2}\}$,
$\mathcal{J}^2 = \{J_i \in \mathcal{J} : u_{i1} > l_{i2}, u_{i2} \leq l_{i1}\} = \{J_i \in \mathcal{J} \setminus \mathcal{J}^0 : u_{i2} \leq l_{i1}\}$,
$\mathcal{J}^* = \{J_i \in \mathcal{J} : u_{i1} > l_{i2}, u_{i2} > l_{i1}\}$.

For each job $J_k \in \mathcal{J}^0$, inequalities $u_{k1} \leq l_{k2}$ and $u_{k2} \leq l_{k1}$ imply inequalities $l_{k1} = u_{k1} = l_{k2} = u_{k2}$. Since both segments of the possible durations of the job J_k on machines M_1 and M_2 become a point, the durations p_{k1} and p_{k2} are fixed and equal for both machines M_1 and M_2: $p_{k1} = p_{k2} =: p_k$. In Reference [41], Theorems 2 and 3 have been proven.

Theorem 2. *There exists a single-element J-solution $S(T) \subset S$, $|S(T)| = 1$, for the problem $F2|l_{ij} \leq p_{ij} \leq u_{ij}|C_{max}$ if and only if*
(a) for any pair of jobs J_i and J_j from the set $\mathcal{J}^1$ (from the set $\mathcal{J}^2$, respectively), either $u_{i1} \leq l_{j1}$ or $u_{j1} \leq l_{i1}$ (either $u_{i2} \leq l_{j2}$ or $u_{j2} \leq l_{i2}$),
(b) $|\mathcal{J}^*| \leq 1$; for job $J_{i*} \in \mathcal{J}^*$, the inequalities $l_{i*1} \geq \max\{u_{i1} : J_i \in \mathcal{J}^1\}$, $l_{i*2} \geq \max\{u_{j2} : J_j \in \mathcal{J}^2\}$ hold; and $\max\{l_{i*1}, l_{i*2}\} \geq p_k$ for each job $J_k \in \mathcal{J}^0$.

Theorem 2 characterizes the simplest case of the problem $F2|l_{ij} \leq p_{ij} \leq u_{ij}|C_{max}$ when one permutation π_k of the jobs $\mathcal{J} = \mathcal{J}_{1,2}$ dominates all other job permutations. The hardest case of this problem is characterized by the following theorem.

Theorem 3. *If* $\max\{l_{ik} : J_i \in \mathcal{J}, M_k \in \mathcal{M}\} < \min\{u_{ik} : J_i \in \mathcal{J}, M_k \in \mathcal{M}\}$, *then* $S(T) = S$.

The J-solution $S(T)$ may be represented in a compact form using the dominance digraph which may be constructed in $O(n^2)$ time. Let $\mathcal{J} \times \mathcal{J}$ denote the Cartesian product of two sets $\mathcal{J}$. One can construct the following binary relation $\mathcal{A}_{\preceq} \subseteq \mathcal{J} \times \mathcal{J}$ over set $\mathcal{J} = \mathcal{J}_{1,2}$.

Definition 2. *For the two jobs* $J_v \in \mathcal{J}$ *and* $J_w \in \mathcal{J}$, *the inclusion* $(J_v, J_w) \in \mathcal{A}_{\preceq}$ *holds if and only if there exists a J-solution* $S(T)$ *for the problem* $F2|l_{ij} \leq p_{ij} \leq u_{ij}|C_{max}$ *such that job* $J_v \in \mathcal{J}$ *is located before job* $J_w \in \mathcal{J}$, $v \neq w$, *in all permutations* π_k, *where* $(\pi_k, \pi_k) \in S(T)$.

The binary relation $(J_v, J_w) \in \mathcal{A}_{\preceq}$ is represented as follows: $J_v \preceq J_w$. Due to Theorem 1, if for the jobs $J_v \in \mathcal{J}$ and $J_w \in \mathcal{J}$ the relation $J_v \preceq J_w$, $v \neq w$, holds, then for the jobs J_v and J_w, at least one of conditions in Inequalities (4) and (5) holds. To construct the binary relation $\mathcal{A}_{\preceq}$ of the jobs on the set $\mathcal{J}$, it is sufficient to check Inequalities (4) and (5) for each pair of jobs J_v and J_w. The binary relation $\mathcal{A}_{\preceq}$ determines the digraph $(\mathcal{J}, \mathcal{A}_{\preceq})$ with vertex set $\mathcal{J}$ and arc set $\mathcal{A}_{\preceq}$. It takes $O(n^2)$ time to construct the digraph $(\mathcal{J}, \mathcal{A}_{\preceq})$. In the general case, the binary relation $\mathcal{A}_{\preceq}$ may be not transitive. In Reference [42], it is proven that, if the binary relation $\mathcal{A}_{\preceq}$ is not transitive, then $\mathcal{J}^0 \neq \emptyset$. We next consider the case with the equality $\mathcal{J}^0 = \emptyset$, i.e., $\mathcal{J} = \mathcal{J}^* \cup \mathcal{J}^1 \cup \mathcal{J}^2$ (the case with $\mathcal{J}^0 \neq \emptyset$ has been considered in Reference [41]). For a pair of jobs $J_v \in \mathcal{J}^1$ and $J_w \in \mathcal{J}^1$ (for a pair of jobs $J_v \in \mathcal{J}^2$ and $J_w \in \mathcal{J}^2$, respectively), it may happen that there exist both J-solution $S(T)$ with job J_v located before job J_w in all permutations π_k, $(\pi_k, \pi_k) \in S(T)$ and J-solution $S'(T)$ with job J_w located before job J_v in all permutations π_l, $(\pi_l, \pi_l) \in S'(T)$.

In Reference [42], the following claim has been proven.

Theorem 4. *The digraph* $(\mathcal{J}, \mathcal{A}_{\preceq})$ *has no circuits if and only if the set* $\mathcal{J} = \mathcal{J}^* \cup \mathcal{J}^1 \cup \mathcal{J}^2$ *includes no pair of jobs* $J_i \in \mathcal{J}^k$ *and* $J_j \in \mathcal{J}^k$ *with* $k \in \{1, 2\}$ *such that* $l_{ik} = u_{ik} = l_{jk} = u_{jk}$.

The binary relation $\mathcal{A}_{\prec} \subset \mathcal{A}_{\preceq} \subseteq \mathcal{J} \times \mathcal{J}$ is defined as follows.

Definition 3. *For the jobs* $J_v \in \mathcal{J}$ *and* $J_w \in \mathcal{J}$, *the inclusion* $(J_v, J_w) \in \mathcal{A}_{\prec}$ *holds if and only if* $J_v \preceq J_w$ *and* $J_w \npreceq J_v$, *or* $J_v \preceq J_w$ *and* $J_w \preceq J_v$ *with* $v < w$.

The relation $(J_v, J_w) \in \mathcal{A}_{\prec}$ is represented as follows: $J_v \prec J_w$. As it is shown in Reference [42], the relation $J_v \prec J_w$ implies that $J_v \preceq J_w$ and that at least one condition in Inequalities (4) or (5) must hold. The relation $J_v \preceq J_w$ implies exactly one of the relations $J_v \prec J_w$ or $J_w \prec J_v$.

Since it is assumed that set $\mathcal{J}^0$ is empty, the binary relation $\mathcal{A}_{\prec}$ is an antireflective, antisymmetric, and transitive relation, i.e., the binary relation $\mathcal{A}_{\prec}$ is a strict order. The strict order $\mathcal{A}_{\prec}$ determines the digraph $\mathcal{G} = (\mathcal{J}, \mathcal{A}_{\prec})$ with arc set $\mathcal{A}_{\prec}$. The digraph $\mathcal{G} = (\mathcal{J}, \mathcal{A}_{\prec})$ has neither a circuit nor a loop. Properties of the dominance digraph $\mathcal{G}$ were studied in Reference [42]. The permutation $\pi_k = (J_{k_1}, J_{k_2}, \dots, J_{k_n})$, $(\pi_k, \pi_k) \in S$, may be considered as a total strict order of all jobs of the set $\mathcal{J}$. The total strict order determined by permutation π_k is a linear extension of the partial strict order $\mathcal{A}_{\prec}$ if each inclusion $(J_{k_v}, J_{k_w}) \in \mathcal{A}_{\prec}$ implies inequality $v < w$. Let $\Pi(\mathcal{G})$ denote a set of permutations $\pi_k \in S_{1,2}$ defining all linear extensions of the partial strict order $\mathcal{A}_{\prec}$. The cases when $\Pi(\mathcal{G}) = S_{1,2}$ and $\Pi(\mathcal{G}) = \{\pi_k\}$ are characterized in Theorems 2 and 3. In the latter case, the strict order $\mathcal{A}_{\prec}$ over set $\mathcal{J}$ can be represented as follows: $J_{k_1} \prec \dots \prec J_{k_i} \prec J_{k_{i+1}} \prec \dots \prec J_{k_{n_{1,2}}}$. In Reference [42], the following claims have been proven.

Theorem 5. *Let $\mathcal{J} = \mathcal{J}^* \cup \mathcal{J}^1 \cup \mathcal{J}^2$. For any scenario $p \in T$, the set $\Pi(\mathcal{G})$ contains a Johnson's permutation for the problem $F2|p|C_{max}$.*

Corollary 1. *If $\mathcal{J} = \mathcal{J}^* \cup \mathcal{J}^1 \cup \mathcal{J}^2$, then there exists a J-solution $S(T)$ for the problem $F2|l_{ij} \leq p_{ij} \leq u_{ij}|C_{max}$ such that $\pi' \in \Pi(\mathcal{G})$ for all pairs of job permutations, $\{(\pi', \pi')\} \in S(T)$.*

In Reference [42], it was studied how to construct a minimal dominant set $S(T) = \{(\pi', \pi')\}$, $\pi' \in \Pi(\mathcal{G})$. Two types of redundant permutations were examined, and the following claim was proven.

Lemma 1. *Let $\mathcal{J} = \mathcal{J}^* \cup \mathcal{J}_1 \cup \mathcal{J}_2$. If permutation $\pi_t \in \Pi(\mathcal{G})$ is redundant in the set $\Pi(\mathcal{G})$, then π_t is a redundant permutation either of type 1 or type 2.*

Testing whether set $\Pi(\mathcal{G})$ contains a redundant permutation of type 1 takes $O(n^2)$ time, and testing whether permutation $\pi_g \in \Pi(\mathcal{G})$ is a redundant permutation of type 2 takes $O(n)$ time. In Reference [42], it is shown how to delete all redundant permutations from the set $\Pi(\mathcal{G})$. Let $\Pi^*(\mathcal{G})$ denote a set of permutations remaining in the set $\Pi(\mathcal{G})$ after deleting all redundant permutations of type 1 and type 2.

Theorem 6. *Assume the following condition:*

$$\max\{l_{i,3-k}, l_{j,3-k}\} < l_{ik} = u_{ik} = l_{jk} = u_{jk} < \min\{u_{i,3-k}, u_{j,3-k}\}. \tag{6}$$

If set $\mathcal{J} = \mathcal{J}^* \cup \mathcal{J}^1 \cup \mathcal{J}^2$ does not contain a pair of jobs $J_i \in \mathcal{J}^k$ and $J_j \in \mathcal{J}^k$, $k \in \{1,2\}$, such that the above condition holds, then $S(T) =< \Pi^*(\mathcal{G}), \Pi^*(\mathcal{G}) >$.

To test conditions of Theorem 6 takes $O(n)$ time. Due to Theorem 6 and Lemma 1, if there are no jobs such that condition (6) holds, then a J-solution can be constructed via deleting redundant permutations from set $\Pi(\mathcal{G})$. Since the set $\Pi^*(\mathcal{G})$ is uniquely determined [42], we obtain Corollary 2.

Corollary 2 ([42]). *If set $\mathcal{J} = \mathcal{J}^* \cup \mathcal{J}^1 \cup \mathcal{J}^2$ does not contain a pair of jobs J_i and J_j such that condition (6) holds, then the binary relation $\mathcal{A}_\prec$ determines a unique J-solution $S(T) =< \Pi^*(\mathcal{G}), \Pi^*(\mathcal{G}) >$ for the problem $F2|l_{ij} \leq p_{ij} \leq u_{ij}|C_{max}$.*

The condition of Theorem 6 is sufficient for the uniqueness of a J-solution $\Pi^*(\mathcal{G}) = S(T)$ for the problem $F2|l_{ij} \leq p_{ij} \leq u_{ij}|C_{max}$. Due to Theorem 1, one can construct a digraph $\mathcal{G} = (\mathcal{J}, \mathcal{A}_\prec)$ in $O(n^2)$ time. The digraph $\mathcal{G} = (\mathcal{J}, \mathcal{A}_\prec)$ determines a set $S(T)$ and may be considered a condensed form of a J-solution for the problem $F2|l_{ij} \leq p_{ij} \leq u_{ij}|C_{max}$. The results presented in this section are used in Section 5 for constructing precedence digraphs for the problem $J2|l_{ij} \leq p_{ij} \leq u_{ij}, n_i \leq 2|C_{max}$.

5. Properties of the Optimal Pairs of Job Permutations

We consider the uncertain job-shop problem $J2|l_{ij} \leq p_{ij} \leq u_{ij}, n_i \leq 2|C_{max}$ and prove sufficient conditions for determining a small dominant set of schedules for this problem. In what follows, we use Definition 4 of the dominant set $DS(T) \subseteq S$ along with Definition 1 of the J-solution $S(T) \subseteq S$.

Definition 4. *A set of the pairs of job permutations $DS(T) \subseteq S$ is called a dominant set (of schedules) for the uncertain problem $J2|l_{ij} \leq p_{ij} \leq u_{ij}, n_i \leq 2|C_{max}$ if, for each scenario $p \in T$, the set $DS(T)$ contains at least one optimal pair of job permutations for the individual deterministic problem $J2|p, n_i \leq 2|C_{max}$ with scenario p.*

Every J-solution (Definition 1) is a dominant set for the problem $J2|l_{ij} \leq p_{ij} \leq u_{ij}, n_i \leq 2|C_{max}$. Before processing jobs of the set $\mathcal{J}$ (before the realization of a schedule $s \in S$), a scheduler does not know exact values of the job durations. Nevertheless, it is needed to choose a pair of permutations of the jobs $\mathcal{J}$, i.e., it is needed to determine orders of jobs for processing them on machine M_1 and machine M_2. When all jobs will be processed on machines $\mathcal{M}$ (a schedule will be realized) and the job durations will take on exact values p_{ij}^*, $l_{ij} \leq p_{ij}^* \leq u_{ij}$, and so a factual scenario $p^* \in T$ will be determined. A schedule s chosen for the realization should be optimal for the obtained factual scenario p^*. In the stability approach, one can use two phases of scheduling for solving an uncertain scheduling problem: the off-line phase and the on-line phase. The off-line phase of scheduling is finished before starting the realization of a schedule. At this phase, a scheduler knows only given segments of the job durations and the aim is to find a pair of job permutations (π', π'') which is optimal for the most scenarios $p \in T$. After constructing a small dominant set of schedules $DS(T)$, a scheduler can choose a pair of job permutations in the set $DS(T)$, which dominates the most pairs of job permutations $(\pi', \pi'') \in S$ for the given scenarios T. Note that making a decision at the off-line phase may be time-consuming since the realization of a schedule is not started.

The on-line phase of scheduling can begin once the earliest job in the schedule (π', π'') starts. At this phase, a scheduler can use additional on-line information on the job duration since, for each operation O_{ij}, the exact value p_{ij}^* becomes known at the time of the completion of this operation. At the on-line phase, the selection of a next job for processing should be quick.

In Section 5.1, we investigate sufficient conditions for a pair of job permutations (π', π'') such that equality $DS(T) = \{(\pi', \pi'')\}$ holds. In Section 5.2, the sufficient conditions allowing to construct a single optimal schedule dominating all other schedules in the set S are proven. If a single-element dominant set $DS(T)$ does not exist, then one should construct two partial strict orders $A_{\prec}^{1,2}$ and $A_{\prec}^{2,1}$ on the set $\mathcal{J}_{1,2}$ and on the set $\mathcal{J}_{2,1}$ of jobs as it is described in Section 4.2. These orders may be constructed in the form of the two precedence digraphs allowing a scheduler to reduce a size of the dominant set $DS(T)$. Section 5.4 presents Algorithm 1 for constructing a semi-active schedule, which is optimal for the problem $F2|l_{ij} \leq p_{ij} \leq u_{ij}|C_{max}$ for all possible scenarios T provided that such a schedule exists. Otherwise, Algorithm 1 constructs the precedence digraphs determining a minimal dominant set of schedules for the problem $F2|l_{ij} \leq p_{ij} \leq u_{ij}|C_{max}$.

5.1. Sufficient Conditions for an Optimal Pair of Job Permutations

In the proofs of several claims, we use a notion of the main machine, which is introduced within the proof of the following theorem.

Theorem 7. *Consider the following conditions in Inequalities (7) or (8):*

$$\sum_{J_i \in \mathcal{J}_{1,2}} u_{i1} \leq \sum_{J_i \in \mathcal{J}_{2,1} \cup \mathcal{J}_2} l_{i2} \ \text{and} \ \sum_{J_i \in \mathcal{J}_{1,2}} l_{i2} \geq \sum_{J_i \in \mathcal{J}_{2,1} \cup \mathcal{J}_1} u_{i1} \tag{7}$$

$$\sum_{J_i \in \mathcal{J}_{2,1}} u_{i2} \leq \sum_{J_i \in \mathcal{J}_{1,2} \cup \mathcal{J}_1} l_{i1} \ \text{and} \ \sum_{J_i \in \mathcal{J}_{2,1}} l_{i1} \geq \sum_{J_i \in \mathcal{J}_{1,2} \cup \mathcal{J}_2} u_{i2} \tag{8}$$

If one of the above conditions holds, then any pair of job permutations $(\pi', \pi'') \in S$ is a single-element dominant set $DS(T) = \{(\pi', \pi'')\}$ for the problem $J2|l_{ij} \leq p_{ij} \leq u_{ij}, n_i \leq 2|C_{max}$ with set $\mathcal{J} = \mathcal{J}_1 \cup \mathcal{J}_2 \cup \mathcal{J}_{1,2} \cup \mathcal{J}_{2,1}$ of the given jobs.

Proof. Let the condition in Inequalities (7) hold. Then, we consider an arbitrary pair of job permutations $(\pi', \pi'') \in S$ with any fixed scenario $p \in T$ and show that this pair of job permutations (π', π'') is optimal for the individual deterministic problem $J2|p, n_i \leq 2|C_{max}$ with scenario p, i.e., $C_{max}(\pi', \pi'') = C_{max}$.

Let $c_1(\pi')$ $(c_2(\pi''))$ denote a completion time of all jobs $\mathcal{J}_1 \cup \mathcal{J}_{1,2} \cup \mathcal{J}_{2,1}$ (jobs $\mathcal{J}_2 \cup \mathcal{J}_{1,2} \cup \mathcal{J}_{2,1}$) on machine M_1 (machine M_2) in the schedule (π', π''), where $\pi' = (\pi_{1,2}, \pi_1, \pi_{2,1})$ and $\pi'' = (\pi_{2,1}, \pi_2, \pi_{1,2})$. For the problem $J2|p, n_i \leq 2|C_{max}$, the maximal completion time of the jobs in schedule (π', π'') may be calculated as follows: $C_{max}(\pi', \pi'') = \max\{c_1(\pi'), c_2(\pi'')\}$.

Machine M_1 (machine M_2) is called a *main machine* for the schedule (π', π'') if equality $C_{max}(\pi', \pi'') = c_1(\pi')$ holds (equality $C_{max}(\pi', \pi'') = c_2(\pi'')$ holds, respectively).

For schedule $(\pi', \pi'') \in S$, the following equality holds:

$$c_1(\pi') = \sum_{J_i \in \mathcal{J}_{1,2} \cup \mathcal{J}_{2,1} \cup \mathcal{J}_1} p_{i1} + I_1; \quad c_2(\pi'') = \sum_{J_i \in \mathcal{J}_{1,2} \cup \mathcal{J}_{2,1} \cup \mathcal{J}_2} p_{i2} + I_2,$$

where I_1 and I_2 denote total idle times of machine M_1 and machine M_2 in the schedule (π', π''), respectively. We next show that, if the condition in Inequalities (7) holds, then machine M_2 is a main machine for schedule (π', π'') and machine M_2 has no idle time, i.e., machine M_2 is completely filled in the segment $[0, c_2(\pi'')]$ for processing jobs from the set $\mathcal{J}_{1,2} \cup \mathcal{J}_{2,1} \cup \mathcal{J}_2$. At the initial time $t = 0$, machine M_2 begins to process jobs from the set $\mathcal{J}_{2,1} \cup \mathcal{J}_2$ without idle times until the time moment $t_1 = \sum_{J_i \in \mathcal{J}_{2,1} \cup \mathcal{J}_2} p_{i2}$.

From the first inequality in (7), we obtain the following relations:

$$\sum_{J_i \in \mathcal{J}_{1,2}} p_{i1} \leq \sum_{J_i \in \mathcal{J}_{1,2}} u_{i1} \leq \sum_{J_i \in \mathcal{J}_{2,1} \cup \mathcal{J}_2} l_{i2} \leq \sum_{J_i \in \mathcal{J}_{2,1} \cup \mathcal{J}_2} p_{i2} = t_1.$$

Therefore, at the time moment t_1, machine M_2 begins to process jobs from the set $\mathcal{J}_{1,2}$ without idle times and we obtain the following equality: $c_2(\pi'') = \sum_{J_i \in \mathcal{J}_{1,2} \cup \mathcal{J}_{2,1} \cup \mathcal{J}_2} p_{i2}$, where $I_2 = 0$ and machine M_2 has no idle time. We next show that machine M_2 is a main machine for the schedule (π', π''). To this end, we consider the following two possible cases.

(a) Let machine M_1 have no idle time.

By summing Inequalities (7), we obtain the following inequality:

$$\sum_{J_i \in \mathcal{J}_{1,2} \cup \mathcal{J}_{2,1} \cup \mathcal{J}_1} u_{i1} \leq \sum_{J_i \in \mathcal{J}_{1,2} \cup \mathcal{J}_{2,1} \cup \mathcal{J}_2} l_{i2}.$$

Thus, the following relations hold:

$$c_1(\pi') = \sum_{J_i \in \mathcal{J}_{1,2} \cup \mathcal{J}_{2,1} \cup \mathcal{J}_1} p_{i1} \leq \sum_{J_i \in \mathcal{J}_{1,2} \cup \mathcal{J}_{2,1} \cup \mathcal{J}_1} u_{i1} \leq \sum_{J_i \in \mathcal{J}_{1,2} \cup \mathcal{J}_{2,1} \cup \mathcal{J}_2} l_{i2} \leq \sum_{J_i \in \mathcal{J}_{1,2} \cup \mathcal{J}_{2,1} \cup \mathcal{J}_2} p_{i2} = c_2(\pi'').$$

Hence, machine M_2 is a main machine for the schedule (π', π'').

(b) Let machine M_1 have an idle time.

An idle time of machine M_1 is only possible if some job J_j from set $\mathcal{J}_{2,1}$ is processed on machine M_2 at the time moment t_2 when this job J_j could be processed on machine M_1.

Obviously, after the time moment $\sum_{J_i \in \mathcal{J}_{2,1}} p_{i2}$ when machine M_2 completes all jobs from set $\mathcal{J}_{2,1}$, machine M_1 can process some jobs from set $\mathcal{J}_{2,1}$ without an idle time. Therefore, the inequality $t_2 + I_1 \leq \sum_{J_i \in \mathcal{J}_{2,1}} p_{i2}$ holds and we obtain the following relations:

$$c_1(\pi') \leq t_2 + I_1 + \sum_{J_i \in \mathcal{J}_{2,1}} p_{i1} \leq \sum_{J_i \in \mathcal{J}_{2,1}} p_{i2} + \sum_{J_i \in \mathcal{J}_{2,1} \cup \mathcal{J}_1} p_{i1} \leq \sum_{J_i \in \mathcal{J}_{2,1}} p_{i2} + \sum_{J_i \in \mathcal{J}_{2,1} \cup \mathcal{J}_1} u_{i1}$$

$$\leq \sum_{J_i \in \mathcal{J}_{2,1}} p_{i2} + \sum_{J_i \in \mathcal{J}_{1,2}} l_{i2} \leq \sum_{J_i \in \mathcal{J}_{2,1}} p_{i2} + \sum_{J_i \in \mathcal{J}_{1,2}} p_{i2} \leq \sum_{J_i \in \mathcal{J}_{2,1} \cup \mathcal{J}_2 \cup \mathcal{J}_{1,2}} p_{i2} = c_2(\pi'').$$

We conclude that, in case *(b)*, machine M_2 is a main machine for the schedule (π', π''). Thus, if the condition in Inequalities (7) holds, then machine M_2 is a main machine for the schedule (π', π'') and machine M_2 has no idle time, i.e., equality $C_{max}(\pi', \pi'') = c_2(\pi'')$ holds and machine M_2 is completely filled in the segment $[0, c_2(\pi'')]$ with processing jobs from the set $\mathcal{J}_{1,2} \cup \mathcal{J}_{2,1} \cup \mathcal{J}_2$.

Thus, the pair of permutations (π', π'') is optimal for scenario $p \in T$. Since scenario p was chosen arbitrarily in the set T, we conclude that the pair of job permutations (π', π'') is a singleton $DS(T) = \{(\pi', \pi'')\}$ for the problem $J2|l_{ij} \leq p_{ij} \leq u_{ij}, n_i \leq 2|C_{max}$ with set $\mathcal{J} = \mathcal{J}_1 \cup \mathcal{J}_2 \cup \mathcal{J}_{1,2} \cup \mathcal{J}_{2,1}$

of the given jobs. As a pair of permutations (π', π'') is an arbitrary pair of job permutations in the set S, any pair of job permutations $(\pi', \pi'') \in S$ is a singleton $DS(T) = \{(\pi', \pi'')\}$ for the problem $J2|l_{ij} \leq p_{ij} \leq u_{ij}, n_i \leq 2|C_{max}$ with job set $\mathcal{J} = \mathcal{J}_1 \cup \mathcal{J}_2 \cup \mathcal{J}_{1,2} \cup \mathcal{J}_{2,1}$.

The case when the condition in Inequalities (8) holds may be analyzed similarly via replacing machine M_1 by machine M_2 and vice versa. $\square$

If conditions of Theorem 7 hold, then in the optimal pair of job permutations (π', π'') existing for the problem $J2|l_{ij} \leq p_{ij} \leq u_{ij}, n_i \leq 2|C_{max}$, the orders of jobs from sets $\mathcal{J}_{1,2} \subseteq \mathcal{J}$ and $\mathcal{J}_{2,1} \subseteq \mathcal{J}$ may be chosen arbitrarily. Theorem 7 implies the following two corollaries.

Corollary 3. *If the following inequality holds:*

$$\sum_{J_j \in \mathcal{J}_{1,2}} u_{i1} \leq \sum_{J_j \in \mathcal{J}_{2,1} \cup \mathcal{J}_2} l_{i2}, \tag{9}$$

then set $< \{\pi_{1,2}\}, S_{2,1} > \subseteq S$, *where $\pi_{1,2}$ is an arbitrary permutation in set $S_{1,2}$, is a dominant set of schedules for the problem $J2|l_{ij} \leq p_{ij} \leq u_{ij}, n_i \leq 2|C_{max}$ with set $\mathcal{J} = \mathcal{J}_1 \cup \mathcal{J}_2 \cup \mathcal{J}_{1,2} \cup \mathcal{J}_{2,1}$ of the given jobs.*

Proof. We consider an arbitrary vector $p \in T$ of the job durations and an arbitrary permutation $\pi_{1,2}$ in the set $S_{1,2}$. The set $S_{2,1}$ contains at least one Johnson's permutation $\pi_{2,1}^*$ for the deterministic problem $F2|p_{2,1}|C_{max}$ with job set $\mathcal{J}_{2,1}$ and scenario $p_{2,1}$ (the components of vector $p_{2,1}$ are equal to the corresponding components of vector p). We consider a pair of job permutations (π', π'') $= ((\pi_{1,2}, \pi_1, \pi_{2,1}^*), (\pi_{2,1}^*, \pi_2, \pi_{1,2})) \in <\{\pi_{1,2}\}, S_{2,1} > \subseteq S$ and show that it is an optimal pair of job permutations for the problem $J2|p, n_i \leq 2|C_{max}$ with job set $\mathcal{J}$ and scenario p. Without loss of generality, both permutations π_1 and π_2 are ordered in increasing order of the indexes of their jobs.

Similar to the proof of Theorem 7, one can show that, if the condition in Inequalities (9) holds, then machine M_2 processes jobs without idle times and equality $c_2(\pi'') = \sum_{J_i \in \mathcal{J}_{1,2} \cup \mathcal{J}_{2,1} \cup \mathcal{J}_2} p_{i2}$ holds, where the value of $c_2(\pi'')$ cannot be reduced. If machine M_1 has no idle time, we obtain equalities

$$C_{max}(\pi', \pi'') = \max\{c_1(\pi'), c_2(\pi'')\} = \max\{ \sum_{J_i \in \mathcal{J}_{1,2} \cup \mathcal{J}_{2,1} \cup \mathcal{J}_1} p_{i1}, \sum_{J_i \in \mathcal{J}_{1,2} \cup \mathcal{J}_{2,1} \cup \mathcal{J}_2} p_{i2}\} = C_{max}.$$

On the other hand, an idle time of machine M_1 is only possible if some job J_j from set $\mathcal{J}_{2,1}$ is processed on machine M_2 at the time moment t_2 when job J_j could be processed on machine M_1. In such a case, the value of $c_1(\pi')$ is equal to the makespan $C_{max}(\pi_{2,1}^*)$ for the problem $F2|p_{2,1}|C_{max}$ with job set $\mathcal{J}_{2,1}$ and scenario $p_{2,1}$. As the permutation $\pi_{2,1}^*$ is a Johnson's permutation, the value of $C_{max}(\pi_{2,1}^*)$ cannot be reduced and we obtain the following equalities:

$$C_{max}(\pi', \pi'') = \max\{c_1(\pi'), c_2(\pi'')\} = \max\{C_{max}(\pi_{2,1}^*), \sum_{J_i \in \mathcal{J}_{1,2} \cup \mathcal{J}_{2,1} \cup \mathcal{J}_2} p_{i2}\} = C_{max}.$$

Thus, the pair of job permutation $(\pi', \pi'') = ((\pi_{1,2}, \pi_1, \pi_{2,1}^*), (\pi_{2,1}^*, \pi_2, \pi_{1,2})) \in <\{\pi_{1,2}\}, S_{2,1} > \subseteq S$ is optimal for the problem $J2|p, n_i \leq 2|C_{max}$ with scenario $p \in T$. The optimal pair of job permutations for the problem $J2|p, n_i \leq 2|C_{max}$ with scenario $p \in T$ belongs to the set $<\{\pi_{1,2}\}, S_{2,1} >$. As vector p is an arbitrary vector in the set T, the set $< \{\pi_{1,2}\}, S_{2,1} >$ contains an optimal pair of job permutations for all scenarios from set T. Due to Definition 4, the set $< \{\pi_{1,2}\}, S_{2,1} > \subseteq S$ is a dominant set of schedules for the problem $J2|l_{ij} \leq p_{ij} \leq u_{ij}, n_i \leq 2|C_{max}$ with job set $\mathcal{J}$. $\square$

Corollary 4. *Consider the following inequality:*

$$\sum_{J_j \in \mathcal{J}_{2,1}} u_{i2} \leq \sum_{J_j \in \mathcal{J}_{1,2} \cup \mathcal{J}_1} l_{i1}.$$

If the above inequality holds, then set $< S_{1,2}, \{\pi_{2,1}\} >$, where $\pi_{2,1}$ is an arbitrary permutation in set $S_{2,1}$, is a dominant set of schedules for the problem $J2|l_{ij} \leq p_{ij} \leq u_{ij}, n_i \leq 2|C_{max}$ with set $\mathcal{J} = \mathcal{J}_1 \cup \mathcal{J}_2 \cup \mathcal{J}_{1,2} \cup \mathcal{J}_{2,1}$ of the given jobs.

This claim may be proven similar to Corollary 3. If the conditions of Corollary 3 (Corollary 4) hold, then the order for processing jobs from set $\mathcal{J}_{1,2} \subseteq \mathcal{J}$ (set $\mathcal{J}_{2,1} \subseteq \mathcal{J}$, respectively) in the optimal schedule $(\pi', \pi'') = ((\pi_{1,2}, \pi_1, \pi_{2,1}), (\pi_{2,1}, \pi_2, \pi_{1,2}))$ for the problem $J2|l_{ij} \leq p_{ij} \leq u_{ij}, n_i \leq 2|C_{max}$ may be arbitrary. Since the orders of jobs from the sets $\mathcal{J}_1$ and $\mathcal{J}_2$ are fixed in the optimal schedule (Remark 1), we need to determine only orders for processing jobs from set $\mathcal{J}_{2,1}$ (set $\mathcal{J}_{1,2}$, respectively). To do this, we will consider two uncertain problems $F2|l_{ij} \leq p_{ij} \leq u_{ij}|C_{max}$ with job set $\mathcal{J}_{1,2} \subseteq \mathcal{J}$ and with the machine route (M_1, M_2) and that with job set $\mathcal{J}_{2,1} \subseteq \mathcal{J}$ and with the opposite machine route (M_2, M_1).

Lemma 2. *If $S'_{1,2} \subseteq S_{1,2}$ is a set of permutations from the dominant set for the problem $F2|l_{ij} \leq p_{ij} \leq u_{ij}|C_{max}$ with job set $\mathcal{J}_{1,2}$, then $< S'_{1,2}, S_{2,1} > \subseteq S$ is a dominant set for the problem $J2|l_{ij} \leq p_{ij} \leq u_{ij}, n_i \leq 2|C_{max}$ with job set $\mathcal{J} = \mathcal{J}_1 \cup \mathcal{J}_2 \cup \mathcal{J}_{1,2} \cup \mathcal{J}_{2,1}$.*

The proof of Lemma 2 and those for other statements in this section are given in Appendix A.

Lemma 3. *Let $S'_{2,1} \subseteq S_{2,1}$ be a set of permutations from the dominant set for the problem $F2|l_{ij} \leq p_{ij} \leq u_{ij}|C_{max}$ with job set $\mathcal{J}_{2,1}$, $S'_{2,1} \subseteq S_{2,1}$. Then, $< S_{1,2}, S'_{2,1} >$ is a dominant set for the problem $J2|l_{ij} \leq p_{ij} \leq u_{ij}, n_i \leq 2|C_{max}$ with job set $\mathcal{J}$.*

The proof of this claim is similar to that for Lemma 2 (see Appendix A).

Theorem 8. *Let $S'_{1,2} \subseteq S_{1,2}$ be a set of permutations from the dominant set for the problem $F2|l_{ij} \leq p_{ij} \leq u_{ij}|C_{max}$ with job set $\mathcal{J}_{1,2}$, and let $S'_{2,1} \subseteq S_{2,1}$ be a set of permutations from the dominant set for the problem $F2|l_{ij} \leq p_{ij} \leq u_{ij}|C_{max}$ with job set $\mathcal{J}_{2,1}$. Then, $< S'_{1,2}, S'_{2,1} > \subseteq S$ is a dominant set for the problem $J2|l_{ij} \leq p_{ij} \leq u_{ij}, n_i \leq 2|C_{max}$ with job set $\mathcal{J} = \mathcal{J}_1 \cup \mathcal{J}_2 \cup \mathcal{J}_{1,2} \cup \mathcal{J}_{2,1}$.*

Theorem 9. *Let a pair of identical permutations $(\pi_{1,2}, \pi_{1,2})$ determine a single-element J-solution for the problem $F2|l_{ij} \leq p_{ij} \leq u_{ij}|C_{max}$ with job set $\mathcal{J}_{1,2}$, and let a pair of identical permutations $(\pi_{2,1}, \pi_{2,1})$ determine a single-element J-solution for the problem $F2|l_{ij} \leq p_{ij} \leq u_{ij}|C_{max}$ with job set $\mathcal{J}_{2,1}$. Then, the pairs of permutations $\{(\pi_{1,2}, \pi_1, \pi_{2,1}) and (\pi_{1,2}, \pi_2, \pi_{2,1})\}$ are a single-element dominant set $DS(T)$ for the problem $J2|l_{ij} \leq p_{ij} \leq u_{ij}, n_i \leq 2|C_{max}$ with job set $\mathcal{J} = \mathcal{J}_1 \cup \mathcal{J}_2 \cup \mathcal{J}_{1,2} \cup \mathcal{J}_{2,1}$.*

The following claim follows directly from Theorem 9.

Corollary 5. *If the conditions of Theorem 9 hold, then there exists a single pair of job permutations, which is an optimal pair of job permutations for the problem $J2|p, n_i \leq 2|C_{max}$ with job set $\mathcal{J}$ and any scenario $p \in T$.*

Theorem 9 implies also the following corollary proven in Appendix A.

Corollary 6. *If the conditions of Theorem 9 hold, then there exists a single pair of job permutations which is a J-solution for the problem $J2|l_{ij} \leq p_{ij} \leq u_{ij}, n_i \leq 2|C_{max}$ with job set $\mathcal{J} = \mathcal{J}_1 \cup \mathcal{J}_2 \cup \mathcal{J}_{1,2} \cup \mathcal{J}_{2,1}$.*

Note that the criterion for a single-element J-solution for the problem $F2|l_{ij} \leq p_{ij} \leq u_{ij}|C_{max}$ is given in Theorem 2.

5.2. Precedence Digraphs Determining a Minimal Dominant Set of Schedules

In Section 4.2, it is assumed that $\mathcal{J}_{1,2} = \mathcal{J}_{1,2}^1 \cup \mathcal{J}_{1,2}^2 \cup \mathcal{J}_{1,2}^*$ and $\mathcal{J}_{2,1} = \mathcal{J}_{2,1}^1 \cup \mathcal{J}_{2,1}^2 \cup \mathcal{J}_{2,1}^*$, i.e., $\mathcal{J}_{1,2}^0 = \mathcal{J}_{2,1}^0 = \varnothing$. Based on the results presented in Section 4.2, we can determine a binary relation $A_{\prec}^{1,2}$ for the problem $F2|l_{ij} \leq p_{ij} \leq u_{ij}|C_{max}$ with job set $\mathcal{J}_{1,2}$ and a binary relation $A_{\prec}^{2,1}$ for this problem with job set $\mathcal{J}_{2,1}$. For job set $\mathcal{J}_{1,2}$, the binary relation $A_{\prec}^{1,2}$ determines the digraph $G_{1,2} = (\mathcal{J}_{1,2}, A_{\prec}^{1,2})$ with the vertex set $\mathcal{J}_{1,2}$ and the arc set $A_{\prec}^{1,2}$. For job set $\mathcal{J}_{2,1}$, the binary relation $A_{\prec}^{2,1}$ determines the digraph $G_{2,1} = (\mathcal{J}_{2,1}, A_{\prec}^{2,1})$ with the vertex set $\mathcal{J}_{2,1}$ and the arc set $A_{\prec}^{2,1}$.

Let us consider the problem $F2|l_{ij} \leq p_{ij} \leq u_{ij}|C_{max}$ with job set $\mathcal{J}_{1,2}$ and the corresponding digraph $G_{1,2} = (\mathcal{J}_{1,2}, A_{\prec}^{1,2})$ (the same results for the problem $F2|l_{ij} \leq p_{ij} \leq u_{ij}|C_{max}$ with job set $\mathcal{J}_{2,1}$ can be derived in a similar way).

Definition 5. *Two jobs, $J_x \in \mathcal{J}_{1,2}$ and $J_y \in \mathcal{J}_{1,2}$, $x \neq y$, are called conflict jobs if they are not in the relation $A_{\prec}^{1,2}$, i.e., $(J_x, J_y) \notin A_{\prec}^{1,2}$ and $(J_y, J_x) \notin A_{\prec}^{1,2}$.*

Due to Definitions 2 and 3, for the conflict jobs $J_x \in \mathcal{J}_{1,2}$ and $J_y \in \mathcal{J}_{1,2}$, $x \neq y$, Inequalities (4) and (5) do not hold either for the case $v = x$ with $w = y$ or for the case $v = y$ with $w = x$.

Definition 6. *The subset $\mathcal{J}_x \subseteq \mathcal{J}_{1,2}$ is called a conflict set of jobs if, for any job $J_y \in \mathcal{J}_{1,2} \setminus \mathcal{J}_x$, either relation $(J_x, J_y) \in A_{\prec}^{1,2}$ or relation $(J_y, J_x) \in A_{\prec}^{1,2}$ holds for each job $J_x \in \mathcal{J}_x$ (provided that any proper subset of the set $\mathcal{J}_x$ does not possess such a property).*

From Definition 6, it follows that the conflict set $\mathcal{J}_x$ is a minimal set (with respect to the inclusion). Obviously, there may exist several conflict sets in the set $\mathcal{J}_{1,2}$. (A conflict set of the jobs $\mathcal{J}_x \subseteq \mathcal{J}_{2,1}$ can be determined similarly.) Let the strict order $A_{\prec}^{1,2}$ for the problem $F2|l_{ij} \leq p_{ij} \leq u_{ij}|C_{max}$ with job set $\mathcal{J}_{1,2}$ be represented as follows:

$$J_1 \prec J_2 \prec \ldots \prec J_k \prec \{J_{k+1}, J_{k+2}, \ldots, J_{k+r}\} \prec J_{k+r+1} \prec J_{k+r+2} \prec \ldots \prec J_{m_{1,2}}, \tag{10}$$

where all jobs between braces are conflict ones and each of these jobs is in relation $A_{\prec}^{1,2}$ with any job located outside the brackets in Relation (10). In such a case, an optimal order for processing jobs from the set $\{J_1, J_2, \ldots, J_k\}$ is determined as follows: $(J_1, J_2, \ldots, J_k)$.

Due to Theorem 5, we obtain that set $\Pi(G_{1,2})$ of the permutations generated by the digraph $G_{1,2}$ contains an optimal Johnson's permutation for each vector $p_{1,2}$ of the durations of jobs from the set $\mathcal{J}_{1,2}$. Thus, due to Definition 1, the singleton $\{(\pi_{1,2}, \pi_{1,2})\}$, where $\pi_{1,2} \in \Pi(G_{1,2})$, is a J-solution for the problem $F2|l_{ij} \leq p_{ij} \leq u_{ij}|C_{max}$ with job set $\mathcal{J}_{1,2}$. Analogously, the singleton $\{(\pi_{2,1}, \pi_{2,1})\}$, where $\pi_{2,1} \in \Pi(G_{2,1})$, is a J-solution for the problem $F2|l_{ij} \leq p_{ij} \leq u_{ij}|C_{max}$ with job set $\mathcal{J}_{2,1}$. We can determine a dominant set of schedules for the problem $J2|l_{ij} \leq p_{ij} \leq u_{ij}, n_i \leq 2|C_{max}$ with job set $\mathcal{J}$ as follows: $<\Pi(G_{1,2}), \Pi(G_{2,1})> \subseteq S$. The following theorems allow us to reduce a dominant set for the problem $J2|l_{ij} \leq p_{ij} \leq u_{ij}, n_i \leq 2|C_{max}$. We use the following notation: $L_2 = \sum_{J_i \in \mathcal{J}_{2,1} \cup \mathcal{J}_2} l_{i2}$.

Theorem 10. *Let the strict order $A_{\prec}^{1,2}$ over set $\mathcal{J}_{1,2} = \mathcal{J}_{1,2}^* \cup \mathcal{J}_{1,2}^1 \cup \mathcal{J}_{1,2}^2$ be determined as follows: $J_1 \prec \ldots \prec J_k \prec \{J_{k+1}, J_{k+2}, \ldots, J_{k+r}\} \prec J_{k+r+1} \prec \ldots \prec J_{m_{1,2}}$. Consider the following inequality:*

$$\sum_{i=1}^{k+r} u_{i1} \leq L_2 + \sum_{i=1}^{k} l_{i2}, \tag{11}$$

If the above inequality holds, then set $S' = < \{\pi\}, \Pi(G_{2,1}) > \subset S$ with $\pi \in \Pi(G_{1,2})$ is a dominant set of schedules for the problem $J2|l_{ij} \leq p_{ij} \leq u_{ij}, n_i \leq 2|C_{max}$ with job set $\mathcal{J}$.

Proof. We consider an arbitrary vector $p \in T$ of the job durations and an arbitrary permutation π from the set $\Pi(G_{1,2})$. The set $\Pi(G_{2,1})$ contains at least one optimal Johnson's permutation $\pi^*_{2,1}$ for the problem $F2|p_{2,1}|C_{max}$ with job set $\mathcal{J}_{2,1}$ and vector $p_{2,1}$ of the job durations (components of this vector are equal to the corresponding components of the vector p).

We consider a pair of job permutations $(\pi', \pi'') = ((\pi, \pi_1, \pi^*_{2,1}), (\pi^*_{2,1}, \pi_2, \pi)) \in S'$ and show that it is an optimal pair of job permutations for the problem $J2|p, n_i \leq 2|C_{max}$ with set $\mathcal{J}$ of the jobs and scenario p. To this end, we show that the value of $C_{max}(\pi', \pi'') = \max\{c_1(\pi'), c_2(\pi'')\}$ cannot be reduced. Indeed, an idle time for machine M_1 is only possible if some job J_j from the set $\mathcal{J}_{2,1}$ is processed on machine M_2 at the same time when job J_j could be processed on machine M_1. In such a case, $c_1(\pi')$ is equal to the makespan $C_{max}(\pi^*_{2,1})$ for the problem $F2|p_{2,1}|C_{max}$ with job set $\mathcal{J}_{2,1}$ and vector $p_{2,1}$ of the job durations. As permutation $\pi^*_{2,1}$ is a Johnson's permutation, the value of

$$c_1(\pi') = \max\{ \sum_{J_i \in \mathcal{J}_{1,2} \cup \mathcal{J}_{2,1} \cup \mathcal{J}_1} p_{i1}, C_{max}(\pi^*_{2,1}) \}$$

cannot be reduced. In the beginning of the permutation π, the jobs of set $\{J_1, J_2, \ldots, J_k\}$ are arranged in the Johnson's order. Thus, if machine M_2 has an idle time while processing these jobs, this idle time cannot be reduced. From Inequality (11), it follows that machine M_2 has no idle time while processing jobs from the conflict set.

In the end of the permutation π, jobs of set $\{J_{k+r+1}, \ldots, J_{m_{1,2}}\}$ are arranged in Johnson's order. Therefore, if machine M_2 has an idle time while processing these jobs, this idle time cannot be reduced. Thus, the value of $c_2(\pi'')$ cannot be reduced by changing the order of jobs in the conflict set.

We obtain the qualities $C_{max}(\pi', \pi'') = \max\{c_1(\pi'), c_2(\pi'')\} = C_{max}$. The pair of job permutations $(\pi', \pi'') = ((\pi, \pi_1, \pi^*_{2,1}), (\pi^*_{2,1}, \pi_2, \pi))$ is optimal for the problem $J2|p, n_i \leq 2|C_{max}$ with scenario $p \in T$. Thus, set $S' =<\{\pi\}, \Pi(G_{2,1})>$ contains an optimal pair of job permutations for the problem $J2|p, n_i \leq 2|C_{max}$ with scenario $p \in T$. As vector p is an arbitrary vector in set T, set S' contains an optimal pair of job permutations for each vector from set T. Due to Definition 4, set S' is a dominant set of schedules for the problem $J2|l_{ij} \leq p_{ij} \leq u_{ij}, n_i \leq 2|C_{max}$ with job set $\mathcal{J}$. $\square$

Theorem 11. *Let the partial strict order $A^{1,2}_{\prec}$ over set $\mathcal{J}_{1,2} = \mathcal{J}^*_{1,2} \cup \mathcal{J}^1_{1,2} \cup \mathcal{J}^2_{1,2}$ be determined as follows: $J_1 \prec \ldots \prec J_k \prec \{J_{k+1}, J_{k+2}, \ldots, J_{k+r}\} \prec J_{k+r+1} \prec \ldots \prec J_{m_{1,2}}$. Consider the following inequality:*

$$u_{k+s,1} \leq L_2 + \sum_{i=1}^{k+s-1} (l_{i2} - u_{i1}) \tag{12}$$

If the above inequality holds for all $s \in \{1, 2, \ldots, r\}$, then the set $S' =< \{\pi\}, S_{2,1} >$, where $\pi = (J_1, \ldots, J_{k-1}, J_k, J_{k+1}, J_{k+2}, \ldots, J_{k+r}, J_{k+r+1}, \ldots, J_{m_{1,2}}) \in \Pi(G_{1,2})$, is a dominant set for the problem $J2|l_{ij} \leq p_{ij} \leq u_{ij}, n_i \leq 2|C_{max}$ with job set $\mathcal{J}$.

Proof. We consider an arbitrary scenario $p \in T$ and a pair of job permutations $(\pi', \pi'') = ((\pi, \pi_1, \pi^*_{2,1}), (\pi^*_{2,1}, \pi_2, \pi)) \in S'$, where $\pi^*_{2,1} \in S_{2,1}$ is a Johnson's permutation of the jobs from the set $\mathcal{J}_{2,1}$ with vector $p_{2,1}$ of the job durations (components of this vector are equal to the corresponding components of vector p). We next show that this pair of job permutations (π', π'') is optimal for the individual deterministic problem $J2|p, n_i \leq 2|C_{max}$ with scenario p, i.e., $C_{max}(\pi', \pi'') = C_{max}$.

If conditions of Theorem 11 hold, then machine M_2 processes jobs from the conflict set $\{J_{k+1}, J_{k+2}, \ldots, J_{k+r}\}$ without idle times. At the initial time $t = 0$, machine M_1 begins to process jobs from the permutation π without idle times. Let a time moment t_1 be as follows: $t_1 = \sum_{i=1}^{k+1} p_{i1}$. At the time moment t_1, job J_{k+1} is ready for processing on machine M_2.

On the other hand, at the time $t = 0$, machine M_2 begins to process jobs from the set $\mathcal{J}_{2,1} \cup \mathcal{J}_2$ without idle times and then jobs from the permutation $(J_1, J_2, \ldots, J_{k+1})$. Let t_2 denote the first time moment when machine M_2 is ready for processing job J_{k+1}. Obviously, the following inequality holds: $t_2 \geq L_2 + \sum_{i=1}^{k+1} p_{i2}$. From the condition in Inequality (12) with $s = 1$, we obtain inequality $\sum_{i=1}^{k+1} u_{i1} \leq L_2 + \sum_{i=1}^{k} l_{i2}$.

Therefore, the following relations hold:

$$t_1 = \sum_{i=1}^{k+1} p_{i1} \leq \sum_{i=1}^{k+1} u_{i1} \leq L_2 + \sum_{i=1}^{k} l_{i2} \leq L_2 + \sum_{i=1}^{k+1} p_{i2} = t_2.$$

Machine M_2 processes job J_{k+1} without an idle time between job J_k and job J_{k+1}.

Analogously, using $s \in \{2, 3, \ldots, r\}$, one can show that machine M_2 processes jobs from the conflict set $\{J_{k+1}, J_{k+2}, \ldots, J_{k+r}\}$ without idle times between jobs J_{k+1} and J_{k+2}, between jobs J_{k+2} and J_{k+3}, and so on to between jobs J_{k+r-1} and J_{k+r}. To end this proof, we have to show that the value of $C_{max}(\pi', \pi'') = \max\{c_1(\pi'), c_2(\pi'')\}$ cannot be reduced.

An idle time for machine M_1 is only possible between some jobs from the set $\mathcal{J}_{2,1}$. However, the permutation $\pi_{2,1}^*$ is a Johnson's permutation of the jobs from the set $\mathcal{J}_{2,1}$ for the vector $p_{2,1}$ of the job durations. Therefore, the value of $c_1(\pi')$ cannot be reduced. On the other hand, in the permutation π, all jobs $J_1, J_2, \ldots, J_k$ and all jobs $J_{k+r+1}, \ldots, J_{m_{1,2}}$ are arranged in Johnson's orders. Therefore, if machine M_2 has an idle time while processing these jobs, this idle time cannot be reduced. It is clear that machine M_2 has no idle time while processing jobs from the conflict set. Thus, the value of $c_2(\pi'')$ cannot be reduced by changing the order of jobs from the conflict set. We obtain the equalities $C_{max}(\pi', \pi'') = \max\{c_1(\pi'), c_2(\pi'')\} = C_{max}$.

It is shown that the pair of job permutations $(\pi', \pi'') = ((\pi, \pi_1, \pi_{2,1}^*), (\pi_{2,1}^*, \pi_2, \pi)) \in S'$ is optimal for the problem $J2|p, n_i \leq 2|C_{max}$ with vector $p \in T$ of job durations. As vector p is an arbitrary one in set T, the set S' contains an optimal pair of job permutations for each scenario from set T. Due to Definition 4, the set S' is a dominant set of schedules for the problem $J2|l_{ij} \leq p_{ij} \leq u_{ij}, n_i \leq 2|C_{max}$ with job set $\mathcal{J}$. $\square$

The proof of the following theorem is given in Appendix A.

Theorem 12. *Let the partial strict order $A_{\prec}^{1,2}$ over set $\mathcal{J}_{1,2} = \mathcal{J}_{1,2}^* \cup \mathcal{J}_{1,2}^1 \cup \mathcal{J}_{1,2}^2$ have the form $J_1 \prec \ldots \prec J_k \prec \{J_{k+1}, J_{k+2}, \ldots, J_{k+r}\} \prec J_{k+r+1} \prec \ldots \prec J_{m_{1,2}}$. If inequalities*

$$\sum_{i=r-s+2}^{r+1} l_{k+i,1} \geq \sum_{j=r-s+1}^{r} u_{k+j,2} \tag{13}$$

hold for all indexes $s \in \{1, 2, \ldots, r\}$, then the set $S' = <\{\pi\}, S_{2,1}>$, where $\pi = (J_1, \ldots, J_{k-1}, J_k, J_{k+1}, J_{k+2}, \ldots, J_{k+r}, J_{k+r+1}, \ldots, J_{m_{1,2}}) \in \Pi(G_{1,2})$, is a dominant set of pairs of permutations for the problem $J2|l_{ij} \leq p_{ij} \leq u_{ij}, n_i \leq 2|C_{max}$ with job set $\mathcal{J}$.

Similarly, one can prove sufficient conditions for the existence of an optimal job permutation for the problem $F2|l_{ij} \leq p_{ij} \leq u_{ij}|C_{max}$ with job set $\mathcal{J}_{2,1}$, when the partial strict order $A_{\prec}^{2,1}$ on the set $\mathcal{J}_{2,1} = \mathcal{J}_{2,1}^* \cup \mathcal{J}_{2,1}^1 \cup \mathcal{J}_{2,1}^2$ has the following form: $J_1 \prec \ldots \prec J_k \prec \{J_{k+1}, J_{k+2}, \ldots, J_{k+r}\} \prec J_{k+r+1} \prec \ldots \prec J_{m_{2,1}}$.

To apply Theorems 11 and 12, one can construct a job permutation that satisfies the strict order $A_{\prec}^{1,2}$. Then, one can check the conditions of Theorems 11 and 12 for the constructed permutation. If the set of jobs $\{J_1, J_2, \ldots, J_k\}$ is empty in the constructed permutation, one needs to check conditions of Theorem 12. If the set of jobs $\{J_{k+r+1}, \ldots, J_{m_{1,2}}\}$ is empty, one needs to check the conditions of Theorem 11. It is needed to construct only one permutation to check Theorem 11 and only one permutation to check Theorem 12.

5.3. Two Illustrative Examples

Example 1. *We consider the uncertain job-shop scheduling problem $J2|l_{ij} \leq p_{ij} \leq u_{ij}, n_i \leq 2|C_{max}$ with lower and upper bounds of the job durations given in Table 1.*

Table 1. Input data for Example 1.

J_i	l_{i1}	u_{i1}	l_{i2}	u_{i2}
J_1	6	7	6	7
J_2	8	9	5	6
J_3	7	9	5	6
J_4	2	3	-	-
J_5	-	-	16	20
J_6	1	3	3	4
J_7	1	3	3	4
J_8	1	3	3	4

These bounds determine the set T of possible scenarios. In Example 1, jobs J_1, J_2, and J_3 have the machine route (M_1, M_2); jobs J_6, J_7, and J_8 have the machine route (M_2, M_1); and job J_4 (job J_5, respectively) has to be processed only on machine M_1 (on machine M_2, respectively). Thus, $\mathcal{J}_{1,2} = \{J_1, J_2, J_3\}$, $\mathcal{J}_{2,1} = \{J_6, J_7, J_8\}$, $\mathcal{J}_1 = \{J_4\}$, $\mathcal{J}_2 = \{J_5\}$.

We check the conditions of Theorem 7 for a single pair of job permutations, which is optimal for all scenarios T. For the given jobs, the condition in Inequalities (7) of Theorem 7 holds due to the following relations:

$$\sum_{J_i \in \mathcal{J}_{1,2}} u_{i1} = u_{1,1} + u_{2,1} + u_{3,1} = 7 + 9 + 9 = 25 \le \sum_{J_i \in \mathcal{J}_{2,1} \cup \mathcal{J}_2} l_{i2} = l_{6,2} + l_{7,2} + l_{8,2} + l_{5,2} = 3 + 3 + 3 + 16 = 25;$$
$$\sum_{J_i \in \mathcal{J}_{1,2}} l_{i2} = l_{1,2} + l_{2,2} + l_{3,2} = 6 + 5 + 5 = 16 \ge \sum_{J_i \in \mathcal{J}_{2,1} \cup \mathcal{J}_1} u_{i1} = u_{6,1} + u_{7,1} + u_{8,1} + u_{4,1} = 3 + 3 + 3 + 3 = 12.$$

Due to Theorem 7, the order of jobs from the set $\mathcal{J}_{1,2} = \{J_1, J_2, J_3\}$ and the order of jobs from the set $\mathcal{J}_{2,1} = \{J_6, J_7, J_8\}$ may be arbitrary in the optimal pair of job permutations for the problem $J2|l_{ij} \le p_{ij} \le u_{ij}, n_i \le 2|C_{max}$ under consideration. Thus, any pair of job permutations $(\pi', \pi'') \in S$ is a single-element dominant set $DS(T) = \{(\pi', \pi'')\}$ for Example 1.

Example 2. *Let us now consider the problem* $J2|l_{ij} \le p_{ij} \le u_{ij}, n_i \le 2|C_{max}$ *with numerical input data given in Table 1 with the following two exceptions:* $l_{5,2} = 2$ *and* $u_{5,2} = 3$.

We check the condition in Inequalities (7) of Theorem 7 and obtain

$$\sum_{J_i \in \mathcal{J}_{1,2}} u_{i1} = u_{1,1} + u_{2,1} + u_{3,1} = 7 + 9 + 9 = 25 \not\le \sum_{J_i \in \mathcal{J}_{2,1} \cup \mathcal{J}_2} l_{i2} = l_{6,2} + l_{7,2} + l_{8,2} + l_{5,2} = 3 + 3 + 3 + 2 = 11. \tag{14}$$

Thus, the condition of Inequalities (7) does not hold for Example 2. We check the condition of Inequalities (8) of Theorem 7 and obtain

$$\sum_{J_i \in \mathcal{J}_{2,1}} u_{i2} = u_{6,2} + u_{7,2} + u_{8,2} = 4 + 4 + 4 = 12 \le \sum_{J_i \in \mathcal{J}_{1,2} \cup \mathcal{J}_1} l_{i1} = l_{1,1} + l_{2,1} + l_{3,1} + l_{4,1} = 6 + 8 + 7 + 2 = 23. \tag{15}$$

However, we see that the condition of Equation (8) does not hold:

$$\sum_{J_i \in \mathcal{J}_{2,1}} l_{i1} = l_{6,1} + l_{7,1} + l_{8,1} = 1 + 1 + 1 = 3 \not\ge \sum_{J_i \in \mathcal{J}_{1,2} \cup \mathcal{J}_2} u_{i2} = u_{1,2} + u_{2,2} + u_{3,2} + u_{5,2} = 7 + 6 + 6 + 3 = 22.$$

From Equation (14), it follows that the condition of Inequalities (9) of Corollary 3 does not hold. On the other hand, due to Equation (15), conditions of Corollary 4 hold. Thus, the order for processing jobs from set $\mathcal{J}_{2,1} \subseteq \mathcal{J}$ in the optimal schedule $(\pi', \pi'') = ((\pi_{1,2}, \pi_1, \pi_{2,1}), (\pi_{2,1}, \pi_2, \pi_{1,2}))$ for the problem $J2|l_{ij} \le p_{ij} \le u_{ij}, n_i \le 2|C_{max}$ may be arbitrary. One can fix permutation $\pi_{2,1}$ with the increasing order of the indexes of their jobs: $\pi_{2,1} = (J_6, J_7, J_8)$. Since the orders of jobs from the sets $\mathcal{J}_1$ and $\mathcal{J}_2$ are fixed in the optimal schedule (Remark 1), i.e., $\pi_1 = (J_4)$ and $\pi_2 = (J_5)$, we need to determine the order for processing jobs in set $\mathcal{J}_{1,2}$. To this end, we consider the problem $F2|l_{ij} \le p_{ij} \le u_{ij}|C_{max}$ with job set $\mathcal{J}_{1,2}$. We see that conditions of Theorem 2 do not hold for the jobs in set $\mathcal{J}_{1,2}$ since $J_1 \in \mathcal{J}_{1,2}^*$, $J_2 \in \mathcal{J}_{1,2}^2$, and $J_3 \in \mathcal{J}_{1,2}^2$; however the following inequalities hold: $u_{2,2} > l_{3,2}$ and $u_{3,2} > l_{2,2}$.

We next construct the binary relation $A^{1,2}_{\prec}$ over set $\mathcal{J}_{1,2}$ based on Definition 3 and Theorem 1. Due to checking Inequalities (4) and (5), we conclude that the inequality in Equation (5) holds for the pair of jobs J_1 and J_2. We obtain the relation $J_1 \prec J_2$. Analogously, we obtain the relation $J_1 \prec J_3$. For the pair of jobs J_2 and J_3, neither Inequality (4) nor Inequality (5) hold. Therefore, the partial strict order $A^{1,2}_{\prec}$ over set $\mathcal{J}_{1,2}$ has the following form: $J_1 \prec \{J_2, J_3\}$. The job set $\{J_2, J_3\}$ is a conflict set of these jobs (Definition 6).

Let us check whether the sufficient conditions given in Section 5.2 hold.

We check the conditions of Theorem 10 for the jobs from set $\mathcal{J}_{1,2}$. For $k = 1$ and $r = 2$, we obtain the following equalities: $L_2 = \sum_{J_i \in \mathcal{J}_{2,1} \cup \mathcal{J}_2} l_{i2} = l_{6,2} + l_{7,2} + l_{8,2} + l_{5,2} = 3 + 3 + 3 + 2 = 11$. The condition of Theorem 10 does not hold since the following relations hold:

$$\sum_{i=1}^{k+r} u_{i1} = u_{1,1} + u_{2,1} + u_{3,1} = 7 + 9 + 9 = 25 \nleq L_2 + \sum_{i=1}^{k} l_{i2} = L_2 + l_{1,2} = 11 + 6 = 17.$$

For checking the conditions of Theorem 11, we need to check both permutations of the jobs from set $\mathcal{J}_{1,2}$, which satisfy the partial strict order $A^{1,2}_{\prec}$: $\Pi(\mathcal{G}_{1,2}) = \{\pi^1_{1,2}, \pi^2_{1,2}\}$, where $\pi^1_{1,2} = \{J_1, J_2, J_3\}$ and $\pi^2_{1,2} = \{J_1, J_3, J_2\}$.

We consider permutation $\pi^1_{1,2}$. As in the previous case, $L_2 = 11$, $k = 1$, $r = 2$, and we must consider two inequalities in the condition in Equaiton (12) with $s = 1$ and $s = 2$. For $s = 1$, we obtain the following:

$$u_{1+1,1} = u_{2,1} = 9 \leq L_2 + \sum_{i=1}^{1+1-1} (l_{i2} - u_{i1}) = L_2 + \sum_{i=1}^{1}(l_{i2} - u_{i1}) = 11 + (l_{1,2} - u_{1,1}) = 11 + (6 - 7) = 10.$$

However, for $s = 2$, we obtain

$$u_{1+2,1} = u_{3,1} = 9 \nleq L_2 + \sum_{i=1}^{1+2-1} (l_{i2} - u_{i1}) = L_2 + \sum_{i=1}^{2}(l_{i2} - u_{i1})$$

$$= 11 + (l_{1,2} - u_{1,1}) + (l_{2,2} - u_{2,1}) = 11 + (6 - 7) + (5 - 9) = 6.$$

Thus, the conditions of Theorem 11 do not hold for permutation $\pi^1_{1,2}$.

We consider permutation $\pi^2_{1,2}$, where $J_{k+1} = J_3$ and $J_{k+2} = J_2$. Again, we must test the two inequalities in Equation (12), where either $s = 1$ or $s = 2$. For $s = 1$, we obtain

$$u_{k+1,1} = u_{3,1} = 9 \leq L_2 + \sum_{i=1}^{k+1-1} (l_{i2} - u_{i1}) = L_2 + \sum_{i=1}^{1}(l_{i2} - u_{i1}) = 11 + (l_{1,2} - u_{1,1}) = 11 + (6 - 7) = 10.$$

However, for $s = 2$, we obtain

$$u_{k+2,1} = u_{2,1} = 9 \nleq L_2 + \sum_{i=1}^{k+2-1} (l_{i2} - u_{i1}) = L_2 + \sum_{i=1}^{k+1}(l_{i2} - u_{i1}) = 11 + (l_{1,2} - u_{1,1}) + (l_{3,2} - u_{3,1})$$

$$= 11 + (6 - 7) + (5 - 9) = 6.$$

Thus, the conditions of Theorem 11 do not hold for permutation $\pi^2_{1,2}$.

Note that we do not check the conditions of Theorem 12 since the conflict set of jobs $\{J_2, J_3\}$ is located at the end of the partial strict order $A^{1,2}_{\prec}$. We conclude that none of the proven sufficient conditions are satisfied for a schedule optimality. Thus, there does not exist a pair of permutations of the jobs in set $\mathcal{J} = \mathcal{J}_{1,2} \cup \mathcal{J}_{2,1} \cup \mathcal{J}_1 \cup \mathcal{J}_2$ which is optimal for any scenario $p \in T$. The J-solution $S(T)$ for Example 2 consists of the following two pairs of job permutations: $\{(\pi'_1, \pi''_1), (\pi'_2, \pi''_2)\} = S(T)$, where

$$\pi'_1 = (\pi^1_{1,2}, \pi_1, \pi_{2,1}) = (J_1, J_2, J_3, J_4, J_6, J_7, J_8), \quad \pi''_1 = (\pi_{2,1}, \pi_2, \pi^1_{1,2}) = (J_6, J_7, J_8, J_5, J_1, J_2, J_3),$$

$$\pi_2' = (\pi_{1,2}^2, \pi_1, \pi_{2,1}) = (J_1, J_3, J_2, J_4, J_6, J_7, J_8), \quad \pi_2'' = (\pi_{2,1}, \pi_2, \pi_{1,2}^2) = (J_6, J_7, J_8, J_5, J_1, J_3, J_2).$$

We next show that none of these two pairs of job permutations is optimal for all scenarios $p \in T$ using the following two scenarios: $p' = (7,6,9,5,9,6,2,0,0,2,1,3,1,3,1,3) \in T$ and $p'' = (7,6,9,6,9,5,2,0,0,2,1,3,1,3,1,3) \in T$. For scenario p', only pair of permutations (π_2', π_2'') is optimal with $C_{\max}(\pi_2', \pi_2'') = 30$ since $C_{\max}(\pi_1', \pi_1'') = 31 > 30$. On the other hand, for scenario p'', only the pair of permutations (π_1', π_1'') is optimal with $C_{\max}(\pi_1', \pi_1'') = 30$ since $C_{\max}(\pi_2', \pi_2'') = 31 > 30$.

Note that the whole set S of the semi-active schedules has the cardinality $|S| = m_{1,2}! \cdot m_{2,1}! = 3! \cdot 3! = 6 \cdot 6 = 36$. Thus, for solving Example 2, one needs to consider only two pairs of job permutations $\{(\pi_1', \pi_1''), (\pi_2', \pi_2'')\} = S(T) \subset S$ instead of 36 semi-active schedules.

5.4. An Algorithm for Checking Conditions for the Existence of a Single-Element Dominant Set

We describe Algorithm 1 for checking the existence of an optimal permutation for the problem $F2|l_{ij} \le p_{ij} \le u_{ij}|C_{max}$ with job set $\mathcal{J}_{1,2}$ if the partial strict order $A_{\prec}^{1,2}$ on the set $\mathcal{J}_{1,2}$ has the following form: $J_1 \prec \ldots \prec J_k \prec \{J_{k+1}, J_{k+2}, \ldots, J_{k+r}\} \prec J_{k+r+1} \prec \ldots \prec J_{m_{1,2}}$. Algorithm 1 considers a set of conflict jobs and checks whether the sufficient conditions given in Section 5.2 hold. For a conflict set of jobs, it is needed to construct two permutations and to check the condition in Inequality (12) for the first permutation and the condition in Inequality (13) for the second one. If at least one of these conditions holds, Algorithm 1 constructs a permutation which is optimal for the problem $F2|l_{ij} \le p_{ij} \le u_{ij}|C_{max}$ with any scenario $p \in T$.

Obviously, testing the conditions of Theorems 11 and 12 takes $O(r)$, where the conflict set contains r jobs. The construction of the permutation of r jobs takes $O(r \log r)$. Therefore, the total complexity of Algorithm 1 is $O(r \log r)$.

Remark 3. *If Algorithm 1 is completed at Step 7 (STOP 1), we suggest to consider a set of conflict jobs $\{J_{k+1}, J_{k+2}, \ldots, J_{k+r}\}$ and construct a Johnson's permutation for the deterministic problem $F2|p'|C_{max}$ with job set $\mathcal{J}' = \{J_{k+1}, J_{k+2}, \ldots J_{k+r}\}$, where vector $p' = (p'_{k+1,1}, p'_{k+1,2} \cdots p'_{k+r,1}, p'_{k+r,2})$ of the durations of conflict jobs $\{J_{k+1}, J_{k+2}, \ldots J_{k+r}\}$ is calculated for each operation O_{ij} of the conflict job $J_i \in \{J_{k+1}, J_{k+2}, \ldots J_{k+r}\}$ on the corresponding machine $M_j \in \mathcal{M}$ as folows:*

$$p'_{ij} = (u_{ij} + l_{ij})/2 \tag{16}$$

Theorem 11 and Theorem 12 imply the following claim.

Corollary 7. *Algorithm 1 constructs a permutation π^* either satisfying conditions of Theorem 11 or Theorem 12 (such permutation π^* is optimal for the problem $F2|l_{ij} \le p_{ij} \le u_{ij}|C_{max}$ with job set $\mathcal{J}_{1,2}$ and any scenario $p \in T$) or establishes that an optimal job permutation for the problem $F2|l_{ij} \le p_{ij} \le u_{ij}|C_{max}$ with any scenario $p \in T$ does not exist.*

The set of jobs $\mathcal{J}_{2,1}$ for the problem $F2|l_{ij} \le p_{ij} \le u_{ij}|C_{max}$ with job set $\mathcal{J} = \mathcal{J}_{2,1}$ can be tested similarly to the set of jobs $\mathcal{J}_{1,2}$.

Algorithm 1: Checking conditions for the existence of a single-element dominant set of schedules for the problem $F2|l_{ij} \leq p_{ij} \leq u_{ij}|C_{max}$

Input: Segments $[l_{ij}, u_{ij}]$ for all jobs $J_i \in \mathcal{J}$ and machines $M_j \in \mathcal{M}$,
a partial strict order $A_{\prec}^{1,2}$ on the set $\mathcal{J}_{1,2} = \mathcal{J}_{1,2}^* \cup \mathcal{J}_{1,2}^1 \cup \mathcal{J}_{1,2}^2$ in the form
$J_1 \prec \ldots \prec J_k \prec \{J_{k+1}, J_{k+2}, \ldots, J_{k+r}\} \prec J_{k+r+1} \prec \ldots \prec J_{m_{1,2}}.$

Output: **EITHER** an optimal job permutation for the problem
$F2|l_{ij} \leq p_{ij} \leq u_{ij}|C_{max}$ with job set $\mathcal{J}_{1,2}$ and any scenario $p \in T$, (see STOP 0)
OR there no permutation $\pi_{1,2}$ of jobs from set $\mathcal{J}_{1,2}$, which is optimal
for all scenarios $p \in T$, (see STOP 1).

Step 1: Set $\delta_s = l_{k+s,2} - u_{k+s,1}$ for all $s \in \{1, 2, \ldots, r\}$
construct a partition of the set of conflicting jobs into two subsets X_1 and X_2,
where $J_{k+s} \in X_1$ if $\delta_s \geq 0$, and $J_{k+s} \in X_2$, otherwise.

Step 2: Construct a permutation $\pi^1 = (J_1, J_2, \ldots, J_k, \pi_1, \pi_2, J_{k+r+1}, \ldots, J_{m_{1,2}})$, where the permutation
π_1 contains jobs from the set X_1 in the non-decreasing order of the values $u_{k+i,1}$ and the
permutation π_2 contains jobs from the set X_2 in the non-increasing order of the values
$l_{k+i,2}$, renumber jobs in the permutations π_1 and π_2 based on their orders.

Step 3: **IF** for the permutation π^1 conditions of Theorem 11 hold **THEN GOTO** step 8.

Step 4: Set $\delta_s = l_{k+s,1} - u_{k+s,2}$ for all $s \in \{1, 2, \ldots, r\}$
construct a partition of the set of conflicting jobs into two subsets
Y_1 and Y_2, where $J_{k+s} \in Y_1$ if $\delta_s \geq 0$, and $J_{k+s} \in Y_2$, otherwise.

Step 5: Construct a permutation $\pi^2 = (J_1, J_2, \ldots, J_k, \pi_2, \pi_1, J_{k+r+1}, \ldots, J_{m_{1,2}})$, where the permutation
π_1 contains jobs from the set Y_1 in the non-increasing order of the values $u_{k+i,2}$, and the
permutation π_2 contains jobs from the set Y_2 in the non-decreasing order of the
values $l_{k+i,1}$, renumber jobs in the permutations π_1 and π_2 based on their orders.

Step 6: **IF** for the permutation π^2 conditions of Theorem 12 hold **THEN GOTO** step 9.

Step 7: **ELSE** there is no a single dominant permutation for problem
$F2|l_{ij} \leq p_{ij} \leq u_{ij}|C_{max}$ with job set $\mathcal{J}_{1,2}$ and any scenario $p \in T$ **STOP 1**.

Step 8: **RETURN** permutation π^1, which is a single dominant permutation
for the problem $F2|l_{ij} \leq p_{ij} \leq u_{ij}|C_{max}$ with job set $\mathcal{J}_{1,2}$ **STOP 0**.

Step 9: **RETURN** permutation π^2, which is a single dominant permutation
for the problem $F2|l_{ij} \leq p_{ij} \leq u_{ij}|C_{max}$ with job set $\mathcal{J}_{1,2}$ **STOP 0**.

---

6. Algorithms for Constructing a Small Dominant Set of Schedules for the Problem $J2|l_{ij} \leq p_{ij} \leq u_{ij}, n_i \leq 2|C_{max}$

In this section, we describe Algorithm 2 for constructing a small dominant set $DS(T)$ of schedules for the problem $J2|l_{ij} \leq p_{ij} \leq u_{ij}, n_i \leq 2|C_{max}$. Algorithm 2 is developed for use at the off-line phase of scheduling (before processing any job from the set $\mathcal{J}$). Based on the initial data, Algorithm 2 checks the conditions of Theorem 7 for a single optimal pair of job permutations for the uncertain problem $J2|l_{ij} \leq p_{ij} \leq u_{ij}, n_i \leq 2|C_{max}$. If the sufficient conditions of Theorem 7 do not hold, Algorithm 2 proceeds to consider the problem $F2|l_{ij} \leq p_{ij} \leq u_{ij}|C_{max}$ with job set $\mathcal{J}_{1,2}$ and the problem $F2|l_{ij} \leq p_{ij} \leq u_{ij}|C_{max}$ with job set $\mathcal{J}_{2,1}$. For each of these problems, the conditions of Theorem 2 are checked. If these conditions do not hold, then strict orders of the jobs $\mathcal{J}$ based on Inequalities (4) and (5) are constructed. In this general case, Algorithm 2 constructs a partial strict order $A_{\prec}^{1,2}$ of the jobs from set $\mathcal{J}_{1,2}$ and a partial strict order $A_{\prec}^{2,1}$ of the jobs from set $\mathcal{J}_{2,1}$. Each of these partial orders may contain one or several conflict sets of jobs. For each such conflict set of jobs, Algorithm 2 checks whether the sufficient conditions given in Section 5.2 hold. Thus, if some sufficient conditions for a schedule optimality presented in Sections 4 and 5 are satisfied, then there exists a pair of permutations of jobs from set $\mathcal{J}$ which is optimal for any scenario $p \in T$. Algorithm 2 constructs such a pair of job permutations $\{(\pi', \pi'')\} = DS(T)$. Otherwise, the precedence digraphs determining a minimal dominant set $DS(T)$ of schedules is constructed by Algorithm 2. The more job pairs are involved in the binary relations $A_{\prec}^{1,2}$ and $A_{\prec}^{2,1}$, the more job permutations will be deleted from set S while constructing a J-solution $S(T) \subseteq S$ for the problems $F2|l_{ij} \leq p_{ij} \leq u_{ij}|C_{max}$ with job sets $\mathcal{J}_{1,2}$ and $\mathcal{J}_{2,1}$.

Algorithm 2: Construction of a small dominant set of schedules for the problem $J2|l_{ij} \leq p_{ij} \leq u_{ij}, n_i \leq 2|C_{max}$

Input:	Lower bounds l_{ij} and upper bounds u_{ij}, $0 < l_{ij} \leq u_{ij}$, of the durations
	of all operations O_{ij} of jobs $J_i \in \mathcal{J}$ processed on machines $M_j \in \mathcal{M} = \{M_1, M_2\}$.

Output: **EITHER** pair of permutations $(\pi', \pi'') = ((\pi_{1,2}, \pi_1, \pi_{2,1}), (\pi_{2,1}, \pi_2, \pi_{1,2}))$,
  where π' is a permutation of jobs from set $\mathcal{J}_{1,2} \cup \mathcal{J}_1 \cup \mathcal{J}_{2,1}$ on machine
  M_1, π'' is a permutation of jobs from set $\mathcal{J}_{1,2} \cup \mathcal{J}_2 \cup \mathcal{J}_{2,1}$ on machine M_2,
  such that $\{(\pi', \pi'')\} = DS(T)$, (see STOP 0),
  OR permutation $\pi_{2,1}$ of jobs from set $\mathcal{J}_{2,1}$ on machines M_1 and M_2 and
  a partial strict order $A_{\prec}^{1,2}$ of jobs from set $\mathcal{J}_{1,2}$,
  OR permutation $\pi_{1,2}$ of jobs from set $\mathcal{J}_{1,2}$ on machines M_1 and M_2 and
  a partial strict order $A_{\prec}^{2,1}$ of jobs from set $\mathcal{J}_{2,1}$,
  OR a partial strict order $A_{\prec}^{1,2}$ of jobs from set $\mathcal{J}_{1,2}$ and
  a partial strict order $A_{\prec}^{2,1}$ of jobs from set $\mathcal{J}_{2,1}$, (see STOP 1).

Step 1: Determine a partition $\mathcal{J} = \mathcal{J}_1 \cup \mathcal{J}_2 \cup \mathcal{J}_{1,2} \cup \mathcal{J}_{2,1}$ of the job set $\mathcal{J}$,
  permutation π_1 of jobs from set $\mathcal{J}_1$ and permutation π_2 of jobs from
  set $\mathcal{J}_2$, arrange the jobs in the increasing order of their indexes.

Step 2: **IF** the first inequality in condition (7) of Theorem 7 holds **THEN BEGIN**
  Construct a permutation $\pi_{1,2}$ of jobs from set $\mathcal{J}_{1,2}$,
   arrange them in the increasing order of their indexes;
  IF the second inequality in condition (7) of Theorem 7 holds
   THEN construct a permutation $\pi_{2,1}$ of jobs from set $\mathcal{J}_{2,1}$,
   arrange them in the increasing order of their indexes **GOTO** Step 10 **END**

Step 3: **IF** the first inequality in condition (8) of Theorem 7 holds **THEN BEGIN**
  Construct a permutation $\pi_{2,1}$ of jobs from set $\mathcal{J}_{2,1}$,
   arrange them in the increasing order of their indexes;
  IF the second inequality in condition (8) of Theorem 7 holds **THEN**
   construct a permutation $\pi_{1,2}$ of jobs from set $\mathcal{J}_{1,2}$,
   arrange the jobs in the increasing order of their indexes **END**

Step 4: **IF** both permutations $\pi_{1,2}$ and $\pi_{2,1}$ are constructed **THEN GOTO** Step 10.

Step 5: **IF** permutation $\pi_{1,2}$ is not constructed **THEN** fulfill Algorithm 3.

Step 6: **IF** permutation $\pi_{2,1}$ is not constructed **THEN** fulfill Algorithm 4.

Step 7: **IF** both permutations $\pi_{1,2}$ and $\pi_{2,1}$ are constructed **THEN GOTO** Step 10.

Step 8: **IF** permutation $\pi_{2,1}$ is constructed **THEN GOTO** Step 11.

Step 9: **IF** permutation $\pi_{1,2}$ is constructed **THEN GOTO** Step 12 **ELSE GOTO** Step 13.

Step 10: **RETURN** pair of permutations (π', π''), where π' is the permutation
  of jobs from set $\mathcal{J}_{1,2} \cup \mathcal{J}_1 \cup \mathcal{J}_{2,1}$ processed on machine M_1 and π'' is
  the permutation of jobs from set $\mathcal{J}_{1,2} \cup \mathcal{J}_2 \cup \mathcal{J}_{2,1}$ processed
  on machine M_2 such that $\{(\pi', \pi'')\} = DS(T)$ **STOP 0**.

Step 11: **RETURN** the permutation $\pi_{2,1}$ of jobs from set $\mathcal{J}_{2,1}$ processed on machines M_1 and M_2,
  the partial strict order $A_{\prec}^{1,2}$ of jobs from set $\mathcal{J}_{1,2}$ **GOTO** Step 14.

Step 12: **RETURN** the permutation $\pi_{1,2}$ of jobs from set $\mathcal{J}_{1,2}$ processed on machines M_1 and M_2,
  the partial strict order $A_{\prec}^{2,1}$ of jobs from set $\mathcal{J}_{2,1}$ **GOTO** Step 14.

Step 13: **RETURN** the partial strict order $A_{\prec}^{1,2}$ of jobs from set $\mathcal{J}_{1,2}$
  and the partial strict order $A_{\prec}^{2,1}$ of jobs from set $\mathcal{J}_{2,1}$

Step 14: **STOP 1**.

Algorithm 2 may be applied for solving the problem $J2|l_{ij} \leq p_{ij} \leq u_{ij}, n_i \leq 2|C_{max}$ exactly or approximately as follows. If at least one of the sufficient conditions proven in Section 5.1 hold, then

Algorithm 2 constructs a pair of job permutations $(\pi', \pi'') = ((\pi_{1,2}, \pi_1, \pi_{2,1}), (\pi_{2,1}, \pi_2, \pi_{1,2}))$, which is optimal for any scenario $p \in T$ (Step 10).

It may happen that the constructed strict order on the set $\mathcal{J}_{1,2}$ or on the set $\mathcal{J}_{2,1}$ is not a linear strict order. If for at least one of the sets $\mathcal{J}_{1,2}$ or $\mathcal{J}_{2,1}$, the constructed partial strict order is not a linear one, a heuristic solution for the problem $J2|l_{ij} \leq p_{ij} \leq u_{ij}, n_i \leq 2|C_{max}$ is constructed similar to that for the problem $F2|l_{ij} \leq p_{ij} \leq u_{ij}|C_{max}$ solved by Algorithm 1 (see Section 5.4). If Algorithm 2 is completed at Steps 11-13 (STOP 1), we consider a set of conflict jobs $\{J_{k+1}, J_{k+2}, \dots, J_{k+r}\}$ and construct a Jackson's pair of job permutation for the deterministic problem $J2|p', n_i \leq 2|C_{max}$ with job set $\mathcal{J} = \{J_{k+1}, J_{k+2}, \dots J_{k+r}\}$, where the vector $p' = (p'_{k+1,1}, p'_{k+1,2}, \dots p'_{k+r,1}, p'_{k+r,2})$ of the durations of conflict jobs $\{J_{k+1}, J_{k+2}, \dots J_{k+r}\}$ is calculated using the equality of Equation (16) for each operation O_{ij} of the conflict job $J_i \in \{J_{k+1}, J_{k+2}, \dots J_{k+r}\}$ on the corresponding machine $M_j \in \mathcal{M}$ (Remark 3).

Algorithm 3: Construction of a strict order $A_{\prec}^{1,2}$ on the set $\mathcal{J}_{1,2}$

Input: Lower bounds l_{ij} and upper bounds $u_{ij}, 0 < l_{ij} \leq u_{ij}$, of the durations
of all operations O_{ij} of jobs $J_i \in \mathcal{J}$ on machines $M_j \in \mathcal{M} = \{M_1, M_2\}$.

Output: **EITHER** permutation $\pi_{1,2}$, which is optimal for the problem
$F2|l_{ij} \leq p_{ij} \leq u_{ij}|C_{max}$ with any scenario $p \in T$ for the jobs $\mathcal{J}_{1,2}$,
OR partial strict order $A_{\prec}^{1,2}$ on the set $\mathcal{J}_{1,2}$.

Step 1: Construct a partition $\mathcal{J}_{1,2} = \mathcal{J}_{1,2}^1 \cup \mathcal{J}_{1,2}^2 \cup \mathcal{J}_{1,2}^*$ of the set $\mathcal{J}_{1,2}$ of the jobs.

Step 2: **IF** conditions of Theorem 2 hold **THEN**

Step 3: Construct permutation $\pi_{1,2} = (\pi_{1,2}^1, J_{1,2}^*, \pi_{1,2}^2)$, where $\pi_{1,2}^1$ is a permutation for
processing jobs from the set $\mathcal{J}_{1,2}^1$ in the non-decreasing order of the values u_{i1},
$\pi_{1,2}^2$ is a permutation for processing jobs from the set $\mathcal{J}_{1,2}^2$
in the non-increasing order of the values u_{i2} **GOTO** Step 7 **ELSE**

Step 4: **FOR** each pair of jobs $J_v \in \mathcal{J}_{1,2}$ and $J_w \in \mathcal{J}_{1,2}, v \neq w$, **DO**
IF at least one of two conditions (4) and (5) holds **THEN**
determine the relation $J_v \prec J_w$
END FOR

Step 5: Renumber jobs in the set $\mathcal{J}_{1,2}$ such that relation $v < w$ holds if $J_v \prec J_w$.

Step 6: **FOR** each conflict set of jobs **DO**
IF condition of Theorem 10 holds **THEN**
Order jobs in the conflict set in the increasing order of their indexes **GOTO** Step 7
ELSE fulfill Algorithm 1
END FOR

Step 7: **IF** the partial strict order $A_{\prec}^{1,2}$ is linear **THEN**
construct a permutation $\pi_{1,2}$ generated by the linear order $A_{\prec}^{1,2}$
STOP.

Algorithm 4 is obtained from the above Algorithm 3 by replacing the set $\mathcal{J}_{1,2}$ of jobs by the set $\mathcal{J}_{2,1}$ of jobs, machine M_1 by machine M_2, and vice versa. Obviously, testing the conditions of Theorems 11 and 12 takes $O(r)$, where conflict set contains r jobs. Construction of permutation of r jobs takes $O(r \log r)$. Therefore, the total complexity of Algorithm 1 is $O(r \log r)$.

Testing the conditions of Theorem 2 takes $O(m_{1,2} \log m_{1,2})$ time. A strict order $A_{\prec}^{1,2}$ on the set $\mathcal{J}_{1,2}$ is constructed by comparing no more than $m_{1,2}(m_{1,2} - 1)$ pairs of jobs in the set $\mathcal{J}_{1,2}$. Thus, it takes $O(m_{1,2}(m_{1,2} - 1))$ time. The complexity of Algorithm 1 is $O(r \log r)$ time provided that the conflict set contains r jobs, where $r \leq m_{1,2}$. Since a strict order $A_{\prec}^{1,2}$ is constructed once in Algorithm 3, we conclude that a total complexity of Algorithm 3 (and Algorithm 4) is $O(n^2)$ time.

In Algorithm 2, testing the condition of Theorem 7 takes $O(\max\{m_{1,2}, m_{2,1}\})$ time. Every Algorithm 3 or Algorithm 4 is fulfilled at most once. Therefore, the complexity of Algorithm 2 is $O(n^2)$ time.

7. Computational Experiments

We describe the conducted computational experiments and discuss the results obtained for randomly generated instances of the problem $J2|l_{ij} \leq p_{ij} \leq u_{ij}, n_i \leq 2|C_{max}$. In the computational experiments, each tested series consisted of 1000 randomly generated instances with the same numbers $n \in \{10, 20, \ldots, 100, 200, \ldots, 1000, 2000, \ldots, 10.000\}$ of jobs in the set $\mathcal{J}$ provided that a maximum relative length δ of the given segment of the possible durations of the operations O_{ij} takes the following values: $\{5\%, 10\%, 15\%, 20\%, 30\%, 40\%, \text{and } 50\%\}$. The lower bounds l_{ij} and upper bounds u_{ij} for possible values of the durations p_{ij} of the operations O_{ij}, $p_{ij} \in [l_{ij}, u_{ij}]$ using the value δ have been determined as follows. First, a value of the lower bound l_{ij} is randomly chosen from the segment $[10, 1000]$ using a uniform distribution. Then, the upper bound u_{ij} is calculated using the following equality:

$$ u_{ij} = l_{ij} \left(1 + \frac{\delta}{100} \right) \tag{17} $$

For example, we assume that $\delta = 5\%$. Then, for the lower bounds $l_{ij} = 50$ and $l_{ij} = 500$, the upper bounds $u_{ij} = 52.5$ and $u_{ij} = 525$ are calculated using Reference (17). If $\delta = 50\%$, then based on the lower bounds $l_{ij} = 50$ and $l_{ij} = 500$ and on Reference (17), we obtain the upper bounds $u_{ij} = 75$ and $u_{ij} = 750$. Thus, rather wide ranges for the tested durations of the jobs $\mathcal{J}$ were considered.

In the experiments, the bounds l_{ij} and u_{ij} were decimal fractions with the maximum possible number of digits after the decimal point. For all tested instances of the problem $J2|l_{ij} \leq p_{ij} \leq u_{ij}, n_i \leq 2|C_{max}$, a strict inequality $l_{ij} < u_{ij}$ was guarantied for each job $J_i \in \mathcal{J}$ and each machine $M_j \in \mathcal{M}$.

We used Algorithms 1 – 4 described in Section 5.4 and Section 6 for solving the problem $J2|l_{ij} \leq p_{ij} \leq u_{ij}, n_i \leq 2|C_{max}$. These algorithms were coded in C# and tested on a PC with Intel Core i7-7700 (TM) 4 Quad, 3.6 GHz, and 32.00 GB RAM. Since Algorithms 1 – 4 are polynomial in number n jobs in set $\mathcal{J}$, the calculations were carried out quickly. In the experiments, we tested 15 classes of randomly generated instances of the problem $J2|l_{ij} \leq p_{ij} \leq u_{ij}, n_i \leq 2|C_{max}$ with different ratios between numbers m_1, m_2, $m_{1,2}$, and $m_{2,1}$ of the jobs in subsets $\mathcal{J}_1$, $\mathcal{J}_2$, $\mathcal{J}_{1,2}$, and $\mathcal{J}_{2,1}$ of the set $\mathcal{J}$. The obtained computational results are presented in Tables A1–A15 for 15 classes of the solved instances. Each tested class of the instances of the problem $J2|l_{ij} \leq p_{ij} \leq u_{ij}, n_i \leq 2|C_{max}$ is characterized by the following ratio of the percentages of the number of jobs in the subsets $\mathcal{J}_1$, $\mathcal{J}_2$, $\mathcal{J}_{1,2}$, and $\mathcal{J}_{2,1}$ of the set $\mathcal{J}$:

$$ \frac{m_1}{n} \cdot 100\% : \frac{m_2}{n} \cdot 100\% : \frac{m_{1,2}}{n} \cdot 100\% : \frac{m_{2,1}}{n} \cdot 100\% \tag{18} $$

Tables A1–A9 present the computational results obtained for classes 1–9 of the tested instances characterized by the following ratios (Equation (18)):

25% : 25% : 25% : 25% (Table A1); 10% : 10% : 40% : 40% (Table A2);
10% : 40% : 10% : 40% (Table A3); 10% : 30% : 10% : 50% (Table A4);
10% : 20% : 10% : 60% (Table A5); 10% : 10% : 10% : 70% (Table A6);
5% : 20% : 5% : 70% (Table A7); 5% : 15% : 5% : 75% (Table A8);
5% : 5% : 5% : 85% (Table A9).

Note that all instances from class 1 of the instances with the ratio from Equation (18), 25% : 25% : 25% : 25%, were optimally solved by Algorithm 1 – 4 for all values of $\delta \in \{5\%, 10\%, 15\%, 20\%, 30\%, 40\%, \text{and } 50\%\}$. We also tested classes 10–15 of the hard instances of the problem $J2|l_{ij} \leq p_{ij} \leq u_{ij}, n_i \leq 2|C_{max}$ characterized by the following ratios (Equation (18)):

3% : 2% : 5% : 90% (Table A10); 2% : 3% : 5% : 90% (Table A11);
2% : 2% : 1% : 95% (Table A12); 1% : 2% : 2% : 95% (Table A13);
1% : 1% : 3% : 95% (Table A14); 1% : 1% : 1% : 97% (Table A15).

All Tables A1–A15 are organized as follows. Number n of given jobs $\mathcal{J}$ in the instances of the problem $J2|l_{ij} \leq p_{ij} \leq u_{ij}, n_i \leq 2|C_{max}$ are presented in column 1. The values of δ (a maximum relative length of the given segment of the job durations) in percentages are presented in the first line of each table. For the fixed value of δ, the obtained computational results are presented in four columns called *Opt*, *NC*, *SC*, and *t*. The column *Opt* determines the percentage of instances from the series of 1000 randomly generated instances which were optimally solved using Algorithms 1 – 4. For each such instance, an optimal pair (π', π'') of the job permutations was constructed in spite of the uncertain durations of the given jobs $\mathcal{J}$. In other words, the equality $C_{max}(\pi', \pi'') = C_{max}(\pi^*, \pi^{**})$ holds, where $(\pi^*, \pi^{**}) \in S$ is a pair of job permutations which is optimal for the deterministic problem $J2|p^*, n_i \leq 2|C_{max}$ associated with the factual scenario $p^* \in T$. The factual scenario $p^* \in T$ for the instance of the uncertain problem $J2|l_{ij} \leq p_{ij} \leq u_{ij}, n_i \leq 2|C_{max}$ is assumed to be unknown until completing the jobs $\mathcal{J}$.

Column NC presents total number of conflict sets of the jobs in the partial strict orders $A^{1,2}_{\prec}$ on the job sets $\mathcal{J}_{1,2}$ and partial strict orders $A^{2,1}_{\prec}$ on the job sets $\mathcal{J}_{2,1}$ constructed by Algorithm 2. The value of NC is equal to the total number of decision points, where Algorithm 2 has to select an order for processing jobs from the corresponding conflict set. To make a *correct decision* for such an order means to construct a permutation of all jobs from the conflict set, which is optimal for the factual scenario (which is unknown before scheduling). In particular, if all conflict sets have received correct decisions in Algorithm 2, then the constructed pair of job permutations will be optimal for the problem $J2|p^*, n_i \leq 2|C_{max}$, where $p^* \in T$ is the factual scenario.

Column SC presents a percentage of the correct decisions made for determining optimal orders of the conflict jobs by Algorithm 2 with Algorithms 3 and 4. Column t presents a total CPU time (in seconds) for solving all 1000 instances of the corresponding series.

Average percentages of the instances which were optimally solved (Opt) are presented in Figure 1 for classes 1–9 of the tested instances and in Figure 2 for classes 10–15 of the hard-tested instances.

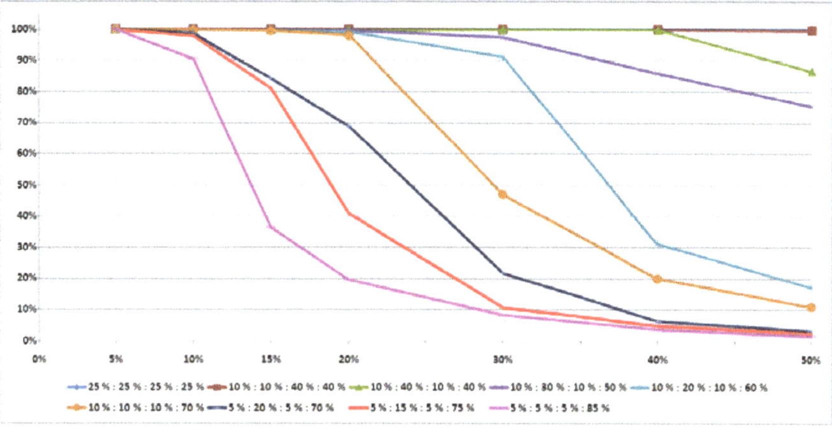

Figure 1. Average percentages of the instances presented in Tables A1–A9, which were optimally solved at the off-line phase of scheduling.

Percentages of the average values of the correct decisions (SC) made for determining optimal orders of the conflict jobs for classes 1–9 are presented in Figure 3. Most instances from these nine classes were optimally solved (Table 2). If the values of δ were no greater than 20%, i.e., $\delta \in \{5\%, 10\%, 15\%, 20\%\}$, then more than 80% of the tested instances were optimally solved in spite of the data uncertainty. If the value δ is increased, the percentage of the optimally solved instances decreased. If the value δ was equal to 50%, then 45% of the tested instances was optimally solved.

For all series of the hard instances presented in Tables A10–A15 (see the third line in Table 2), only a few instances were optimally solved. If $\delta = 5\%$, then 70% of the tested instances was optimally solved. If value δ belongs to the set $\{20\%, 30\%, 40\%, 50\%\}$, then only 1% of the tested instances was optimally solved. There were no hard-tested instances optimally solved for the value of $\delta = 50\%$.

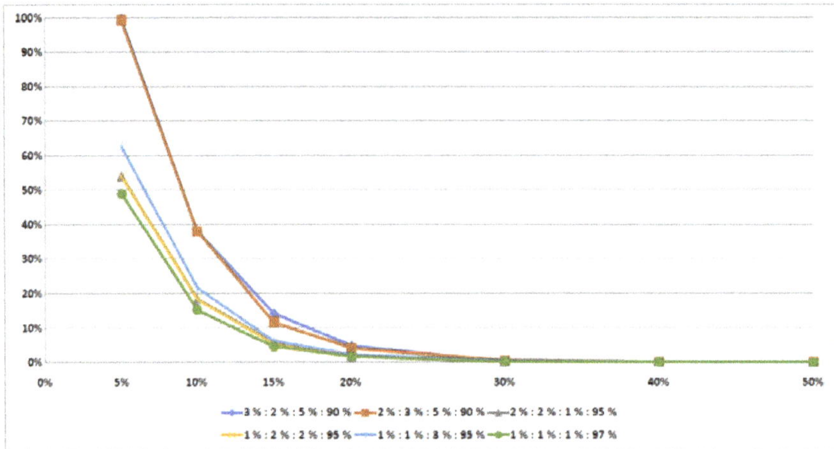

Figure 2. Average percentages of the instances presented in Tables A10–A15, which were optimally solved at the off-line phase of scheduling.

Table 2. Average percentage of the instances which were optimally solved.

$\delta\%$	5%	10%	15%	20%	30%	40%	50%	Average
Instances from Tables A1–A9	99.93	98.48	88.93	80.66	63.97	50.18	44.10	75.18
Instances from Tables A10–A15	69.78	24.89	7.96	2.73	0.20	0.03	0.00	15.08

Percentages of the average values of the correct decisions made for determining optimal orders of the conflict jobs by Algorithm 2, Algorithm 3 and Algorithm 4 for the hard classes 10–15 of the tested instances are presented in Figure 4. Note that there is a correlation between values of *Opt* and *SC* presented in Figures 1 and 3 for classes 1–9 of the tested instances and those presented in Figures 2 and 4 for classes 10–15 of the hard-tested instances.

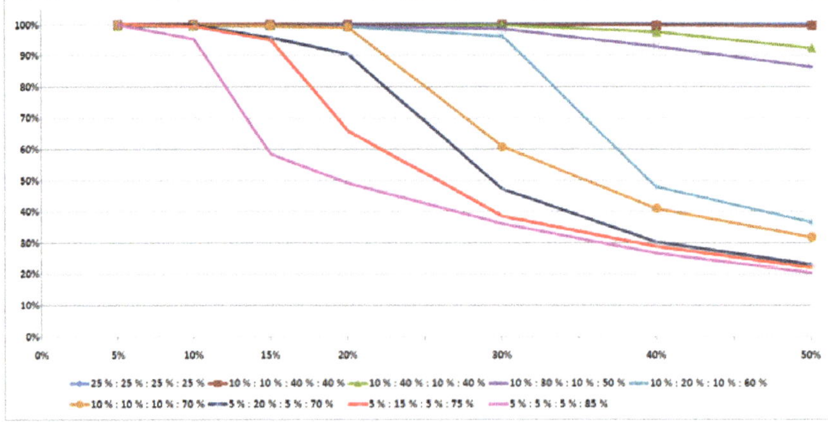

Figure 3. Average percentages of the correct decisions made for constructing permutations of the conflict jobs for the instances presented in Tables A1–A9.

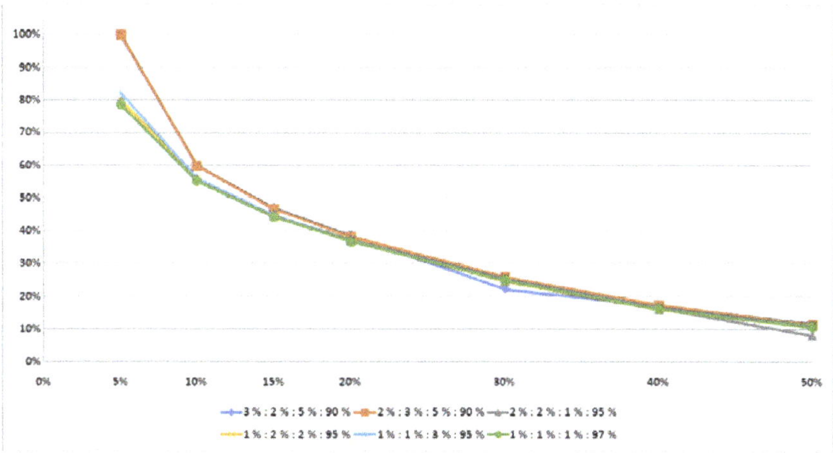

Figure 4. Average percentages of the correct decisions made for constructing permutations of the conflict jobs for the hard instances presented in Tables A10–A15.

8. Concluding Remarks and Future Works

The uncertain flow-shop scheduling problem $F2|l_{ij} \le p_{ij} \le u_{ij}|C_{max}$ and its generalization the job-shop problem $J2|l_{ij} \le p_{ij} \le u_{ij}, n_i \le 2|C_{max}$ attract the attention of researchers since these problems are applicable in many real-life scheduling systems. The optimal scheduling decisions for these problems allow the plant to reduce the costs of productions due to a better utilization of the available machines and other resources. In Section 5, we proved several properties of the optimal pairs (π', π'') of job permutations (Theorems 7–12). Using these properties, we derived Algorithms 1–4 for constructing optimal pairs (π', π'') of job permutations or a small dominant set of schedules for the problem $J2|l_{ij} \le p_{ij} \le u_{ij}, n_i \le 2|C_{max}$. If it is impossible to construct a single pair (π', π'') of job permutations, which dominates all other pairs of job permutations for all possible scenarios T, then Algorithm 2 determines the partial strict order $A_{\prec}^{1,2}$ on the job set $\mathcal{J}_{1,2}$ (Algorithm 3) and the partial strict order $A_{\prec}^{2,1}$ on the job set $\mathcal{J}_{2,1}$ (Algorithm 4). The precedence digraphs $(\mathcal{J}_{1,2}, A_{\prec}^{1,2})$ and $(\mathcal{J}_{2,1}, A_{\prec}^{2,1})$ determine a minimal dominant set of schedules for the problem $J2|l_{ij} \le p_{ij} \le u_{ij}, n_i \le 2|C_{max}$.

From the conducted extensive computational experiments, it follows that pairs of job permutations constructed using Algorithm 2 are close to the optimal pairs of job permutations, which may be determined after completing all jobs $\mathcal{J}$ when factual operation durations become known. We tested 15 classes of the randomly generated instances $J2|l_{ij} \le p_{ij} \le u_{ij}, n_i \le 2|C_{max}$. Most instances from tested classes 1–9 were optimally solved at the off-line phase of scheduling. If the values of δ were no greater than 20%, i.e., $\delta \in \{5\%, 10\%, 15\%, 20\%\}$, then more than 80% of the tested instances was optimally solved in spite of the uncertainty of the input data. If $\delta = 50\%$, then 45% of the tested instances was optimally solved. However, less than 5% of the instances with $\delta \ge 20\%$ from hard classes 10–15 were optimally solved at the off-line phase of scheduling (Figure 2). There were no tested hard instances optimally solved for the value $\delta = 50\%$.

In future research, the on-line phase of scheduling will be studied for the problem $J2|l_{ij} \le p_{ij} \le u_{ij}, n_i \le 2|C_{max}$. To this end, it will be useful to find sufficient conditions for existing a dominant pair of job permutations at the on-line phase of scheduling. The additional information on the factual value of the job duration becomes available once the processing of the job on the corresponding machine is completed. Using this additional information, a scheduler can determine a smaller dominant set DS of schedules, which is based on sufficient conditions for schedule dominance. The smaller DS enables a scheduler to quickly make an on-line scheduling decision whenever additional information on processing the job becomes available. To solve the problem $J2|l_{ij} \le p_{ij} \le u_{ij}, n_i \le 2|C_{max}$ at the on-line phase, a scheduler needs to use fast (better polynomial) algorithms. The investigation of the on-line phase of scheduling for the uncertain job-shop problem is under development.

We suggest to investigate properties of the optimality box and optimality region for a pair (π', π'') of the job permutations and to develop algorithms for constructing a pair (π', π'') of the job permutations that have the largest optimality box (or the largest optimality region). We also suggest to apply the stability approach for solving the uncertain flow-shop and job-shop scheduling problems with $|\mathcal{M}| > 2$ available machines.

Author Contributions: Methodology, Y.N.S.; software, V.D.H.; validation, Y.N.S., N.M.M. and V.D.H.; formal analysis, Y.N.S. and N.M.M.; investigation, Y.N.S. and N.M.M; writing—original draft preparation, Y.N.S. and N.M.M.; writing—review and editing, Y.N.S.; visualization, N.M.M.; supervision, Y.N.S. All authors have read and agreed to the published version of the manuscript.

Funding: This research received no external funding.

Acknowledgments: We are thankful for useful remarks and suggestions provided by the editors and three anonymous reviewers on the earlier draft of our paper.

Conflicts of Interest: The authors declare no conflict of interest.

Appendix A. Proofs of the Statements

Appendix A.1. Proof of Lemma 2

We choose an arbitrary vector p in the set T, $p \in T$, and show that set $< S'_{1,2}, S_{2,1} >$ contains at least one optimal pair of job permutations for the problem $J2|p, n_i \leq 2|C_{max}$ with scenario $p \in T$.

Let $(\pi^*, \pi^{**}) = ((\pi^*_{1,2}, \pi_1, \pi^*_{2,1}), (\pi^*_{2,1}, \pi_2, \pi^*_{1,2}))$ be a Jackson's pair of job permutations for the problem $J2|p, n_i \leq 2|C_{max}$ with scenario $p \in T$, i.e., $C_{max}(\pi^*, \pi^{**}) = C_{max}$. Without loss of generality, one can assume that jobs in both permutations π_1 and π_2 are ordered in increasing order of their indexes. It is clear that $\pi^*_{2,1} \in S_{2,1}$. If inclusion $\pi^*_{1,2} \in S'_{1,2}$ holds as well, then $(\pi^*, \pi^{**}) \in <S'_{1,2}, S_{2,1}>$ and set $< S'_{1,2}, S_{2,1} >$ contains an optimal pair of job permutations for the problem $J2|p, n_i \leq 2|C_{max}$ with scenario $p \in T$. We now assume that $\pi^*_{1,2} \notin S'_{1,2}$. The set $S'_{1,2}$ contains at least one optimal permutation for the problem $F2|p_{1,2}|C_{max}$ with job set $\mathcal{J}_{1,2}$ and scenario $p_{1,2}$ (the components of vector $p_{1,2}$ are equal to the corresponding components of vector p). We denote this permutation as $\pi'_{1,2}$. Remember that permutation $\pi'_{1,2}$ may be not a Johnson's permutation for the problem $F2|p_{1,2}|C_{max}$ with job set $\mathcal{J}_{1,2}$ and scenario $p_{1,2}$. We consider a pair of job permutations $(\pi', \pi^{**}) = ((\pi'_{1,2}, \pi_1, \pi^*_{2,1}), (\pi^*_{2,1}, \pi_2, \pi'_{1,2})) \in <S'_{1,2}, S_{2,1}>$ and show that equality $C_{max}(\pi', \pi^{**}) = C_{max}$ holds. We consider the following two possible cases.

(j) $C_{max}(\pi', \pi^{**}) = c_1(\pi')$.

If equality $c_1(\pi') = \sum_{J_i \in \mathcal{J}_{1,2} \cup \mathcal{J}_{2,1} \cup \mathcal{J}_1} p_{i1}$ holds, then $c_1(\pi') \leq c_1(\pi^*)$.

We now assume that inequality $c_1(\pi') > \sum_{J_i \in \mathcal{J}_{1,2} \cup \mathcal{J}_{2,1} \cup \mathcal{J}_1} p_{i1}$ holds. Then, machine M_1 has an idle time. As it is mentioned in the proof of Theorem 7, an idle time for machine M_1 is only possible if some job J_j from the set $\mathcal{J}_{2,1}$ is processed on machine M_2 at the time moment t_2 when job J_j could be processed on machine M_1. Thus, the value of $c_1(\pi')$ is equal to the makespan $C_{max}(\pi^*_{2,1})$ for the problem $F2|p_{2,1}|C_{max}$ with job set $\mathcal{J}_{2,1}$ and scenario $p_{2,1}$ (the components of vector $p_{2,1}$ are equal to the corresponding components of vector p). As jobs from the set $\mathcal{J}_{2,1}$ are processed as in the permutation $\pi^*_{2,1}$, which is a Johnson's permutation, the value of $c_1(\pi')$ cannot be reduced and so $c_1(\pi') \leq c_1(\pi^*)$. We obtain the following relations: $C_{max}(\pi', \pi^{**}) = c_1(\pi') \leq c_1(\pi^*) \leq \max\{c_1(\pi^*), c_2(\pi^{**})\} = C_{max}(\pi^*, \pi^{**}) = C_{max}$. Thus, equality $C_{max}(\pi', \pi^{**}) = C_{max}$ holds.

(jj) $C_{max}(\pi', \pi^{**}) = c_2(\pi^{**})$.

Similarly to case (j), we obtain the following equality:

$$c_2(\pi^{**}) = \max\{ \sum_{J_i \in \mathcal{J}_{1,2} \cup \mathcal{J}_{2,1} \cup \mathcal{J}_2} p_{i2}, C_{max}(\pi'_{1,2})\},$$

where $C_{max}(\pi'_{1,2})$ is the makespan for the problem $F2|p_{1,2}|C_{max}$ with job set $\mathcal{J}_{1,2}$ and vector $p_{1,2}$ of the job durations (it is assumed that $\pi'_{1,2}$ is an optimal permutation for this problem). Thus, the value of $c_2(\pi^{**})$ cannot be reduced and equality $C_{max}(\pi', \pi^{**}) = C_{max}$ holds.

In both considered cases, the pair of job permutations (π', π^{**}) is an optimal schedule for the problem $J2|p, n_i \leq 2|C_{max}$ with scenario $p \in T$. Therefore, an optimal pair of job permutations for the problem $J2|p, n_i \leq 2|C_{max}$ with scenario $p \in T$ belongs to the set $< S'_{1,2}, S_{2,1} >$. As vector p is an arbitrary vector in set T, the set $< S'_{1,2}, S_{2,1} >$ contains an optimal pair of job permutations for each scenario from set T. Due to Definition 4, the set $< S'_{1,2}, S_{2,1} >$ is a dominant set of schedules for the problem $J2|l_{ij} \leq p_{ij} \leq u_{ij}, n_i \leq 2|C_{max}$ with job set $\mathcal{J}$.

Appendix A.2. Proof of Theorem 8

We consider an arbitrary vector $p \in T$ of the job durations from set T and relevant vectors $p_{1,2}$ and $p_{2,1}$ of the durations of jobs from set $\mathcal{J}_{1,2}$ and set $\mathcal{J}_{2,1}$, respectively. Set $S'_{1,2}$ contains an optimal permutation $\pi'_{1,2}$ for the problem $F2|p_{1,2}|C_{max}$ with job set $\mathcal{J}_{1,2}$ and with vector $p_{1,2}$ of the job durations. Set $S'_{2,1}$ contains an optimal permutation $\pi'_{2,1}$ for the problem $F2|p_{2,1}|C_{max}$ with job set $\mathcal{J}_{2,1}$ and with vector $p_{2,1}$ of the job durations. We next show that the pair of job permutations $(\pi', \pi'') = ((\pi'_{1,2}, \pi_1, \pi'_{2,1}), (\pi'_{2,1}, \pi_2, \pi'_{1,2}))$ is an optimal pair of job permutations for the problem $J2|p, n_i \leq 2|C_{max}$ with scenario $p \in T$ (the jobs in permutations π_1 and π_2 are ordered in increasing order of their indexes). From the proofs of Lemmas 2 and 3, we obtain the value of $C_{max}(\pi', \pi'') = \max\{c_1(\pi'), c_2(\pi'')\}$

$$= \max\{\max\{ \sum_{J_i \in \mathcal{J}_{1,2} \cup \mathcal{J}_{2,1} \cup \mathcal{J}_1} p_{i1}, C_{max}(\pi'_{2,1})\}, \max\{ \sum_{J_i \in \mathcal{J}_{1,2} \cup \mathcal{J}_{2,1} \cup \mathcal{J}_2} p_{i2}, C_{max}(\pi'_{1,2})\}\},$$

which cannot be reduced. Therefore, $C_{max}(\pi', \pi'') = C_{max}$. An optimal pair of job permutations for the problem $J2|p, n_i \leq 2|C_{max}$ with vector $p \in T$ of the job durations belongs to the set $< S'_{1,2}, S'_{2,1} >$. As vector p is arbitrary in set T, the set $< S'_{1,2}, S'_{2,1} >$ contains an optimal pair of job permutations for all vectors from set T. Due to Definition 4, the set $< S'_{1,2}, S'_{2,1} > \subseteq S$ is a dominant set of schedules for the problem $J2|l_{ij} \leq p_{ij} \leq u_{ij}, n_i \leq 2|C_{max}$ with job set $\mathcal{J}$.

Appendix A.3. Proof of Theorem 9

We consider an arbitrary scenario $p \in T$. Due to Definition 1, the permutation $\pi_{1,2}$ is a Johnson's permutation for the problem $F2|p_{1,2}|C_{max}$ with job set $\mathcal{J}_{1,2}$ and scenario $p_{1,2}$ (the components of this vector are equal to the corresponding components of vector p). Due to Definition 4, the singleton $\{(\pi_{1,2}, \pi_{1,2})\}$ is a minimal dominant set of schedules for the problem $F2|l_{ij} \leq p_{ij} \leq u_{ij}|C_{max}$ with job set $\mathcal{J}_{1,2}$.

Similarly, the singleton $\{(\pi_{2,1}, \pi_{2,1})\}$ is a minimal dominant set of schedules for the problem $F2|l_{ij} \leq p_{ij} \leq u_{ij}|C_{max}$ with job set $\mathcal{J}_{2,1}$. We consider permutations π_1 and π_2 of the jobs $\mathcal{J}_1$ and $\mathcal{J}_2$, respectively (due to Remark 1, the jobs in permutations π_1 and π_2 are ordered in increasing order of their indexes). Due to Theorem 8, the pair of permutations $((\pi_{1,2}, \pi_1, \pi_{2,1}), (\pi_{1,2}, \pi_2, \pi_{2,1}))$ is a single-element dominant set (DS(T)) for the problem $J2|l_{ij} \leq p_{ij} \leq u_{ij}, n_i \leq 2|C_{max}$ with job set $\mathcal{J} = \mathcal{J}_1 \cup \mathcal{J}_2 \cup \mathcal{J}_{1,2} \cup \mathcal{J}_{2,1}$.

Appendix A.4. Proof of Corollary 6

In the proof of Theorem 9, it is shown that the pair of job permutations $((\pi_{1,2}, \pi_1, \pi_{2,1}), (\pi_{1,2}, \pi_2, \pi_{2,1}))$ is a single-element dominant set of schedules for the problem $J2|l_{ij} \leq p_{ij} \leq u_{ij}, n_i \leq 2|C_{max}$ with job set $\mathcal{J} = \mathcal{J}_1 \cup \mathcal{J}_2 \cup \mathcal{J}_{1,2} \cup \mathcal{J}_{2,1}$. We next show that the pair of permutations $((\pi_{1,2}, \pi_1, \pi_{2,1}), (\pi_{1,2}, \pi_2, \pi_{2,1}))$ satisfies to Definition 1, i.e., this pair of permutations is a Jackson's pair of job permutations for the problem $J2|l_{ij} \leq p_{ij} \leq u_{ij}, n_i \leq 2|C_{max}$ with job set $\mathcal{J}$ (the minimality condition is obvious). Indeed, due to conditions of Theorem 9, the permutation $\pi_{1,2}$ is a Johnson's permutation for the problem $F2|l_{ij} \leq p_{ij} \leq u_{ij}|C_{max}$ with job set $\mathcal{J}_{1,2}$ and the permutation $\pi_{2,1}$ is a Johnson's permutation for the problem $F2|l_{ij} \leq p_{ij} \leq u_{ij}|C_{max}$ with job set $\mathcal{J}_{2,1}$. Therefore, pair $((\pi_{1,2}, \pi_1, \pi_{2,1}), (\pi_{1,2}, \pi_2, \pi_{2,1}))$ is a Jackson's pair of permutations for the problem $J2|l_{ij} \leq p_{ij} \leq u_{ij}, n_i \leq 2|C_{max}$ with job set $\mathcal{J}$. Due to Definition 1, the pair of job permutations $((\pi_{1,2}, \pi_1, \pi_{2,1}), (\pi_{1,2}, \pi_2, \pi_{2,1}))$ is a single-element J-solution for the problem $J2|l_{ij} \leq p_{ij} \leq u_{ij}, n_i \leq 2|C_{max}$ with job set $\mathcal{J} = \mathcal{J}_1 \cup \mathcal{J}_2 \cup \mathcal{J}_{1,2} \cup \mathcal{J}_{2,1}$.

Appendix A.5. Proof of Theorem 12

We consider any fixed scenario $p \in T$ and a pair of job permutations $(\pi', \pi'') = ((\pi, \pi_1, \pi_{2,1}^*), (\pi_{2,1}^*, \pi_2, \pi)) \in S'$, where $\pi_{2,1}^* \in S_{2,1}$ is a Johnson's permutation of the jobs from the set $\mathcal{J}_{2,1}$ with vector $p_{2,1}$ of the job durations (components of this vector are equal to the corresponding components of vector p). We next show that this pair of job permutations (π', π'') is optimal for the individual problem $J2|p, n_i \leq 2|C_{max}$ with scenario p, i.e., $C_{max}(\pi', \pi'') = C_{max}$.

At time $t = 0$, machine M_1 begins to process jobs from the permutation π without idle times. We denote $t_1 = \sum_{i=1}^{k+r+1} p_{i1}$. At time moment t_1, job J_{k+r+1} is ready for processing on machine M_2. From the condition of Inequality (13) with $s = 1$, it follows that, even if machine M_2 has an idle time before processing job J_{k+r+1}, machine M_2 is available for processing this job at time t_1. If in addition, the condition of Inequality (13) holds with $s \in \{2, 3, \ldots, r\}$, then machine M_2 may also have idle times between processing jobs from the conflict set $\{J_{k+1}, J_{k+2}, \ldots, J_{k+r}\}$. However, machine M_2 is available for processing job J_{k+r+1} from the time moment $t_1 = \sum_{i=1}^{k+r+1} p_{i1}$.

In permutation π, jobs $J_{k+r+1}, \ldots, J_{m_{1,2}}$ are arranged in Johnson's order. Therefore, if machine M_2 has an idle time while processing these jobs, this idle time cannot be reduced.

Thus, the value of $c_2(\pi'')$ cannot be reduced by changing the order of jobs from the conflict set. Note that an idle time for machine M_1 is only possible between some jobs from the set $\mathcal{J}_{2,1}$. Since the permutation $\pi_{2,1}^*$ is a Johnson's permutation of the jobs from set $\mathcal{J}_{2,1}$ with scenario $p_{2,1}$, the value of $c_1(\pi')$ cannot be reduced. Thus, we obtain $C_{max}(\pi', \pi'') = \max\{c_1(\pi'), c_2(\pi'')\} = C_{max}$ and the pair of permutations $(\pi', \pi'') = ((\pi, \pi_1, \pi_{2,1}^*), (\pi_{2,1}^*, \pi_2, \pi)) \in S'$ is optimal for the problem $J2|p, n_i \leq 2|C_{max}$ with scenario $p \in T$. As the vector p is an arbitrary vector in the set T, set S' contains an optimal pair of job permutations for each vector from the set T. Due to Definition 4, set S' is a dominant set of schedules for the problem $J2|l_{ij} \leq p_{ij} \leq u_{ij}, n_i \leq 2|C_{max}$ with job set $\mathcal{J}$.

Appendix B. Tables with Computations Results

Table A1. Computational results for randomly generated instances with ratio 25% : 25% : 25% : 25% of the number of jobs in the subsets.

$\delta\%$	5%				10%				15%				20%				30%				40%				50%			
n	Opt	NC	SC	t	Opt	NC	SC	t	Opt	NC	SC	t	Opt	NC	SC	t	Opt	NC	SC	t	Opt	NC	SC	t	Opt	NC	SC	t
20	100	6	100	0	100	19	100	0	100	35	100	0	100	70	100	0	100	139	100	0	100	250	100	0	100	339	100	0
40	100	0	-	0	100	4	100	0	100	20	100	0	100	33	100	0	100	101	100	0	100	136	100	0	100	333	100	0
50	100	7	100	0	100	3	100	0	100	16	100	0	100	8	100	0	100	50	100	0	100	114	100	0	100	224	100	0
70	100	0	-	0	100	0	-	0	100	2	100	0	100	3	100	0	100	11	100	0	100	71	100	0	100	149	100	0
80	100	0	-	0	100	0	-	0	100	0	-	0	100	3	100	0	100	0	-	0	100	35	100	0	100	122	100	0
100	100	0	-	0	100	0	-	0	100	0	-	0	100	0	-	0	100	0	-	0	100	19	100	0	100	84	100	0
200	100	0	-	0	100	0	-	0	100	0	-	0	100	0	-	0	100	0	-	0	100	0	-	0	100	5	100	0
300	100	0	-	0	100	0	-	0	100	0	-	0	100	0	-	0	100	0	-	0	100	0	-	0	100	0	-	0
400	100	0	-	0	100	0	-	0	100	0	-	0	100	0	-	0	100	0	-	0	100	0	-	0	100	0	-	0
500	100	0	-	0	100	0	-	0	100	0	-	0	100	0	-	0	100	0	-	0	100	0	-	0	100	0	-	0
600	100	0	-	0	100	0	-	0	100	0	-	0	100	0	-	0	100	0	-	0	100	0	-	0	100	0	-	0
700	100	0	-	0	100	0	-	0	100	0	-	0	100	0	-	0	100	0	-	0	100	0	-	0	100	0	-	0
800	100	0	-	0	100	0	-	0	100	0	-	0	100	0	-	0	100	0	-	0	100	0	-	0	100	0	-	0
900	100	0	-	0	100	0	-	0	100	0	-	0	100	0	-	0	100	0	-	0	100	0	-	0	100	0	-	0
1000	100	0	-	0	100	0	-	0	100	0	-	0	100	0	-	0	100	0	-	0	100	0	-	0	100	0	-	0
2000	100	0	-	0	100	0	-	0	100	0	-	0	100	0	-	0	100	0	-	0	100	0	-	0	100	0	-	0
3000	100	0	-	0	100	0	-	0	100	0	-	0	100	0	-	0	100	0	-	0	100	0	-	0	100	0	-	0
4000	100	0	-	0	100	0	-	0	100	0	-	0	100	0	-	0	100	0	-	0	100	0	-	0	100	0	-	0
5000	100	0	-	0	100	0	-	0	100	0	-	0	100	0	-	0	100	0	-	0	100	0	-	0	100	0	-	0
6000	100	0	-	0	100	0	-	0	100	0	-	0	100	0	-	0	100	0	-	0	100	0	-	0	100	0	-	0
7000	100	0	-	0	100	0	-	0	100	0	-	0	100	0	-	0	100	0	-	0	100	0	-	0	100	0	-	0
8000	100	0	-	0	100	0	-	0	100	0	-	0	100	0	-	0	100	0	-	0	100	0	-	0	100	0	-	0
9000	100	0	-	0	100	0	-	0	100	0	-	0	100	0	-	0	100	0	-	0	100	0	-	0	100	0	-	0
10,000	100	0	-	0	100	0	-	0	100	0	-	0	100	0	-	0	100	0	-	0	100	0	-	0	100	0	-	0
Aver.	100	0.54	100	0	100	1.08	100	0	100	3.04	100	0	100	4.88	100	0	100	12.54	100	0	100	26.04	100	0	100	52.33	100	0

Table A2. Computational results for randomly generated instances with ratio 10% : 10% : 40% : 40% of the number of jobs in the subsets.

$\delta\%$	5%				10%				15%				20%				30%				40%				50%			
n	Opt	NC	SC	t	Opt	NC	SC	t	Opt	NC	SC	t	Opt	NC	SC	t	Opt	NC	SC	t	Opt	NC	SC	t	Opt	NC	SC	t
10	100	71	100	0	100	153	100	0	99.6	241	98.34	0	99.4	320	98.13	0	98.1	481	96.05	0	95.8	618	93.20	0	91.9	713	88.64	0
20	100	235	100	0	100	531	100	0	100	811	100	0	100	1032	100	0	100	1341	100	0	99.9	1450	99.93	0	99.5	1424	99.65	0
30	100	460	100	0	100	887	100	0	100	1334	100	0	100	1643	100	0	100	1912	100	0	99.9	1893	99.95	0	100	1808	100	0
40	100	636	100	0	100	1185	100	0	100	1659	100	0	100	2068	100	0	100	2352	100	0	100	2162	100	0	99.9	1953	99.95	0
50	100	824	100	0	100	1496	100	0	100	2074	100	0	100	2411	100	0	100	2546	100	0	100	2211	100	0	100	2009	100	0
60	100	893	100	0	100	1542	100	0	100	2222	100	0	100	2619	100	0	100	2758	100	0	100	2440	100	0	100	2109	100	0
70	100	841	100	0	100	1589	100	0	100	2285	100	0	100	2775	100	0	100	2935	100	0	100	2477	100	0	100	2106	100	0
80	100	981	100	0	100	1570	100	0	100	2342	100	0	100	2896	100	0	100	2995	100	0	100	2567	100	0	100	2249	100	0
90	100	878	100	0	100	1660	100	0	100	2310	100	0	100	2905	100	0	100	3103	100	0	100	2598	100	0	100	2273	100	0
100	100	826	100	0	100	1633	100	0	100	2368	100	0	100	3056	100	0	100	3114	100	0	100	2585	100	0	100	2321	100	0
200	100	411	100	0	100	1145	100	0	100	1999	100	0	100	3065	100	0	100	3392	100	0	100	2709	100	0	100	2250	100	0
300	100	181	100	0	100	721	100	0	100	1708	100	0	100	2888	100	0	100	3365	100	0	100	2579	100	0	100	2117	100	0
400	100	51	100	0	100	302	100	0	100	981	100	0	100	2466	100	0	100	3263	100	0	100	2469	100	0	100	1966	100	0
500	100	11	100	0	100	240	100	0	100	813	100	0	100	2307	100	0	100	3138	100	0	100	2362	100	0	100	1838	100	0
600	100	0	-	0	100	88	100	0	100	499	100	0	100	2076	100	0	100	2951	100	0	100	2202	100	0	100	1692	100	0
700	100	0	-	0	100	45	100	0	100	528	100	0	100	1894	100	0	100	2779	100	1	100	2015	100	1	100	1585	100	1
800	100	0	-	0	100	36	100	0	100	294	100	0	100	1707	100	0	100	2656	100	0	100	1866	100	1	100	1485	100	1
900	100	0	-	0	100	0	-	0	100	318	100	0	100	1442	100	0	100	2392	100	1	100	1677	100	1	100	1420	100	1
1000	100	0	-	0	100	0	-	0	100	196	100	0	100	1275	100	0	100	2255	100	1	100	1630	100	1	100	1298	100	1
2000	100	0	-	0	100	0	-	0	100	3	100	0	100	441	100	0	100	1452	100	3	100	1137	100	3	100	1044	100	2
3000	100	0	-	0	100	0	-	0	100	0	-	0	100	160	100	0	100	1127	100	6	100	1025	100	5	100	1011	100	4
4000	100	0	-	0	100	0	-	0	100	0	-	0	100	86	100	0	100	1032	100	9	100	1005	100	8	100	1000	100	7
5000	100	0	-	0	100	0	-	0	100	0	-	0	100	34	100	0	100	1011	100	14	100	1000	100	12	100	1000	100	10
6000	100	0	-	0	100	0	-	0	100	0	-	0	100	23	100	0	100	1002	100	21	100	1001	100	17	100	1001	100	14
7000	100	0	-	0	100	0	-	0	100	0	-	0	100	8	100	0	100	1000	100	28	100	1000	100	23	100	1000	100	19
8000	100	0	-	0	100	0	-	0	100	0	-	0	100	6	100	0	100	1001	100	37	100	1000	100	31	100	1000	100	25
9000	100	0	-	0	100	0	-	0	100	0	-	0	100	3	100	0	100	1000	100	48	100	1000	100	39	100	1000	100	32
10,000	100	0	-	0	100	0	-	0	100	0	-	0	100	4	100	1	100	1000	100	61	100	1000	100	49	100	1000	100	40
Aver.	100	261	100	0	100	529	100	0	99.99	892	99.92	0	99.98	1486	99.93	0.04	99.93	2120	99.86	8.21	99.84	1774	99.75	6.82	99.69	1560	99.58	5.61

Table A3. Computational results for randomly generated instances with ratio 10% : 40% : 10% : 40% of the number of jobs in the subsets.

$\delta\%$	5%				10%				15%				20%				30%				40%				50%			
n	Opt	NC	SC	t	Opt	NC	SC	t	Opt	NC	SC	t	Opt	NC	SC	t	Opt	NC	SC	t	Opt	NC	SC	t	Opt	NC	SC	t
10	98.7	104	87.5	0	97.3	207	86.96	0	96.9	287	89.20	0	94.5	359	84.68	0	92.5	543	86.19	0	88.5	597	80.74	0	81.5	713	74.05	0
20	99.7	466	99.36	0	99.6	830	99.52	0	99.3	1057	99.24	0	98.5	1244	98.79	0	95.3	1433	96.65	0	91	1467	93.73	0	84.6	1501	89.61	0
30	100	1070	100	0	99.8	1597	99.87	0	99.7	1879	99.84	0	98.2	2007	99.10	0	96.7	2053	98.30	0	91.4	1945	95.58	0	82.1	1790	89.83	0
40	100	1700	100	0	100	2355	100	0	99.8	2660	99.92	0	99.7	2628	99.89	0	97.6	2462	99.03	0	92.4	2135	96.35	0	85.5	1979	92.52	0
50	100	2425	100	0	99.7	3174	99.91	0	99.9	3197	99.97	0	99.8	3048	99.93	0	97.8	2664	99.17	0	92.5	2283	96.54	0	82.2	2075	91.42	0
60	100	3218	100	0	100	3808	100	0	100	3702	100	0	99.9	3394	99.97	0	98.5	2881	99.41	0	91.9	2442	96.60	0	80.5	2172	90.75	0
70	100	3911	100	0	100	4385	100	0	99.9	4063	99.98	0	99.9	3648	99.97	0	98.7	2959	99.56	0	91.9	2517	96.62	0	78	2146	89.47	0
80	100	4817	100	0	100	4902	100	0	99.8	4370	99.95	0	99.9	3829	99.97	0	98.5	3103	99.52	0	92.2	2627	96.99	0	77.3	2257	89.77	0
90	100	5518	100	0	100	5398	100	0	100	4656	100	0	100	3910	100	0	97.7	3154	99.21	0	92.5	2675	97.12	0	79.5	2249	90.84	0
100	100	6195	100	0	100	5697	100	0	99.9	4853	99.98	0	100	4047	100	0	98.2	3207	99.44	0	91.3	2706	96.78	0	75.6	2328	89.48	0
200	100	10,620	100	0	100	7608	100	0	100	5717	100	0	100	4645	100	0	98.9	3320	99.64	0	90.8	2717	96.61	0	72.7	2281	87.90	0
300	100	13,110	100	0	100	8259	100	0	100	6070	100	0	100	4782	100	0	99.5	3369	99.85	0	94.3	2605	97.81	0	74.5	2117	87.81	0
400	100	14,309	100	1	100	8634	100	0	100	6113	100	0	100	4650	100	0	99.8	3247	99.94	0	94.3	2460	97.68	0	73.1	2002	86.56	0
500	100	14,935	100	0	100	8658	100	0	100	6102	100	0	100	4630	100	0	99.9	3137	99.97	0	95.1	2297	97.87	0	78.5	1808	88.11	0
600	100	15,780	100	0	100	8832	100	0	100	6021	100	0	100	4492	100	0	100	2911	100	0	97.2	2153	98.70	0	77.8	1705	86.98	0
700	100	15,971	100	0	100	8753	100	0	100	5789	100	0	100	4379	100	0	100	2786	100	0	97.7	1996	98.85	0	82.4	1613	89.09	0
800	100	16,439	100	0	100	8806	100	0	100	5793	100	0	100	4176	100	0	100	2533	100	0	98.8	1846	99.35	0	84	1487	89.24	0
900	100	16,268	100	0	100	8574	100	1	100	5608	100	1	100	4005	100	1	100	2379	100	0	98.8	1717	99.30	0	89.1	1366	92.02	0
1000	100	16,614	100	1	100	8419	100	1	100	5400	100	1	100	3807	100	1	100	2279	100	1	99.6	1655	99.76	1	90.9	1302	93.01	0
2000	100	15,539	100	2	100	6906	100	2	100	3715	100	2	100	2422	100	2	100	1401	100	1	100	1135	100	1	98.4	1040	98.46	1
3000	100	13,884	100	4	100	5259	100	4	100	2599	100	4	100	1624	100	3	100	1109	100	3	100	1021	100	3	99.8	1006	99.80	3
4000	100	12,302	100	7	100	3911	100	7	100	1874	100	6	100	1291	100	6	100	1044	100	5	100	1004	100	4	99.8	1001	99.80	4
5000	100	10,421	100	13	100	2935	100	11	100	1485	100	10	100	1126	100	9	100	1008	100	8	100	1000	100	6	100	1000	100	5
6000	100	8822	100	17	100	2299	100	16	100	1262	100	14	100	1043	100	13	100	1004	100	10	100	1000	100	9	100	1000	100	8
7000	100	7426	100	24	100	1855	100	22	100	1145	100	20	100	1026	100	17	100	1000	100	14	100	1001	100	11	100	1000	100	10
8000	100	6346	100	33	100	1569	100	30	100	1084	100	26	100	1007	100	23	100	1000	100	18	100	1000	100	15	100	1000	100	13
9000	100	5378	100	42	100	1362	100	38	100	1038	100	33	100	1002	100	30	100	1000	100	24	100	1000	100	19	100	1000	100	17
10,000	100	4529	100	54	100	1237	100	48	100	1028	100	42	100	1000	100	38	100	1000	100	29	100	1000	100	24	100	1000	100	20
Aver.	99.94	8861	99.53	7.07	99.87	4865	99.51	6.43	99.83	3520	99.57	5.68	99.66	2829	99.37	5.11	98.91	2142	99.14	4.04	95.79	1786	97.61	3.32	86.71	1569	92.38	2.89

Table A4. Computational results for randomly generated instances with ratio 10% : 30% : 10% : 50% of the number of jobs in the subsets.

$\delta\%$	5%				10%				15%				20%				30%				40%				50%			
n	Opt	NC	SC	t	Opt	NC	SC	t	Opt	NC	SC	t	Opt	NC	SC	t	Opt	NC	SC	t	Opt	NC	SC	t	Opt	NC	SC	t
10	99.4	194	96.91	0	97.4	334	92.22	0	95.5	474	90.51	0	94.3	566	89.93	0	88.3	749	84.38	0	79.1	849	75.03	0	74.1	934	71.95	0
20	99.9	767	99.87	0	99.4	1232	99.51	0	98.4	1489	98.86	0	97.5	1680	98.39	0	94.7	1762	96.99	0	85.6	1770	91.64	0	75.1	1724	85.15	0
30	99.7	1532	99.80	0	99.5	2150	99.77	0	99.2	2399	99.67	0	99	2499	99.52	0	96	2367	98.31	0	86.8	2200	93.77	0	71.7	1941	85.27	0
40	99.8	2422	99.92	0	99.7	3151	99.90	0	99.8	3295	99.94	0	98.9	2972	99.63	0	95.2	2581	98.06	0	83	2326	92.48	0	69.5	2055	85.06	0
50	100	3422	100	0	99.9	4013	99.98	0	99.9	3844	99.97	0	99.5	3451	99.86	0	95.3	2857	98.35	0	82.1	2486	92.76	0	64.2	2202	83.47	0
60	100	4425	100	0	100	4681	100	0	99.9	4189	99.98	0	99.4	3750	99.84	0	94.8	2981	98.26	0	83.7	2566	93.53	0	64	2238	83.60	0
70	100	5338	100	0	100	5181	100	0	100	4569	100	0	99.3	4027	99.83	0	94.3	3183	98.21	0	79.6	2594	92.14	0	61.1	2284	82.75	0
80	100	6169	100	0	100	5770	100	0	99.9	4915	99.98	0	99.6	4112	99.90	0	95.6	3257	98.62	0	80.5	2625	92.42	0	58	2260	81.15	0
90	100	6998	100	0	100	6018	100	0	100	4984	100	0	99.6	4213	99.91	0	94.6	3332	98.38	0	78.5	2680	91.87	0	52.3	2270	78.72	0
100	100	7714	100	0	100	6298	100	0	100	5197	100	0	99.6	4358	99.91	0	94.5	3367	98.31	0	75.8	2642	90.61	0	54.6	2299	79.99	0
200	100	12,228	100	0	100	7920	100	0	100	5951	100	0	100	4748	100	0	95.7	3330	98.71	0	73.5	2665	90.02	0	45.4	2233	75.41	0
300	100	14,375	100	0	100	8735	100	0	100	6096	100	0	100	4723	100	0	96.4	3285	98.90	0	69.6	2464	87.66	0	43.4	2064	72.53	0
400	100	15,022	100	0	100	8762	100	0	100	6135	100	0	100	4712	100	0	96.4	3036	98.81	0	70.5	2286	87.05	0	48.1	1820	71.43	0
500	100	15,705	100	0	100	8823	100	0	100	5876	100	0	100	4497	100	0	97.5	2842	99.12	0	73.5	2100	87.38	0	55.1	1686	73.37	0
600	100	16,442	100	0	100	8712	100	0	100	5753	100	0	100	4252	100	0	97.6	2674	99.10	0	74.8	1941	87.02	0	62.6	1530	75.56	0
700	100	15,910	100	0	100	8670	100	0	100	5609	100	0	100	3976	100	0	99.1	2510	99.64	0	76.5	1776	86.77	0	69.3	1420	78.31	0
800	100	16,215	100	1	100	8419	100	1	100	5492	100	1	100	3773	100	1	99.3	2271	99.69	1	81.9	1648	89.02	1	75.9	1319	81.73	1
900	100	16,347	100	1	100	8268	100	1	100	5254	100	1	100	3597	100	1	99.2	2173	99.63	1	84.8	1575	90.35	1	80.7	1245	84.50	1
1000	100	16,355	100	1	100	8133	100	1	100	5064	100	1	100	3369	100	1	99.7	1998	99.85	1	86.8	1426	90.74	1	84.9	1189	87.30	1
2000	100	14,679	100	3	100	5955	100	3	100	3095	100	3	100	1972	100	2	100	1243	100	2	98.6	1056	98.67	2	97.7	1017	97.74	2
3000	100	12,643	100	6	100	4207	100	6	100	2036	100	5	100	1354	100	5	100	1038	100	4	99.9	1003	99.90	4	100	1001	100	3
4000	100	10,375	100	12	100	2927	100	11	100	1467	100	10	100	1152	100	9	100	1011	100	7	100	1000	100	6	100	1000	100	6
5000	100	8524	100	19	100	2140	100	18	100	1205	100	15	100	1032	100	14	100	1003	100	11	100	1000	100	9	100	1000	100	8
6000	100	6942	100	28	100	1724	100	26	100	1095	100	23	100	1014	100	20	100	1000	100	16	100	1000	100	14	100	1000	100	12
7000	100	5463	100	40	100	1398	100	35	100	1050	100	32	100	1007	100	28	100	1002	100	23	100	1000	100	18	100	1000	100	16
8000	100	4543	100	54	100	1240	100	48	100	1028	100	43	100	1002	100	45	100	1000	100	30	100	1000	100	24	100	1000	100	21
9000	100	3751	100	69	100	1135	100	62	100	1005	100	55	100	1001	100	48	100	1000	100	38	100	1000	100	32	100	1000	100	26
10,000	100	3056	100	86	100	1078	100	77	100	1003	100	68	100	1000	100	60	100	1000	100	48	100	1000	100	40	100	1000	100	33
Aver.	99.96	8841	99.87	11.43	99.85	4896	99.69	10.32	99.74	3556	99.60	9.18	99.53	2850	99.53	8.36	97.29	2138	98.62	6.50	85.90	1774	92.89	5.43	75.28	1562	86.25	4.64

Table A5. Computational results for randomly generated instances with ratio 10% : 20% : 10% : 60% of the number of jobs in the subsets.

δ%	5%				10%				15%				20%				30%				40%				50%			
n	Opt	NC	SC	t	Opt	NC	SC	t	Opt	NC	SC	t	Opt	NC	SC	t	Opt	NC	SC	t	Opt	NC	SC	t	Opt	NC	SC	t
10	98.8	263	95.44	0	97.5	491	94.91	0	94.5	657	91.48	0	91	791	88.50	0	81.5	1014	81.36	0	74.9	1102	76.59	0	66	1186	70.49	0
20	99.7	1034	99.71	0	98.8	1601	99.19	0	98.8	1904	99.32	0	97.6	1984	98.74	0	89.9	2113	95.13	0	80.4	1968	89.94	0	63	1845	79.73	0
30	99.9	2131	99.95	0	99.8	2778	99.93	0	99	3059	99.67	0	98.3	2878	99.37	0	89.6	2608	95.82	0	74.3	2270	88.46	0	60.3	2060	80.34	0
40	100	3224	100	0	99.9	3747	99.97	0	99.8	3698	99.95	0	98.3	3392	99.50	0	90.9	2877	96.63	0	71.8	2469	88.42	0	52	2174	77.60	0
50	100	4370	100	0	100	4704	100	0	99.6	4174	99.90	0	99	3701	99.73	0	89.4	2988	96.32	0	66.6	2566	86.83	0	47.1	2228	75.94	0
60	100	5473	100	0	100	5368	100	0	99.9	4608	99.98	0	98.2	3987	99.55	0	89.3	3098	96.51	0	67.2	2643	87.48	0	42.6	2279	74.59	0
70	100	6454	100	0	100	5985	100	0	99.9	4968	99.98	0	99.4	4125	99.85	0	87.5	3214	96.02	0	62.6	2669	85.69	0	38.6	2291	72.85	0
80	100	7498	100	0	99.9	6235	99.98	0	99.8	5194	99.94	0	98.8	4333	99.70	0	87.3	3372	96.14	0	61.4	2716	85.64	0	33.6	2260	70.40	0
90	99.9	8281	99.99	0	100	6560	100	0	99.8	5243	99.96	0	98.9	4502	99.76	0	87.8	3388	96.40	0	61.4	2757	85.78	0	32	2361	71.03	0
100	100	9169	100	0	100	7056	100	0	99.7	5507	99.95	0	99.2	4586	99.83	0	83.7	3360	94.97	0	58.2	2689	84.27	0	27.7	2287	68.21	0
200	100	13,366	100	0	100	8131	100	0	99.9	6029	99.98	0	99	4814	99.79	0	83.1	3329	94.89	0	43.9	2541	77.80	0	10.3	2172	58.52	0
300	100	14,999	100	0	100	8869	100	0	100	6010	100	0	98.9	4675	99.76	0	82	3127	94.18	0	32.2	2329	70.85	0	4.6	1870	48.66	0
400	100	15,704	100	0	100	8848	100	0	100	6048	100	0	99.7	4490	99.93	0	82.5	2899	93.96	0	28.3	2120	66.08	0	1.3	1710	42.28	0
500	100	15,775	100	0	100	8720	100	0	100	5825	100	0	99.7	4290	99.93	0	83.2	2638	93.63	0	21.6	1885	58.30	0	0.7	1541	35.56	0
600	100	16,336	100	0	100	8420	100	1	100	5582	100	0	100	3938	100	0	87.7	2420	94.88	1	18	1727	52.52	0	0	1408	28.98	0
700	100	16,298	100	1	100	8466	100	1	100	5360	100	1	100	3733	100	1	88.4	2203	94.73	1	15	1574	46.00	1	0	1282	22.00	1
800	100	16,707	100	1	100	8030	100	1	100	5023	100	1	99.9	3479	99.97	1	88.7	2077	94.56	1	12.9	1457	40.15	1	0.2	1207	17.32	1
900	100	16,135	100	1	100	7936	100	1	100	4808	100	1	100	3265	100	1	89.5	1934	94.52	1	10.8	1368	34.80	1	0	1172	14.68	1
1000	100	16,015	100	1	100	7665	100	1	100	4528	100	1	100	3049	100	1	91.6	1737	95.16	1	9.6	1314	31.20	1	0	1144	12.59	1
2000	100	13,921	100	4	100	5101	100	4	100	2549	100	4	100	1622	100	3	98.6	1138	98.77	3	1.2	1024	3.52	3	0	1002	0.20	3
3000	100	11,344	100	9	100	3400	100	9	100	1636	100	8	100	1210	100	7	99.8	1014	99.80	6	0.4	1003	0.70	5	0	1000	0	5
4000	100	8769	100	17	100	2283	100	16	100	1245	100	14	100	1054	100	13	100	1003	100	10	0.1	1000	0.1	9	0	1000	0	8
5000	100	6948	100	28	100	1691	100	25	100	1102	100	27	100	1023	100	21	100	1001	100	16	0	1000	0	14	0	1000	0	11
6000	100	5409	100	42	100	1362	100	38	100	1041	100	34	100	1003	100	30	100	1000	100	27	0	1000	0	20	0	1000	0	17
7000	100	4121	100	59	100	1214	100	53	100	1016	100	47	100	1001	100	42	100	1000	100	34	0	1000	0	27	0	1000	0	23
8000	100	3368	100	80	100	1093	100	71	100	1006	100	62	100	1000	100	66	100	1000	100	43	0	1000	0	36	0	1000	0	30
9000	100	2646	100	102	100	1048	100	90	100	1000	100	80	100	1000	100	71	100	1000	100	56	0	1000	0	47	0	1000	0	39
10,000	100	2248	100	126	100	1024	100	119	100	1002	100	100	100	1000	100	89	100	1000	100	70	0	1000	0	63	0	1000	0	48
Aver.	99.94	8857	99.82	16.82	99.85	4923	99.79	15.36	99.67	3565	99.65	13.57	99.14	2854	99.43	12.36	91.14	2127	96.23	9.64	31.17	1757	47.90	8.14	17.14	1553	36.50	6.71

Table A6. Computational results for randomly generated instances with ratio 10% : 10% : 10% : 70% of the number of jobs in the subsets.

δ%	5%				10%				15%				20%				30%				40%				50%			
n	Opt	NC	SC	t	Opt	NC	SC	t	Opt	NC	SC	t	Opt	NC	SC	t	Opt	NC	SC	t	Opt	NC	SC	t	Opt	NC	SC	t
10	98.6	371	96.23	0	96.2	612	93.63	0	93.6	853	92.15	0	89.5	1015	89.06	0	82.1	1231	84.73	0	67.1	1301	73.10	0	54.3	1348	63.65	0
20	99.8	1432	99.86	0	99.3	1988	99.65	0	97.2	2273	98.77	0	95.5	2345	98.04	0	85.9	2222	93.61	0	67.5	2119	84.29	0	50.7	1952	74.13	0
30	99.7	2713	99.89	0	99.6	3341	99.88	0	99.2	3397	99.76	0	96.9	3204	99.03	0	85.9	2777	94.63	0	65.2	2344	84.98	0	43.5	2139	72.98	0
40	99.9	4056	99.98	0	99.7	4439	99.93	0	99.2	4015	99.80	0	97.1	3722	99.19	0	85	2983	94.90	0	58.4	2525	82.89	0	36.5	2191	70.61	0
50	100	5231	100	0	99.9	5130	99.98	0	99.7	4593	99.93	0	97.7	3952	99.42	0	80.4	3165	93.52	0	55	2595	82.35	0	28.6	2200	67.18	0
60	100	6574	100	0	99.9	5804	99.98	0	99.3	4934	99.84	0	97.1	4182	99.26	0	84.1	3283	94.94	0	49.9	2656	80.80	0	25.6	2255	66.39	0
70	99.9	7444	99.99	0	100	6365	100	0	99.4	5115	99.88	0	97.8	4328	99.49	0	80.4	3261	93.90	0	48.3	2706	80.60	0	21	2330	65.75	0
80	100	8505	100	0	100	6737	100	0	99.7	5415	99.94	0	96.5	4422	99.21	0	75.3	3304	92.37	0	42.7	2667	78.29	0	16.8	2258	63.02	0
90	100	9185	100	0	99.8	7333	99.97	0	99.8	5623	99.96	0	97.7	4555	99.50	0	76.4	3417	92.98	0	37.6	2696	76.34	0	13.3	2288	61.32	0
100	100	9909	100	0	99.9	7305	99.99	0	99.6	5571	99.93	0	98.2	4546	99.60	0	74.4	3449	92.49	0	35.8	2695	75.92	0	11.8	2314	61.62	0
200	100	13,806	100	0	100	8387	100	0	99.8	6146	99.97	0	96.2	4736	99.18	0	63.5	3261	88.75	0	16.1	2527	66.68	0	2.7	2006	51.40	0
300	100	15,550	100	0	100	8870	100	0	99.9	6084	99.98	0	97.3	4563	99.41	0	53.6	3067	84.84	0	6.9	2215	57.97	0	0.6	1765	43.63	0
400	100	15,856	100	0	100	8573	100	0	99.9	5852	99.98	0	96.9	4304	99.28	0	48.3	2737	81.11	0	3.4	2049	52.76	0	0	1596	37.34	0
500	100	16,158	100	0	100	8576	100	0	100	5760	100	0	97.9	4067	99.48	1	44.9	2471	77.70	0	1.8	1727	43.14	0	0.1	1402	28.74	0
600	100	16,216	100	1	100	8425	100	1	99.9	5416	99.98	1	98.8	3724	99.68	1	42.9	2217	74.24	1	1.6	1539	36.00	1	0	1279	21.81	1
700	100	16,338	100	1	100	8142	100	1	100	5055	100	1	99.1	3432	99.74	1	40.4	2059	71.01	1	1.4	1420	30.56	1	0	1197	16.46	1
800	100	16,548	100	1	100	7909	100	1	100	4744	100	1	99	3206	99.69	1	36.8	1821	65.29	1	0.4	1319	24.49	1	0	1133	11.74	1
900	100	16,000	100	1	100	7494	100	1	100	4477	100	1	99	2929	99.66	1	30.5	1716	59.50	1	0.1	1264	20.97	1	0	1098	8.93	1
1000	100	15,806	100	1	100	7294	100	1	100	4123	100	1	99.3	2694	99.74	1	36.1	1535	58.37	1	0.1	1177	15.12	1	0	1057	5.39	1
2000	100	13,179	100	6	100	4424	100	5	100	2200	100	5	100	1431	100	4	24.3	1059	28.52	4	0	1007	0.70	3	0	1002	0.20	3
3000	100	9960	100	13	100	2746	100	12	100	1407	100	11	100	1103	100	10	20.2	1007	20.75	8	0	1000	0	7	0	1001	0.10	7
4000	100	7402	100	24	100	1843	100	22	100	1130	100	20	100	1027	100	17	16.2	1000	16.2	15	0	1000	0	12	0	1000	0	10
5000	100	5616	100	40	100	1450	100	36	100	1042	100	31	100	1008	100	28	12.2	1000	12.2	23	0	1000	0	19	0	1000	0	15
6000	100	4234	100	59	100	1204	100	53	100	1019	100	56	100	1000	100	42	9.8	1000	9.8	34	0	1000	0	28	0	1000	0	23
7000	100	3236	100	82	100	1083	100	64	100	1007	100	65	100	1000	100	58	7.7	1000	7.7	47	0	1000	0	38	0	1000	0	31
8000	100	2511	100	121	100	1040	100	98	100	1002	100	90	100	1000	100	76	7.6	1000	7.6	61	0	1000	0	51	0	1000	0	43
9000	100	2059	100	140	100	1015	100	124	100	1001	100	110	100	1000	100	96	6.2	1000	6.2	79	0	1000	0	65	0	1000	0	52
10,000	100	1728	100	174	100	1011	100	154	100	1001	100	159	100	1000	100	120	5	1000	5	97	0	1000	0	80	0	1000	0	65
Aver.	99.93	8844	99.85	23.71	99.80	4948	99.75	20.46	99.51	3581	99.64	19.71	98.13	2839	99.20	16.32	47.00	2109	60.82	13.32	19.98	1734	41.00	11	10.91	1529	31.87	9.07

Table A7. Computational results for randomly generated instances with ratio 5% : 20% : 5% : 70% of the number of jobs in the subsets.

$\delta\%$	5%				10%				15%				20%				30%				40%				50%			
n	Opt	NC	SC	t	Opt	NC	SC	t	Opt	NC	SC	t	Opt	NC	SC	t	Opt	NC	SC	t	Opt	NC	SC	t	Opt	NC	SC	t
20	98.8	1353	99.04	0	96.7	1993	98.29	0	93.3	2325	97.08	0	86.9	2343	94.15	0	71.4	2210	86.38	0	52.1	2091	76.09	0	34.8	1941	65.33	0
40	99.2	4101	99.80	0	98.7	4426	99.71	0	93.5	3918	98.34	0	88.8	3685	96.93	0	60.9	3007	86.63	0	33.2	2539	73.02	0	19	2189	62.27	0
60	99.1	6628	99.86	0	98.7	5712	99.75	0	93.1	4893	98.51	0	82.5	4106	95.64	0	48.8	3176	83.44	0	21.1	2584	68.89	0	10.9	2267	60.17	0
80	99.5	8614	99.93	0	98.3	6701	99.75	0	94.7	5335	98.93	0	79.5	4447	95.21	0	38.9	3272	80.96	0	14.5	2738	68.04	0	5.7	2254	57.59	0
100	99.9	10,165	99.99	0	98.7	7202	99.81	0	92.6	5740	98.69	0	76.5	4634	94.80	0	29.4	3338	78.16	0	10.5	2711	66.58	0	2.7	2224	55.94	0
200	100	13,856	100	0	98.6	8509	99.84	0	87.5	6083	97.93	0	59.5	4691	91.20	0	13.4	3333	73.69	0	4.8	2546	62.33	0	0.7	2041	51.20	0
300	100	15,201	100	0	99.3	8705	99.92	0	82.4	6187	97.07	0	44.5	4566	87.69	0	7.3	3025	69.16	0	2.4	2287	57.24	0	0.1	1833	45.50	0
400	99.9	15,924	99.99	0	98.4	8964	99.82	0	75.3	5888	95.77	0	32.3	4338	84.23	0	9.2	2727	66.59	0	1.9	1970	50.20	0	0.1	1592	37.25	0
500	100	16,186	100	0	98	8588	99.76	0	71	5652	94.80	0	28.6	3987	81.89	1	12.6	2468	64.55	0	1.5	1794	45.09	0	0	1379	27.48	0
600	100	16,531	100	1	97.5	8437	99.70	1	65.2	5391	93.51	1	26.8	3660	79.92	1	16.4	2184	61.72	1	0.6	1556	36.05	1	0	1287	22.30	1
700	100	16,251	100	1	98.7	8282	99.84	1	63.8	4967	92.65	1	24.6	3441	78.00	1	17.1	2087	60.28	2	0.3	1452	31.34	1	0	1186	15.68	1
800	100	16,462	100	1	98.3	7937	99.79	1	62.1	4736	91.91	1	26.5	3192	76.94	1	19.3	1806	55.26	1	0	1338	25.26	1	0	1131	11.58	1
900	100	16,099	100	1	97	7613	99.61	1	58.8	4439	90.70	1	28.7	2885	75.22	1	23.5	1694	54.84	1	0.4	1237	19.48	1	0	1118	10.55	1
1000	100	15,750	100	1	96.6	7157	99.52	1	59.2	4186	90.18	1	29.7	2708	74.04	1	23	1551	50.35	1	0.1	1211	17.51	1	0	1080	7.41	1
2000	100	13,055	100	6	97.8	4521	99.51	5	65.8	2164	84.20	5	76.3	1416	83.26	5	23.5	1063	28.03	4	0	1012	1.19	3	0	1003	0.30	3
3000	100	10,038	100	13	99.5	2766	99.82	12	86.4	1403	90.31	11	94	1109	94.59	10	17.9	1007	18.47	8	0	1000	0	7	0	1000	0	6
4000	100	7568	100	25	99.6	1823	99.78	22	96.3	1118	96.69	20	98.5	1021	98.53	18	15.9	1001	15.98	14	0	1000	0	12	0	1000	0	10
5000	100	5613	100	40	100	1430	100	35	97.4	1056	97.54	32	99.5	1008	99.50	29	11.4	1000	11.4	23	0	1000	0	19	0	1000	0	16
6000	100	4157	100	59	100	1187	100	54	99.5	1014	99.51	48	99.9	1002	99.90	46	10.4	1000	10.4	34	0	1000	0	33	0	1000	0	23
7000	100	3108	100	85	100	1076	100	74	99.7	1007	99.70	66	100	1000	100	60	8.6	1000	8.6	47	0	1000	0	39	0	1000	0	31
8000	100	2581	100	110	100	1051	100	98	99.7	1004	99.70	88	100	1000	100	78	6.4	1000	6.4	65	0	1000	0	50	0	1000	0	41
9000	100	2029	100	140	100	1014	100	131	100	1000	100	115	100	1000	100	99	6.7	1000	6.7	79	0	1000	0	63	0	1000	0	52
10,000	100	1672	100	175	100	1010	100	166	100	1000	100	138	100	1000	100	122	5.9	1000	5.9	99	0	1000	0	80	0	1000	0	66
Aver.	99.84	9693	99.94	28.61	98.71	5048	99.75	26.17	84.23	3500	95.81	22.96	68.85	2706	90.51	20.57	21.65	1954	47.13	16.48	6.23	1612	30.36	13.52	3.22	1414	23.07	11

Table A8. Computational results for randomly generated instances with ratio 5% : 15% : 5% : 75% of the number of jobs in the subsets.

δ%	5%				10%				15%				20%				30%				40%				50%			
n	Opt	NC	SC	t	Opt	NC	SC	t	Opt	NC	SC	t	Opt	NC	SC	t	Opt	NC	SC	t	Opt	NC	SC	t	Opt	NC	SC	t
20	99	1475	99.32	0	96.5	2242	98.44	0	93.4	2450	97.18	0	84.7	2514	93.52	0	68.7	2340	86.07	0	49.2	2103	74.99	0	30.8	1952	63.73	0
40	99.4	4487	99.84	0	98.2	4516	99.53	0	93.5	4245	98.35	0	85.9	3850	96.21	0	52.9	3112	84.38	0	30.1	2565	71.85	0	13.7	2149	59.19	0
60	100	6924	100	0	97.6	5942	99.58	0	91.7	4936	98.30	0	80.3	4225	95.24	0	43.5	3242	82.02	0	17.7	2620	67.82	0	5.9	2282	58.15	0
80	99.8	9113	99.98	0	98.7	6863	99.78	0	92.2	5512	98.55	0	75.3	4487	94.25	0	32.8	3350	79.37	0	12.3	2686	66.57	0	3.1	2301	57.32	0
100	99.9	10,503	99.99	0	98.3	7707	99.78	0	91.4	5884	98.45	0	73	4606	93.90	0	25.9	3411	77.84	0	4.4	2744	64.72	0	1.1	2292	56.68	0
200	100	14,104	100	0	97.7	8731	99.73	0	83.2	6090	97.14	0	53	4744	89.80	0	10.2	3300	72.42	0	0.9	2486	59.90	0	0	2009	50.02	0
300	99.8	15,767	99.99	0	98.2	8987	99.80	0	77.5	6167	96.25	0	36.8	4562	85.88	0	2.3	2969	66.86	0	0	2144	53.36	0	0	1739	42.50	0
400	99.9	15,948	99.99	0	97.5	8704	99.71	0	69	5775	94.53	0	25.9	4239	82.33	0	2.1	2675	63.18	0	0	1939	48.38	0	0	1522	34.30	0
500	99.9	16,456	99.99	1	96.7	8566	99.61	1	63.5	5460	93.28	1	20.8	3926	79.67	1	1.3	2393	58.71	1	0	1695	40.94	0	0	1357	26.31	0
600	100	16,327	100	1	96.6	8386	99.59	1	58.8	5149	91.96	1	17	3616	76.94	1	1.5	2102	53.09	1	0	1510	33.77	1	0	1280	21.88	1
700	100	16,349	100	1	95.8	8152	99.48	1	52.1	4831	90.04	1	14.2	3287	73.81	1	0.6	1911	47.83	1	0	1373	27.17	1	0	1190	15.97	1
800	100	16,329	100	1	94.8	7753	99.33	1	52	4584	89.49	1	15.7	3044	72.17	1	0.4	1732	42.44	1	0	1311	23.72	1	0	1123	10.95	1
900	100	16,036	100	1	96.4	7430	99.52	2	50.3	4206	88.11	1	17.4	2760	70.04	1	0.4	1550	35.68	1	0	1223	18.23	1	0	1084	7.75	1
1000	100	15,802	100	2	94.9	7018	99.27	2	50.3	3859	87.12	2	18.8	2588	68.62	2	0.2	1460	31.64	1	0	1155	13.42	1	0	1057	5.39	1
2000	100	12,622	100	6	94.3	4221	98.63	6	65.2	2006	82.65	6	36.5	1340	52.61	5	0	1044	4.21	4	0	1009	0.89	4	0	1000	0	3
3000	100	9378	100	15	97.3	2524	98.93	14	84.7	1308	88.30	13	36.7	1080	41.39	12	0	1008	0.79	11	0	1000	0	8	0	1000	0	7
4000	100	6813	100	28	99.7	1677	99.82	25	96.4	1092	96.70	23	40.8	1012	41.50	20	0	1000	0	16	0	1000	0	14	0	1000	0	11
5000	100	4874	100	48	99.8	1318	99.85	42	98.6	1031	98.64	38	38.1	1004	38.35	33	0	1000	0	27	0	1000	0	21	0	1000	0	18
6000	100	3737	100	69	100	1178	100	62	99.4	1010	99.41	55	37.8	1003	37.99	49	0	1000	0	39	0	1000	0	31	0	1000	0	26
7000	100	2782	100	96	100	1062	100	84	99.8	1004	99.80	77	35.9	1000	35.9	68	0	1000	0	54	0	1000	0	41	0	1000	0	36
8000	100	2280	100	128	100	1024	100	112	100	1000	100	101	33.2	1000	33.2	90	0	1000	0	71	0	1000	0	57	0	1000	0	52
9000	100	1802	100	163	100	1006	100	142	99.9	1001	99.9	127	33.9	1000	33.9	112	0	1000	0	94	0	1000	0	73	0	1000	0	61
10,000	100	1536	100	199	100	1010	100	176	100	1000	100	160	31	1000	31	141	0	1000	0	114	0	1000	0	91	0	1000	0	75
Aver.	99.9	9628	99.96	33	97.78	5044	99.58	29.17	81.00	3461	94.96	26.39	40.99	2691	66.01	23.35	10.56	1939	38.55	18.96	4.98	1590	28.95	15	2.37	1406	22.18	12.78

Table A9. Computational results for randomly generated instances with ratio 5% : 5% : 5% : 85% of the number of jobs in the subsets.

$\delta\%$	5%				10%				15%				20%				30%				40%				50%			
n	Opt	NC	SC	t	Opt	NC	SC	t	Opt	NC	SC	t	Opt	NC	SC	t	Opt	NC	SC	t	Opt	NC	SC	t	Opt	NC	SC	t
20	98.8	1896	99.31	0	94.9	2585	97.95	0	90.4	2763	96.20	0	82.2	2777	93.30	0	61.1	2532	83.37	0	41.9	2220	72.03	0	24.2	2063	61.66	0
40	99.5	5138	99.90	0	97.4	5066	99.49	0	92.9	4468	98.30	0	79.9	3928	94.63	0	46.8	3040	81.71	0	21.4	2588	68.66	0	8.7	2231	58.09	0
60	99.8	7963	99.97	0	98	6409	99.69	0	91.5	5209	98.33	0	72.5	4332	93.44	0	35.5	3301	80.01	0	11.5	2681	66.21	0	3.2	2349	58.49	0
80	99.7	9889	99.97	0	97.8	7272	99.68	0	88.7	5564	97.90	0	62.9	4545	91.44	0	25	3415	77.39	0	6.7	2718	65.16	0	0.6	2273	55.70	0
100	99.4	11,142	99.95	0	97.5	7878	99.67	0	86.9	5749	97.67	0	64.1	4732	92.16	0	17.4	3326	74.74	0	3.5	2705	64.07	0	0.2	2257	55.38	0
200	99.8	14,582	99.99	0	96.9	8778	99.64	0	73.5	6081	95.54	0	38.2	4773	86.80	0	2.9	3190	69.28	0	0.2	2407	58.33	0	0	1915	47.78	0
300	100	15,862	100	0	95.8	8856	99.51	0	63.9	5907	93.84	0	21.9	4440	82.27	0	0.6	2808	64.32	0	0	2070	51.64	0	0	1651	39.37	0
400	100	16,410	100	0	93.8	8685	99.25	0	51.1	5715	91.32	0	11.6	4129	78.40	0	0.1	2496	59.90	0	0	1777	43.73	0	0	1438	30.39	0
500	100	16,610	100	1	92.8	8433	99.13	1	43.5	5286	89.14	1	8.4	3665	74.92	1	0	2170	53.98	1	0	1514	33.95	1	0	1280	21.88	1
600	100	16,129	100	1	90.5	8109	98.79	1	37.3	4975	87.32	1	4.5	3355	71.51	1	0	1966	49.14	1	0	1407	28.86	1	0	1178	15.11	1
700	100	16,180	100	1	90.6	7795	98.78	1	34.7	4557	85.63	1	2.6	3089	68.37	1	0	1757	43.09	1	0	1289	22.34	1	0	1117	10.47	1
800	100	15,972	100	1	90.8	7404	98.76	1	29.3	4281	83.44	1	2.1	2723	63.90	1	0	1607	37.77	1	0	1206	17.08	1	0	1090	8.26	1
900	100	15,602	100	2	88.3	6867	98.28	2	24.7	3918	80.78	2	0.7	2547	61.01	2	0	1448	30.94	1	0	1145	12.66	1	0	1044	4.21	1
1000	100	15,428	100	2	86.6	6570	97.95	2	19.3	3586	77.44	2	0.3	2279	56.16	2	0	1314	23.90	2	0	1093	8.51	2	0	1031	3.01	1
2000	100	11,672	100	8	83	3610	95.29	8	6.2	1716	45.34	7	0	1220	18.03	7	0	1029	2.82	6	0	1010	0.99	5	0	1000	0	4
3000	100	8320	100	19	79.4	2171	90.51	18	1.9	1186	17.28	17	0	1044	4.21	15	0	1001	0.10	12	0	1000	0	10	0	1000	0	8
4000	100	5648	100	37	80.9	1472	87.02	34	1.5	1050	6.19	30	0	1007	0.70	27	0	1000	0	22	0	1000	0	18	0	1000	0	15
5000	100	4110	100	61	82.8	1186	85.50	56	0.3	1019	2.16	49	0	1002	0.20	44	0	1000	0	35	0	1000	0	29	0	1000	0	23
6000	100	2976	100	91	83.8	1091	85.15	80	0.3	1003	0.60	72	0	1000	0	91	0	1000	0	51	0	1000	0	44	0	1000	0	34
7000	100	2280	100	125	89	1036	89.38	110	0.1	1001	0.20	99	0	1000	0	88	0	1000	0	70	0	1000	0	57	0	1000	0	53
8000	100	1803	100	164	88.8	1010	88.91	145	0	1000	0	136	0	1000	0	115	0	1000	0	93	0	1000	0	75	0	1000	0	62
9000	100	1517	100	209	89.5	1003	89.53	184	0.1	1000	0.1	164	0	1000	0	147	0	1000	0	117	0	1000	0	96	0	1000	0	79
10,000	100	1290	100	270	90.5	1003	90.53	230	0	1000	0	204	0	1000	0	182	0	1000	0	147	0	1000	0	119	0	1000	0	98
Aver.	99.87	9496	99.96	43.13	90.41	4969	95.15	37.96	36.44	3393	58.47	34.17	19.65	2634	49.19	31.48	8.23	1887	36.19	24.35	3.70	1558	26.70	20	1.60	1388	20.43	16.61

Table A10. Computational results for randomly generated instances with ratio 3% : 2% : 5% : 90% of the number of jobs in the subsets.

δ%	5%				10%				15%				20%				30%				40%				50%			
n	Opt	NC	SC	t	Opt	NC	SC	t	Opt	NC	SC	t	Opt	NC	SC	t	Opt	NC	SC	t	Opt	NC	SC	t	Opt	NC	SC	t
100	99.7	11511	0.9997	0	95.4	8016	0.994	0	79	5914	0.963	0	50.5	4697	0.892	0	0.6	3217	0.688	0	1	2712	0.631	0	0	2282	0.559	0
200	99.7	15,060	0.9998	0	94	8798	0.993	0	60.4	6244	0.935	0	24.3	4606	0.833	0	0.1	2812	0.643	0	0	2352	0.574	0	0	1897	0.472	0
300	99.6	15,656	0.9997	0	88	8797	0.986	0	41	5866	0.898	0	10.2	4394	0.792	0	0	2395	0.582	1	0	2002	0.499	0	0	1572	0.364	0
400	99.5	16,208	0.9997	0	81.6	8503	0.978	1	26	5641	0.868	1	3.5	3960	0.756	1	0	2142	0.533	1	0	1721	0.418	0	0	1376	0.273	1
500	99.7	16,397	0.9998	1	77.2	8360	0.973	1	20.3	5140	0.844	1	1.4	3621	0.726	1	0	1868	0.464	1	0	1555	0.357	1	0	1242	0.195	1
600	99.6	16,083	0.9998	1	71.9	7983	0.964	1	14.2	4801	0.821	1	0.7	3294	0.697	1	0	1688	0.408	1	0	1374	0.272	1	0	1171	0.146	1
700	99.6	15,873	0.9997	1	69.1	7623	0.959	1	11.2	4438	0.800	1	0.4	2956	0.663	1	0	1525	0.344	1	0	1221	0.181	1	0	1093	0.085	1
800	99.6	15,935	0.9997	2	64.6	7140	0.950	2	8.1	4116	0.776	2	0.5	2671	0.627	2	0	1403	0.287	2	0	1178	0.151	1	0	1066	0.062	1
900	99.6	15,490	0.9997	2	63.6	6760	0.945	2	4.9	3735	0.745	2	0.1	2368	0.578	2	0	1324	0.245	2	0	1110	0.099	2	0	1045	0.043	1
1000	99.7	15,121	0.9998	2	60.5	6413	0.937	3	4.4	3405	0.719	2	0	2154	0.536	2	0	1019	0.019	7	0	1091	0.083	2	0	1027	0.026	2
2000	99.9	11,277	0.9999	10	30.5	3309	0.790	9	0	1624	0.384	9	0	1195	0.163	8	0	1003	0.003	14	0	1004	0.004	7	0	1000	0	5
3000	99.8	7873	0.9997	25	19.2	1949	0.585	21	0	1154	0.133	21	0	1027	0.026	17	0	1000	0	25	0	1001	0.001	12	0	1000	0	10
4000	99.7	5275	0.9994	43	15.6	1380	0.388	39	0	1042	0.040	35	0	1007	0.007	31	0	1000	0	40	0	1000	0	22	0	1000	0	17
5000	99.9	3641	0.9997	70	9.8	1141	0.209	64	0	1008	0.008	57	0	1001	0.001	50	0	1000	0	59	0	1000	0	33	0	1000	0	27
6000	100	2667	1	102	9.1	1041	0.127	98	0	1001	0.001	82	0	1000	0	73	0	1000	0	62	0	1000	0	48	0	1000	0	39
7000	100	2045	1	141	7.6	1024	0.098	127	0	1001	0.001	118	0	1000	0	100	0	1000	0	92	0	1000	0	73	0	1000	0	54
8000	100	1633	1	186	6.2	1005	0.067	166	0	1000	0	147	0	1000	0	163	0	1000	0	105	0	1000	0	85	0	1000	0	71
9000	100	1407	1	236	4.5	1003	0.048	210	0	1000	0	186	0	1000	0	165	0	1000	0	133	0	1000	0	109	0	1000	0	89
10,000	100	1248	1	294	2.5	1000	0.025	274	0	1000	0	231	0	1000	0	206	0	1000	0	168	0	1000	0	137	0	1000	0	112
Aver.	99.77	10021	1.00	58.74	45.84	4803	0.63	53.63	14.18	3112	0.47	47.16	4.82	2313	0.38	43.32	0.04	1495	0.22	37.58	0.05	1333	0.17	28.11	0	1198	0.12	22.74

Table A11. Computational results for randomly generated instances with ratio 2% : 3% : 5% : 90% of the number of jobs in the subsets.

δ%	5%				10%				15%				20%				30%				40%				50%			
n	Opt	NC	SC	t	Opt	NC	SC	t	Opt	NC	SC	t	Opt	NC	SC	t	Opt	NC	SC	t	Opt	NC	SC	t	Opt	NC	SC	t
100	99	11,556	0.999	0	94	7914	0.992	0	75	5859	0.956	0	46.8	4631	0.881	0	9.3	3418	0.731	0	0.9	2679	0.624	0	0	2206	0.544	0
200	99.3	14,862	1.000	0	90.1	8619	0.988	0	51.3	6177	0.919	0	18.5	4644	0.820	0	0.4	3093	0.676	0	0	2324	0.567	0	0	1889	0.468	0
300	99.4	15,924	1.000	0	85.8	8892	0.984	0	35.3	5901	0.888	0	6.7	4360	0.783	0	0.2	2749	0.635	0	0	2034	0.508	0	0	1611	0.379	0
400	99.6	16,312	1.000	0	77.8	8622	0.974	1	22.4	5652	0.861	1	2.6	3936	0.752	1	0	2412	0.585	1	0	1733	0.422	1	0	1371	0.270	0
500	99.7	16,305	1.000	1	72.6	8415	0.967	1	13.2	5135	0.829	1	1.2	3653	0.728	1	0	2122	0.528	1	0	1528	0.346	1	0	1232	0.188	1
600	99.2	16,297	1.000	1	66.1	7956	0.956	1	9.5	4917	0.814	1	0.5	3199	0.688	1	0	1850	0.459	1	0	1388	0.280	1	0	1179	0.152	1
700	99.7	16,252	1.000	1	58.2	7472	0.943	1	7.6	4480	0.793	2	0	2895	0.655	1	0	1660	0.398	1	0	1233	0.189	1	0	1080	0.074	1
800	99.5	16,075	1.000	2	51.8	7186	0.933	2	3.3	4088	0.763	2	0	2662	0.624	2	0	1506	0.336	1	0	1178	0.151	1	0	1061	0.057	1
900	99.4	15,617	1.000	2	48.6	6771	0.923	2	2.6	3828	0.744	2	0	2398	0.583	2	0	1379	0.275	2	0	1141	0.124	3	0	1036	0.035	1
1000	99.4	15,297	1.000	2	42.5	6381	0.909	3	2.2	3435	0.715	4	0	2114	0.527	2	0	1352	0.260	2	0	1092	0.084	3	0	1025	0.024	2
2000	98.8	11,349	0.999	10	20.1	3332	0.760	9	0	1646	0.392	9	0	1182	0.154	8	0	1024	0.023	7	0	1000	0	6	0	1001	0.001	5
3000	98.1	7777	0.998	23	5.4	1938	0.512	21	0	1166	0.142	19	0	1038	0.037	17	0	1002	0.002	14	0	1000	0	13	0	1000	0	10
4000	97.4	5297	0.995	45	3.9	1391	0.309	47	0	1037	0.036	35	0	1003	0.003	31	0	1000	0	26	0	1000	0	20	0	1000	0	17
5000	99.1	3802	0.998	70	2.2	1131	0.135	64	0	1009	0.009	56	0	1000	0	50	0	1000	0	46	0	1000	0	33	0	1000	0	27
6000	99	2686	0.996	102	0.8	1047	0.053	99	0	1003	0.003	82	0	1000	0	84	0	1000	0	58	0	1000	0	47	0	1000	0	39
7000	99.6	1996	0.998	142	0.5	1011	0.016	126	0	1001	0.001	112	0	1000	0	100	0	1000	0	87	0	1000	0	66	0	1000	0	54
8000	99.9	1641	0.999	185	0.2	1004	0.006	178	0	1000	0	150	0	1000	0	132	0	1000	0	106	0	1000	0	86	0	1000	0	70
9000	99.8	1387	0.999	236	0.2	1005	0.007	211	0	1000	0	186	0	1000	0	180	0	1000	0	134	0	1000	0	109	0	1000	0	89
10,000	99.9	1247	0.999	293	0	1001	0.001	260	0	1000	0	231	0	1000	0	205	0	1000	0	165	0	1000	0	134	0	1000	0	111
Aver.	99.25	10,088	1.00	58.68	37.94	4794	0.60	54	11.71	3123	0.47	47	4.02	2301	0.38	43	0.52	1609	0.26	34.32	0.05	1333	0.17	27.63	0	1194	0.12	22.58

Table A12. Computational results for randomly generated instances with ratio 2% : 2% : 1% : 95% of the numbers of jobs in the subsets.

$\delta\%$	5%				10%				15%				20%				30%				40%				50%			
n	Opt	NC	SC	t	Opt	NC	SC	t	Opt	NC	SC	t	Opt	NC	SC	t	Opt	NC	SC	t	Opt	NC	SC	t	Opt	NC	SC	t
100	97.4	12,124	0.998	0	86.5	7871	0.982	0	58.4	5867	0.926	0	27.3	4689	0.840	0	2.8	3377	0.705	0	0.2	2727	0.630	0	0	1898	0.473	0
200	97.1	15,322	0.998	0	74.2	8612	0.969	0	30.8	6247	0.886	0	6.1	4774	0.799	0	0	3086	0.674	0	0	2326	0.568	0	0	1586	0.369	0
300	95.5	15,872	0.997	0	57	8728	0.950	0	10.8	5787	0.843	0	0.9	4379	0.771	0	0	2664	0.624	0	0	1969	0.492	0	0	1348	0.258	0
400	95	16,354	0.997	1	40.9	8608	0.930	1	5.4	5518	0.827	1	0	3918	0.743	1	0	2334	0.570	1	0	1685	0.407	1	0	1232	0.188	1
500	94.1	16,103	0.996	1	30.9	8194	0.914	1	1.1	5012	0.801	1	0	3606	0.720	1	0	2076	0.518	1	0	1439	0.305	1	0	1135	0.119	1
600	91.4	16,241	0.995	1	21	7792	0.898	1	0.7	4690	0.787	1	0	3096	0.677	1	0	1750	0.429	1	0	1322	0.244	2	0	1067	0.063	1
700	92.3	16,138	0.995	1	16.7	7403	0.886	2	0	4265	0.765	2	0	2837	0.647	2	0	1604	0.377	1	0	1201	0.167	1	0	1058	0.055	1
800	89.6	15,785	0.993	2	10.7	7081	0.873	2	0.2	3909	0.744	2	0	2544	0.606	2	0	1464	0.317	2	0	1150	0.130	1	0	1039	0.038	2
900	87.3	15,368	0.992	2	6.9	6505	0.856	2	0	3595	0.721	2	0	2321	0.569	2	0	1347	0.258	2	0	1094	0.086	2	0	1019	0.019	2
1000	84.5	15,200	0.990	3	5.2	6203	0.847	3	0	3266	0.694	3	0	2082	0.520	3	0	1254	0.203	2	0	1069	0.065	2	0	1000	0	6
2000	48.3	10,867	0.952	11	0.1	3141	0.682	11	0	1536	0.349	10	0	1163	0.140	9	0	1015	0.015	7	0	1002	0.002	6	0	1000	0	11
3000	27.9	7207	0.900	26	0	1847	0.459	24	0	1124	0.110	21	0	1011	0.011	19	0	1001	0.001	16	0	1000	0	13	0	1000	0	21
4000	14.1	4913	0.825	49	0	1304	0.233	46	0	1019	0.019	41	0	1003	0.003	35	0	1000	0	28	0	1000	0	23	0	1000	0	30
5000	6.9	3353	0.722	80	0	1125	0.111	71	0	1002	0.002	63	0	1001	0.001	57	0	1000	0	48	0	1000	0	36	0	1000	0	47
6000	3.4	2359	0.591	118	0	1037	0.036	104	0	1003	0.003	92	0	1000	0	81	0	1000	0	65	0	1000	0	53	0	1000	0	44
7000	1.9	1895	0.482	175	0	1017	0.017	142	0	1000	0	126	0	1001	0.001	112	0	1000	0	90	0	1000	0	73	0	1000	0	65
8000	0.8	1509	0.343	210	0	1003	0.003	184	0	1000	0	164	0	1000	0	146	0	1000	0	138	0	1000	0	95	0	1000	0	78
9000	0.7	1308	0.241	269	0	1001	0.001	235	0	1000	0	210	0	1000	0	187	0	1000	0	150	0	1000	0	121	0	1000	0	164
10,000	0.5	1155	0.139	330	0	1001	0.001	290	0	1000	0	257	0	1000	0	230	0	1000	0	183	0	1000	0	151	0	1000	0	123
Aver.	54.14	9951	0.80	67.32	18.43	4709	0.56	58.89	5.65	3044	0.45	52.42	1.81	2286	0.37	46.74	0.15	1577	0.25	38.68	0.01	1315	0.16	30.58	0	1125	0.08	31.42

Table A13. Computational results for randomly generated instances with ratio 1% : 2% : 2% : 95% of the number of jobs in the subsets.

$\delta\%$	5%				10%				15%				20%				30%				40%				50%			
n	Opt	NC	SC	t	Opt	NC	SC	t	Opt	NC	SC	t	Opt	NC	SC	t	Opt	NC	SC	t	Opt	NC	SC	t	Opt	NC	SC	t
100	97.3	11,883	0.998	0	85.4	8060	0.981	0	58.5	5861	0.928	0	30.4	4718	0.846	0	3.2	3449	0.714	0	0.5	2646	0.621	0	0	2167	0.537	0
200	96.1	15,255	0.997	0	72.7	8727	0.968	0	29	6058	0.881	0	4.9	4698	0.795	0	0	3070	0.673	0	0	2301	0.565	0	0	1828	0.451	0
300	96.9	16,030	0.998	0	57.9	8787	0.950	0	10.6	5869	0.845	0	1.7	4304	0.770	0	0	2730	0.632	0	0	1956	0.488	0	0	1545	0.353	0
400	93.7	15,829	0.996	1	44.7	8623	0.934	1	3.9	5462	0.823	1	0	3931	0.745	1	0	2349	0.574	1	0	1629	0.386	1	0	1389	0.280	1
500	92.8	16,212	0.995	1	31.2	8178	0.915	1	1.4	5161	0.806	1	0.2	3590	0.721	1	0	2038	0.508	1	0	1472	0.321	1	0	1241	0.194	1
600	94	16,414	0.996	1	22.9	7766	0.899	1	0.8	4733	0.789	1	0	3212	0.687	1	0	1806	0.446	1	0	1310	0.237	1	0	1160	0.138	1
700	91.4	15,944	0.994	1	14	7296	0.880	2	0.2	4171	0.760	2	0	2799	0.643	1	0	1562	0.360	1	0	1235	0.190	1	0	1077	0.071	1
800	89.7	15,772	0.993	2	9.3	7079	0.870	2	0.1	3918	0.745	2	0	2510	0.602	2	0	1452	0.311	2	0	1135	0.119	1	0	1064	0.060	1
900	85.4	15,267	0.990	2	6.2	6618	0.858	2	0	3584	0.720	2	0	2259	0.557	2	0	1314	0.239	2	0	1106	0.096	2	0	1025	0.024	2
1000	84	15,005	0.989	3	3.3	6135	0.841	3	0	3251	0.692	3	0	2074	0.517	3	0	1263	0.208	2	0	1066	0.062	2	0	1014	0.014	2
2000	48.7	10,938	0.953	13	0	3157	0.683	10	0	1501	0.334	9	0	1145	0.127	9	0	1007	0.007	7	0	1001	0.001	6	0	1000	0	6
3000	27.6	7287	0.900	28	0	1792	0.441	24	0	1115	0.103	21	0	1021	0.021	22	0	1001	0.001	16	0	1000	0	13	0	1000	0	11
4000	13.7	4925	0.825	49	0	1290	0.225	44	0	1029	0.028	39	0	1006	0.006	35	0	1000	0	28	0	1000	0	24	0	1000	0	21
5000	7.9	3378	0.727	80	0	1102	0.093	72	0	1006	0.006	63	0	1000	0	56	0	1000	0	45	0	1000	0	36	0	1000	0	34
6000	4.4	2449	0.610	117	0	1049	0.047	104	0	1000	0	91	0	1000	0	81	0	1000	0	67	0	1000	0	53	0	1000	0	43
7000	2.1	1878	0.479	160	0	1019	0.019	141	0	1001	0.001	126	0	1000	0	112	0	1000	0	89	0	1000	0	74	0	1000	0	60
8000	0.9	1514	0.345	210	0	1004	0.004	185	0	1000	0	169	0	1000	0	147	0	1000	0	117	0	1000	0	96	0	1000	0	78
9000	0.4	1298	0.233	286	0	1001	0.001	236	0	1000	0	208	0	1000	0	186	0	1000	0	149	0	1000	0	121	0	1000	0	100
10,000	0.2	1165	0.143	330	0	1000	0	290	0	1000	0	303	0	1000	0	230	0	1000	0	184	0	1000	0	149	0	1000	0	123
Aver.	54.06	9918	0.80	67.58	18.29	4720	0.56	58.84	5.50	3034	0.45	54.79	1.96	2277	0.37	46.79	0.17	1581	0.25	37.47	0.03	1308	0.16	30.58	0	1185	0.11	25.53

Table A14. Computational results for randomly generated instances with ratio 1% : 1% : 3% : 95% of the number of jobs in the subsets.

$\delta\%$	5%				10%				15%				20%				30%				40%				50%			
n	Opt	NC	SC	t	Opt	NC	SC	t	Opt	NC	SC	t	Opt	NC	SC	t	Opt	NC	SC	t	Opt	NC	SC	t	Opt	NC	SC	t
100	98.6	11,377	0.999	0	88.5	7897	0.985	0	61.8	5894	0.933	0	30	4695	0.846	0	3.5	3376	0.708	0	0.4	2676	0.620	0	0	2222	0.545	0
200	98	15,012	0.999	0	76.8	8729	0.972	0	30.8	6003	0.881	0	8.4	4701	0.800	0	0	3170	0.682	0	0	2327	0.569	0	0	1856	0.460	0
300	97.4	15,847	0.998	0	67.8	8727	0.963	0	14.3	5842	0.851	0	1.5	4258	0.767	0	0	2746	0.634	0	0	1959	0.489	0	0	1558	0.358	0
400	96.7	16,261	0.998	1	49.6	8486	0.939	1	6.1	5500	0.828	1	0.4	3922	0.745	1	0	2339	0.572	1	0	1724	0.420	1	0	1330	0.248	0
500	95.9	16,266	0.997	1	38.1	8240	0.923	1	2.7	5142	0.810	1	0	3540	0.716	1	0	2060	0.515	1	0	1457	0.314	1	0	1230	0.187	1
600	95	15,911	0.997	1	29.3	7829	0.909	1	1.3	4759	0.791	1	0	3151	0.682	1	0	1830	0.454	1	0	1325	0.245	1	0	1126	0.112	1
700	92.6	15,922	0.995	1	23.1	7315	0.894	2	0.3	4317	0.769	2	0	2806	0.643	2	0	1643	0.391	1	0	1208	0.172	1	0	1093	0.085	1
800	94.4	15,727	0.996	2	15.8	7005	0.879	2	0.4	3906	0.745	2	0	2508	0.601	2	0	1469	0.319	2	0	1154	0.133	1	0	1051	0.049	1
900	92.4	15,311	0.995	2	12.7	6628	0.868	2	0	3606	0.721	2	0	2262	0.557	2	0	1380	0.275	2	0	1102	0.093	2	0	1030	0.029	2
1000	90.2	14,979	0.993	3	8.1	6215	0.852	3	0	3231	0.690	3	0	2011	0.503	3	0	1255	0.203	3	0	1060	0.057	2	0	1020	0.020	2
2000	67.9	10,778	0.970	11	0.5	3068	0.676	10	0	1567	0.362	9	0	1147	0.128	9	0	1013	0.013	7	0	1000	0	6	0	1000	0	5
3000	51.3	7398	0.934	27	0	1775	0.437	24	0	1129	0.114	22	0	1024	0.023	19	0	1000	0	17	0	1000	0	13	0	1000	0	18
4000	34.2	4903	0.866	49	0	1306	0.234	48	0	1017	0.017	39	0	1003	0.003	35	0	1000	0	28	0	1000	0	23	0	1000	0	19
5000	22.9	3312	0.767	79	0	1113	0.102	71	0	1009	0.009	63	0	1000	0	56	0	1000	0	45	0	1000	0	36	0	1000	0	30
6000	17.6	2440	0.662	116	0	1030	0.029	103	0	1001	0.001	92	0	1000	0	82	0	1000	0	65	0	1000	0	53	0	1000	0	44
7000	15.4	1838	0.540	160	0	1016	0.016	141	0	1000	0	125	0	1000	0	111	0	1000	0	89	0	1000	0	73	0	1000	0	60
8000	10	1470	0.388	208	0	1007	0.007	185	0	1000	0	165	0	1000	0	146	0	1000	0	117	0	1000	0	95	0	1000	0	79
9000	8.4	1339	0.316	266	0	1000	0	236	0	1000	0	208	0	1000	0	185	0	1000	0	148	0	1000	0	121	0	1000	0	100
10,000	7.8	1171	0.213	331	0	1000	0	292	0	1000	0	322	0	1000	0	229	0	1000	0	200	0	1000	0	162	0	1000	0	124
Aver.	62.46	9856	0.82	66.21	21.59	4705	0.56	59.05	6.19	3049	0.45	55.63	2.12	2265	0.37	46.53	0.18	1594	0.25	38.26	0.02	1315	0.16	31.11	0	1185	0.11	25.63

Table A15. Computational results for randomly generated instances with ratio 1% : 1% : 1% : 97% of the number of jobs in the subsets.

δ%	5%				10%				15%				20%				30%				40%				50%			
n	Opt	NC	SC	t	Opt	NC	SC	t	Opt	NC	SC	t	Opt	NC	SC	t	Opt	NC	SC	t	Opt	NC	SC	t	Opt	NC	SC	t
100	96.9	12,063	0.997	0	84.9	7901	0.980	0	52.5	5921	0.916	0	26.1	4713	0.837	0	2.4	3366	0.705	0	0	2618	0.612	0	0	2180	0.534	0
200	96.1	15,300	0.997	0	68.1	8663	0.962	0	22.1	6051	0.868	0	4.1	4665	0.791	0	0	3116	0.677	0	0	2355	0.574	0	0	1837	0.456	0
300	93.6	15,913	0.996	0	50.4	8790	0.942	0	6.9	5883	0.840	0	0.8	4267	0.765	0	0	2741	0.634	0	0	1939	0.483	0	0	1554	0.356	0
400	93.4	16,711	0.996	1	32.3	8577	0.919	1	3.1	5548	0.822	1	0	3935	0.745	1	0	2357	0.576	1	0	1667	0.400	1	0	1328	0.247	1
500	90.6	16,157	0.994	1	20.6	8180	0.902	1	1.2	5054	0.803	1	0	3430	0.708	1	0	2048	0.512	1	0	1447	0.309	1	0	1198	0.165	1
600	89.5	16,069	0.993	1	13.2	7871	0.889	1	0.1	4747	0.788	1	0	3089	0.676	1	0	1808	0.447	1	0	1326	0.246	1	0	1128	0.113	1
700	87.1	16,038	0.992	2	8.4	7320	0.874	2	0	4260	0.763	2	0	2710	0.630	1	0	1641	0.389	1	0	1200	0.167	1	0	1093	0.085	1
800	84.6	15,770	0.990	2	4.5	7035	0.863	2	0	3760	0.733	2	0	2510	0.602	2	0	1458	0.314	2	0	1148	0.129	1	0	1040	0.038	1
900	78.4	15,360	0.986	2	3.8	6525	0.851	3	0	3467	0.711	2	0	2233	0.552	2	0	1354	0.261	2	0	1091	0.083	2	0	1024	0.023	2
1000	73.7	15,050	0.982	3	2.1	6045	0.837	3	0	3186	0.686	3	0	2041	0.510	3	0	1236	0.191	2	0	1073	0.068	2	0	1016	0.016	2
2000	30.1	10,811	0.935	12	0	3034	0.670	11	0	1538	0.350	10	0	1147	0.128	9	0	1012	0.012	8	0	1003	0.003	7	0	1000	0	6
3000	11.8	7095	0.875	28	0	1790	0.441	25	0	1110	0.099	22	0	1020	0.020	20	0	1000	0	16	0	1000	0	13	0	1000	0	11
4000	3.5	4711	0.795	51	0	1274	0.215	61	0	1030	0.029	41	0	1001	0.001	36	0	1000	0	29	0	1000	0	24	0	1000	0	21
5000	0.7	3274	0.696	84	0	1104	0.094	74	0	1003	0.003	67	0	1000	0	59	0	1000	0	47	0	1000	0	41	0	1000	0	32
6000	0.3	2335	0.573	121	0	1039	0.038	107	0	1001	0.001	95	0	1000	0	85	0	1000	0	68	0	1000	0	55	0	1000	0	46
7000	0.1	1747	0.428	166	0	1009	0.009	147	0	1000	0	130	0	1000	0	117	0	1000	0	93	0	1000	0	76	0	1000	0	63
8000	0	1487	0.328	218	0	1005	0.005	194	0	1000	0	170	0	1000	0	151	0	1000	0	123	0	1000	0	100	0	1000	0	81
9000	0	1281	0.219	280	0	1000	0	266	0	1000	0	217	0	1000	0	194	0	1000	0	155	0	1000	0	126	0	1000	0	105
10,000	0	1164	0.141	344	0	1000	0	305	0	1000	0	268	0	1000	0	239	0	1000	0	298	0	1000	0	156	0	1000	0	130
Aver.	48.97	9912	0.78	69.26	15.17	4693	0.55	63.32	4.52	3029	0.44	54.32	1.63	2251	0.37	48.47	0.13	1586	0.25	44.58	0	1309	0.16	31.95	0	1179	0.11	26.53

References

1. Pinedo, M. *Scheduling: Theory, Algorithms, and Systems*; Prentice-Hall: Englewood Cliffs, NJ, USA, 2002.
2. Elmaghraby, S.; Thoney, K.A. Two-machine flowshop problem with arbitrary processing time distributions. *IIE Trans.* **2000**, *31*, 467–477. [CrossRef]
3. Kamburowski, J. Stochastically minimizing the makespan in two-machine flow shops without blocking. *Eur. J. Oper. Res.* **1999**, *112* 304–309. [CrossRef]
4. Ku, P.S.; Niu, S.C. On Johnson's two-machine flow-shop with random processing times. *Oper. Res.* **1986**, *34*, 130–136. [CrossRef]
5. Allahverdi, A. Stochastically minimizing total flowtime in flowshops with no waiting space. *Eur. J. Oper. Res.* **1999**, *113*, 101–112. [CrossRef]
6. Allahverdi, A.; Mittenthal, J. Two-machine ordered flowshop scheduling under random breakdowns. *Math. Comput. Model.* **1999**, *20*, 9–17. [CrossRef]
7. Portougal, V.; Trietsch, D. Johnson's problem with stochastic processing times and optimal service level. *Eur. J. Oper. Res.* **2006**, *169*, 751–760. [CrossRef]
8. Daniels, R.L.; Kouvelis, P. Robust scheduling to hedge against processing time uncertainty in single stage production. *Manag. Sci.* **1995**, *41*, 363–376. [CrossRef]
9. Sabuncuoglu, I.; Goren, S. Hedging production schedules against uncertainty in manufacturing environment with a review of robustness and stability research. *Int. J. Comput. Integr. Manuf.* **2009**, *22*, 138–157. [CrossRef]
10. Sotskov, Y.N.; Werner, F. *Sequencing and Scheduling with Inaccurate Data*; Nova Science Publishers: Hauppauge, NY, USA, 2014.
11. Pereira, J. The robust (minmax regret) single machine scheduling with interval processing times and total weighted completion time objective. *Comput. Oper. Res.* **2016**, *66*, 141–152. [CrossRef]
12. Kasperski, A.; Zielinski, P. A 2-approximation algorithm for interval data minmax regret sequencing problems with total flow time criterion. *Oper. Res. Lett.* **2008**, *36*, 343–344. [CrossRef]
13. Wu, Z.; Yu, S.; Li, T. A meta-model-based multi-objective evolutionary approach to robust job shop scheduling. *Mathematics* **2019**, *7*, 529. [CrossRef]
14. Grabot, B.; Geneste, L. Dispatching rules in scheduling: A fuzzy approach. *Int. J. Prod. Res.* **1994**, *32*, 903–915. [CrossRef]
15. Özelkan, E.C.; Duckstein, L. Optimal fuzzy counterparts of scheduling rules. *Eur. J. Oper. Res.* **1999**, *113*, 593–609. [CrossRef]
16. Kasperski, A.; Zielinski, P. Possibilistic minmax regret sequencing problems with fuzzy parameteres. *IEEE Trans. Fuzzy Syst.* **2011**, *19*, 1072–1082. [CrossRef]
17. Lai, T.-C.; Sotskov, Y.N.; Sotskova, N.Y.; Werner, F. Optimal makespan scheduling with given bounds of processing times. *Math. Comput. Model.* **1997**, *26*, 67–86.
18. Lai, T.-C.; Sotskov, Y.N. Sequencing with uncertain numerical data for makespan minimization. *J. Oper. Res. Soc.* **1999**, *50*, 230–243. [CrossRef]
19. Lai, T.-C.; Sotskov, Y.N.; Sotskova, N.Y. ; Werner, F. Mean flow time minimization with given bounds of processing times. *Eur. J. Oper. Res.* **2004**, *159*, 558–573. [CrossRef]
20. Sotskov, Y.N.; Egorova, N.M.; Lai, T.-C. Minimizing total weighted flow time of a set of jobs with interval processing times. *Math. Comput. Model.* **2009**, *50*, 556–573. [CrossRef]
21. Cheng, T.C.E.; Shakhlevich, N.V. Proportionate flow shop with controllable processing times. *J. Sched.* **1999**, *27*, 253–265. [CrossRef]
22. Cheng, T.C.E.; Kovalyov, M.Y.; Shakhlevich, N.V. Scheduling with controllable release dates and processing times: Makespan minimization. *Eur. J. Oper. Res.* **2006**, *175*, 751–768. [CrossRef]
23. Jansen, K.; Mastrolilli, M.; Solis-Oba, R. Approximation schemes for job shop scheduling problems with controllable processing times. *Eur. J. Oper. Res.* **2005**, *167*, 297–319. [CrossRef]
24. Graham, R.L.; Lawler, E.L.; Lenstra, J.K.; Rinnoy Kan, A.H.G. Optimization and approximation in deterministic sequencing and scheduling. *Ann. Discret. Appl. Math.* **1979**, *5*, 287–326.
25. Brucker, P. *Scheduling Algorithms*; Springer: Berlin, Germany, 1995.
26. Tanaev, V.S.; Sotskov, Y.N.; Strusevich, V.A. *Scheduling Theory: Multi-Stage Systems*; Kluwer Academic Publishers: Dordrecht, The Netherlands, 1994.

27. Jackson, J.R. An extension of Johnson's results on job lot scheduling. *Nav. Res. Logist. Q.* **1956**, *3*, 201–203. [CrossRef]
28. Johnson, S.M. Optimal two and three stage production schedules with set up times included. *Nav. Res. Logist. Q.* **1954**, *1*, 61–68. [CrossRef]
29. Allahverdi, A.; Sotskov, Y.N. Two-machine flowshop minimum-lenght scheduling problem with random and bounded processing times. *Int. Trans. Oper. Res.* **2003**, *10*, 65–76. [CrossRef]
30. Matsveichuk, N.M.; Sotskov, Y.N. A stability approach to two-stage scheduling problems with uncertain processing times. In *Sequencing and Scheduling with Inaccurate Data*; Sotskov, Y.N., Werner, F., Eds.; Nova Science Publishers: Hauppauge, NY, USA, 2014; pp. 377–407.
31. Shafranski Y.M. On the existence of globally optimal schedules for the Bellman-Johnson problem for two machines under uncertainty. *Informatika* **2009**, *3*, 100–110. (In Russian)
32. Lai, T.-C.; Sotskov, Y.N.; Egorova, N.G.; Werner, F. The optimality box in uncertain data for minimising the sum of the weighted job completion times. *Int. J. Prod. Res.* **2018**, *56*, 6336–6362. [CrossRef]
33. Sotskov, Y.N.; Egorova, N.G. Single machine scheduling problem with interval processing times and total completion time objective. *Algorithms* **2018**, *11*, 66. [CrossRef]
34. Sotskov, Y.N.; Lai, T.-C. Minimizing total weighted flow time under uncertainty using dominance and a stability box. *Comput. Oper. Res.* **2012**, *39*, 1271–1289. [CrossRef]
35. Sotskov, Y.N.; Egorova, N.G. Minimizing total flow time under uncertainty using optimality and stability boxes. In *Sequencing and Scheduling with Inaccurate Data*; Sotskov, Y.N., Werner, F., Eds.; Nova Science Publishers: Hauppauge, NY, USA, 2014; pp. 345–376.
36. Matsveichuk, N.M.; Sotskov, Y.N.; Egorova, N.G.; Lai, T.-C. Schedule execution for two-machine flow-shop with interval processing times. *Math. Comput. Model.* **2009**, *49*, 991–1011. [CrossRef]
37. Sotskov, Y.N.; Allahverdi, A.; Lai, T.-C. Flowshop scheduling problem to minimize total completion time with random and bounded processing times. *J. Oper. Res. Soc.* **2004**, *55*, 277–286. [CrossRef]
38. Allahverdi, A.; Aldowaisan, T.; Sotskov, Y.N. Two-machine flowshop scheduling problem to minimize makespan or total completion time with random and bounded setup times. *Int. J. Math. Math. Sci.* **2003**, *39*, 2475–2486. [CrossRef]
39. Kouvelis, P.; Yu, G. *Robust Discrete Optimization and Its Application*; Kluwer Academic Publishers: Boston, MA, USA, 1997.
40. Kouvelis, P.; Daniels, R.L.; Vairaktarakis, G. Robust scheduling of a two-machine flow shop with uncertain processing times. *IEEE Trans.* **2011**, *32*, 421–432. [CrossRef]
41. Ng, C.T.; Matsveichuk, N.M.; Sotskov, Y.N.; Cheng, T.C.E. Two-machine flow-shop minimum-length scheduling with interval processing times. *Asia-Pac. J. Oper. Res.* **2009**, *26*, 1–20. [CrossRef]
42. Matsveichuk, N.M.; Sotskov, Y.N.; Werner, F. The dominance digraph as a solution to the two-machine flow-shop problem with interval processing times. *Optimization* **2011**, *60*, 1493–1517. [CrossRef]

Article

Simple Constructive, Insertion, and Improvement Heuristics Based on the Girding Polygon for the Euclidean Traveling Salesman Problem

Víctor Pacheco-Valencia [1], José Alberto Hernández [2], José María Sigarreta [3] and Nodari Vakhania [1,*]

1   Centro de Investigación en Ciencias UAEMor, Universidad Autónoma del Estado de Morelos, Cuernavaca 62209, Mexico; vhpacval@gmail.com
2   Facultad de Contaduría, Administración e Informática UAEMor, Cuernavaca 62209, Mexico; jose_hernandez@uaem.mx
3   Facultad de Matemáticas UAGro, Universidad Autónoma de Guerrero, Acapulco 39650, Mexico; josemariasigarretaalmira@hotmail.com
*   Correspondence: nodari@uaem.mx

Received: 15 November 2019; Accepted: 17 December 2019; Published: 21 December 2019

Abstract: The Traveling Salesman Problem (TSP) aims at finding the shortest trip for a salesman, who has to visit each of the locations from a given set exactly once, starting and ending at the same location. Here, we consider the Euclidean version of the problem, in which the locations are points in the two-dimensional Euclidean space and the distances are correspondingly Euclidean distances. We propose simple, fast, and easily implementable heuristics that work well, in practice, for large real-life problem instances. The algorithm works on three phases, the constructive, the insertion, and the improvement phases. The first two phases run in time $O(n^2)$ and the number of repetitions in the improvement phase, in practice, is bounded by a small constant. We have tested the practical behavior of our heuristics on the available benchmark problem instances. The approximation provided by our algorithm for the tested benchmark problem instances did not beat best known results. At the same time, comparing the CPU time used by our algorithm with that of the earlier known ones, in about 92% of the cases our algorithm has required less computational time. Our algorithm is also memory efficient: for the largest tested problem instance with 744,710 cities, it has used about 50 MiB, whereas the average memory usage for the remained 217 instances was 1.6 MiB.

Keywords: heuristic algorithm; traveling salesman problem; computational experiment; time complexity

---

1. Introduction

The Traveling Salesman Problem (TSP) is one of the most studied strongly NP-hard combinatorial optimization problems. Given an $n \times n$ matrix of distances between n objects, call them cities, one looks for a shortest possible feasible tour which can be seen as a permutation of the given n objects: a feasible tour visits each of the n cities exactly once except the first visited city with which the tour ends. The cost of a tour is the sum of the distances between each pair of the neighboring cities in that tour. This problem can also be described in graph terms. We have an undirected weighted complete graph $G = (V, E)$, where V is the set of $n = |V|$ vertices (cities) and E is the set of the $n^2 - n$ edges $(i, j) = (j, i)$, $i \neq j$. A non-negative weight of an edge (i, j), $w(i, j)$ is the distance between vertices i and j. There are two basic sets of restrictions that define feasible solution (a tour that has to start and complete at the same vertex and has to contain all the vertices from set V exactly once). A feasible tour T can be represented as:

$$T = (i_1, i_2, \cdots, i_{n-1}, i_n, i_1); \; i_k \in V, \tag{1}$$

and its cost is

$$C(T) = \sum_{k=1}^{n-1} w(i_k, i_{k+1}) + w(i_n, i_1). \tag{2}$$

The objective is to find an *optimal* tour, a feasible one with the minimum cost $\min_T C(T)$.

Some special cases of the problem have been commonly considered. For instance, in the symmetric version, the distance matrix is symmetric (i.e., for each edge (i,j), $w(i,j) = w(j,i)$); in another setting, the distances between the cities are Euclidean distances (i.e., set V can be represented as points in the two-dimensional Euclidean space). Clearly, the Euclidean TSP is also a symmetric TSP but not vice versa. The Euclidean TSP has a straightforward immediate application in the real-life scenario when a salesman wishes to visit the cities using the shortest possible tour. Because in the Euclidean version the cities are points in plane, for each pair of points, the triangle inequality holds, which makes the problem a bit more accessible in the sense that simple geometric rules can be used for calculating the cost of a tour or the cost of the inclusion of a new point in a partial tour, unlike the general setting. Nevertheless, the Euclidean TSP remains strongly NP-hard; see Papadimitriou [1] and Garey et al. [2].

The exact solution methods for TSP can only solve problem instances with a moderate number of cities; hence, approximation algorithms are of a primary interest. There exist a vast amount of approximation heuristic algorithms for TSP. The literature on TSP is very wide-ranging, and it is not our goal to overview all the important relevant work here (we refer the reader, e.g., to a book by Lawler et al. [3] and an overview chapter by Jünger [4]).

The literature distinguishes two basic types of approximation algorithms for TSP: tour construction and loop improvement algorithms. The construction heuristics create a feasible tour in one pass so that the taken decisions are not reconsidered later. A feasible solution delivered by a construction heuristic can be used in a loop improvement heuristic as an initial feasible solution (though such initial solution can be constructed randomly). Given the current feasible tour, iteratively, an improvement algorithm, based on some local optimality criteria, makes some changes in that tour resulting in a new feasible solution with less cost. Well-known examples of tour improvement algorithms are *2-Opt* Croes *2-Opt*, its generalizations *3-Opt* and *k-Opt*, and the algorithm by Lin and Kernighan [5], to mention a few.

The most successful algorithms we have found in the literature for large-scale TSP instances are Ant Colony Optimization (ACO) meta heuristics, with which we compare our results. On one hand, these algorithms give a good approximation. On the other hand, the traditional ACO-based algorithms tend to require a considerable computer memory, which is necessary to keep an $n \times n$ pheromone matrix. Typically, the time complexity of the selection of each next move using ACO is also costly. These drawbacks are addressed in some recent ACO-based algorithms in which, at each iteration of the calculation of the pheromone levels, the intermediate data are reduced storing only a limited number of the most promising tours in computer memory. With Partial ACO (PACO), only some part of a known good tour is altered. A PACO-based heuristic was proposed in Chitty [6] and the experimental results for four problem instances from library Art Gallery were reported. Effective Strategies + ACO (ESACO) uses pheromone values directly in the 2-opt local search for the solution improvement and reduces the pheromone matrix, yielding linear space complexity (see, for example, Ismkman [7]). Parallel Cooperative Hybrid Algorithm ACO (PACO-3Opt) uses a multi-colony of ants to prevent a possible stagnation (see, for example, Gülcü et al. [8]). In a very recent Restricted Pheromone Matrix Method (RPMM) [9], the pheromone matrix is reduced with a linear memory complexity, resulting in an essentially lower memory consumption. Another recent successful ACO-based Dynamic Flying ACO (DFACO) heuristic was proposed by Dahan et al. [10]. Besides these ACO-based heuristics, we have compared our heuristics with other two meta-heuristics. One of them is a parallel algorithm based on the nearest neighborhood search suggested by Al-Adwan et al. [11], and the other one, proposed by Zhong et al. [12], is a Discrete Pigeon-Inspired Optimization (DPIO) metaheuristic. We have also implemented directly the Nearest Neighborhood (NN) algorithm for the comparison purposes (see Section 4 and Appendix A).

In Table A1 in Appendix A, we give a summary of the above heuristics including the information on the type and the number of the instances for which these algorithms were tested and the number of the runs of each of these algorithms. Unlike these heuristics, the heuristic that we propose here is deterministic, in the sense that, for any input, it delivers the same solution each time it is invoked; hence, there is no need in the repeated runs of our algorithm. We have tested the performance of our algorithm on 218 benchmark problem instances (the number of the reported instances for the algorithms from Table A1 vary from 6 to 36). The relative error of our algorithm for the tested instances did not beat the earlier known best results; however, for some instances, our error was better than that of the above-mentioned algorithms (see Table 9 at the end of Section 3). The error percentage provided by our algorithm has varied from 0% to 17%, with an average relative error of 7.16%. The standard error deviation over all the tested instances was 0.03.

In terms of the CPU time, our algorithm was faster than ones from Table A1 except for six instances from Art Gallery RPMM [9] and Partial-ACO [6], and for two instances from TSPLIB DPIO [12] were faster (see Table 10). Among all the comparisons we made, in about 92% of the cases, our algorithm has required less computational time. We have halted the execution of our algorithm for the two of the above-mentioned largest problem instances in 15 days, and for the next largest instance *ara238025* with 238,025 cities our algorithm has halted in about 36 h. The average CPU time for the remained instances were 19.2 min. The standard CPU time deviation for these instances was 89.3 min (for all the instances, including the above-mentioned three largest ones, it was 2068.4 min).

Our algorithm consumes very little computer memory. For the largest problem instance with 744,710 cities, it has used only about 50 MiB (mebibytes). The average memory usage for the remained 217 instances was 1.6 MiB (the average for all the instances including the above largest one was 1.88 MiB). The standard deviation of the usage of the memory is 4.6 MiB. Equation (3) below (see also Figure 15 in Section 3) shows the dependence of the memory required by our algorithm on the total number of cities n. As we can observe, this dependence is linear:

$$RAM = 0.0000685n + 0.563 \ MiB. \tag{3}$$

Our algorithm consists of the constructive, the insertion and the improvement phases, we call it the Constructive, Insertion, and Improvement algorithm, the *CII*-algorithm, for short. The constructive heuristics of Phase 1 deliver a partial tour that includes solely the points of the girding polygon. The insertion heuristic of Phase 2 completes the partial tour of Phase 1 to a complete feasible tour using the cheapest insertion strategy: iteratively, the current partial tour is augmented with a new point, one yielding the minimal increase in the cost in an auxiliary, specially formed tour. We use simple geometry in the decision-making process at Phases 2 and 3. The tour improvement heuristic of Phase 3 improves iteratively the tour of Phase 2 based on the local optimality conditions: it uses two heuristic algorithms which carry out some local rearrangement of the current tour. At Phase 1, the girding polygon for the points of set V and an initial, yet infeasible (partial) tour including the vertices of that polygon is constructed in time $O(n^2)$. The initial tour of Phase 1 is iteratively extended with the new points from the internal area of the polygon at Phase 2. Phase 2 also runs in time $O(n^2)$ and basically uses the triangle inequality for the selection of each newly added point. Phase 3 uses two heuristic algorithms. The first one, called *2-Opt*, is a local search algorithm proposed by Croes [13]. The second one is based on the procedure of Phase 2. The two heuristics are repeatedly applied in the iterative improvement cycle until a special approximation condition is satisfied. The number of repetitions in the improvement cycle, in practice, is bounded by a small constant. In particular, the average number of the repetitions for all the tested instances was about 9 (the maximum of 49 repetitions was attained for one of the moderate sized instances *lra498378*, and for the largest instance *lrb744710* with 744,710 points, Phase 3 was repeated 18 times).

The rest of the paper is organized as follows. In Section 2, we describe the *CII*-algorithm and show its time complexity. In Section 3, we give the implementation details and the results of our computational experiments, and, in Section 4, we give some concluding remarks and possible directions for the future work. The tables presented in Appendix A contain the complete data of our computational results.

2. Methods

We start this section with a brief aggregated description of our algorithm and in the following subsections we describe its three phases (Figure 1).

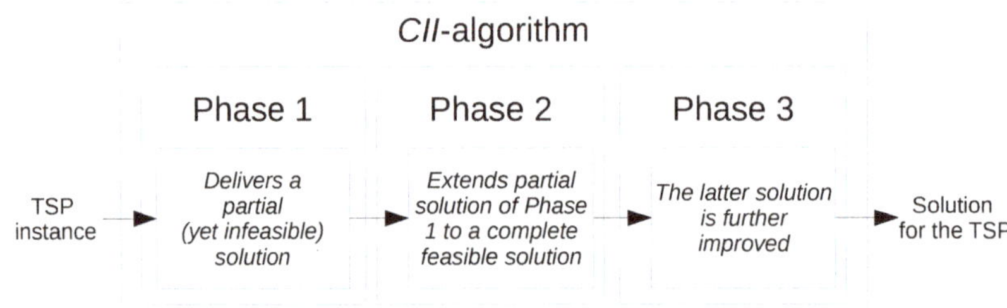

Figure 1. Block diagram of the *CII*-algorithm: (**a**) Phase 1 delivers a partial (yet infeasible) solution, (**b**) Phase 2 extends the partial solution of Phase 1 to a complete feasible solution, and, (**c**) at Phase 3, the latter solution is further improved.

2.1. Phase 1

2.1.1. Procedure to Locate the Extreme Points

At Phase 1, we construct the girding polygon for the points of set V and construct an initial yet infeasible (partial) tour that includes the points of that polygon. The construction of this polygon employs four *extreme* points v^1, v^2, v^3 and v^4; the *uppermost, leftmost, lowermost,* and *rightmost*, respectively [14], with ones from set V defined as follows. First, we define the sets of points T', L', B' and R' with $T' = \{i \mid y_i \text{ is maximum, } i \in V\}$, $L' = \{i \mid x_i \text{ is minimum, } i \in V\}$, $B' = \{i \mid y_i \text{ is minimum, } i \in V\}$, and $R' = \{i \mid x_i \text{ is maximum, } i \in V\}$. Then,

$$v^1 = j \mid x_j \text{ is maximum; } j \in T', \tag{4}$$

$$v^2 = j \mid y_j \text{ is maximum; } j \in L', \tag{5}$$

$$v^3 = j \mid x_j \text{ is minimum; } j \in B', \tag{6}$$

and

$$v^4 = j \mid y_j \text{ is minimum; } j \in R'. \tag{7}$$

See the next procedure for the extreme points in Table 1.

Table 1. Procedure *extreme_points*.

PROCEDURE $extreme\_points(V = \{i_1, i_2, \cdots, i_n\})$	
1	$y_{max} := y_{i_1}$ *//Initializing variables*
2	$x_{min} := x_{i_1}$
3	$y_{min} := x_{i_1}$
4	$x_{max} := y_{i_1}$
5	**FOR** $j := 2$ **TO** n **DO**
6	**IF** $y_{i_j} > y_{max}$ **THEN** $y_{max} := y_{i_j}$
7	**IF** $x_{i_j} < x_{min}$ **THEN** $x_{min} := x_{i_j}$
8	**IF** $y_{i_j} < y_{min}$ **THEN** $y_{min} := y_{i_j}$
9	**IF** $x_{i_j} > x_{max}$ **THEN** $x_{max} := x_{i_j}$
10	$T' = L' = B' = R' := \varnothing$
11	**FOR** $j := 1$ **TO** n **DO**
12	**IF** $y_{i_j} = y_{max}$ **THEN** $T' := T' \cup \{i_j\}$
13	**IF** $x_{i_j} = x_{min}$ **THEN** $L' := L' \cup \{i_j\}$
14	**IF** $y_{i_j} = y_{min}$ **THEN** $B' := B' \cup \{i_j\}$
15	**IF** $x_{i_j} = x_{max}$ **THEN** $R' := R' \cup \{i_j\}$
16	$v^1 := t'_1$ $// \; T' = \{t'_1, t'_2, \cdots, t'_{|T'|}\}, \; |T'| \leq n$
17	$v^2 := l'_1$ $// \; L' = \{l'_1, l'_2, \cdots, l'_{|L'|}\}, \; |L'| \leq n$
18	$v^3 := b'_1$ $// \; B' = \{b'_1, b'_2, \cdots, b'_{|B'|}\}, |B'| \leq n$
19	$v^4 := r'_1$ $// \; R' = \{r'_1, r'_2, \cdots, r'_{|R'|}\}, |R'| \leq n$
20	**FOR** $j := 2$ **TO** $|T'|$ **DO**
21	**IF** $x_{t'_j} > x_{v^1}$ **THEN** $v^1 := t'_j$
22	**FOR** $j := 2$ **TO** $|L'|$ **DO**
23	**IF** $x_{l'_j} > x_{v^2}$ **THEN** $v^2 := l'_j$
24	**FOR** $j := 2$ **TO** $|B'|$ **DO**
25	**IF** $x_{b'_j} > x_{v^3}$ **THEN** $v^3 := b'_j$
26	**FOR** $j := 2$ **TO** $|R'|$ **DO**
27	**IF** $x_{r'_j} > x_{v^4}$ **THEN** $v^4 := r'_j$
28	**RETURN** v^1, v^2, v^3, v^4

Lemma 1. *The time complexity of Procedure extreme_points is $O(n)$.*

Proof of Lemma 1. In this and in the following proofs, we only consider those lines in the formal descriptions in which the number of elementary operations, denote it by $f(n)$, depends on n (ignoring the lines yielding a constant number of operations). In lines 5–9, there is a loop with $n - 1$ cycles, hence $\{f(n) = n - 1\}$. In lines 11–15, there is a loop with n cycles, hence $\{f(n) = n\}$ In lines 20–21, 22–23, 24–25 and 26–27; there are four loops, each one with at most has n cycles, so $\{f(n) = 4n\}$. Hence, the total cost is $O(n)$. $\square$

2.1.2. Procedure for the Construction of the Girding Polygon

Before we describe the procedure, let us define function $\theta(i, j)$, returning the angle formed between the edge (i, j) and the positive direction of the x-axis (Equation (8) and Figure 2):

$$\theta(i, j) = \begin{cases} \arccos \dfrac{x_j - x_i}{w(i, j)} & if \; \arcsin \dfrac{y_j - y_i}{w(i, j)} \geq 0, \\[3mm] -\arccos \dfrac{x_j - x_i}{w(i, j)} & if \; \arcsin \dfrac{y_j - y_i}{w(i, j)} < 0. \end{cases} \tag{8}$$

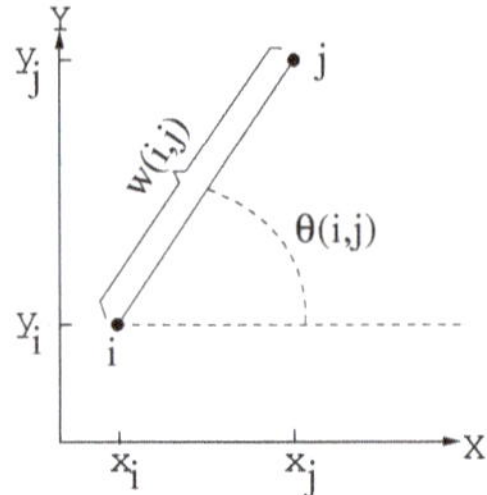

Figure 2. Angle $\theta(i, j)$.

The girding Polygon $P = P(V)$ is a convex geometric figure in a two-dimensional plane, such that any point in V either belongs to that polygon or to the area of that polygon Vakhania et al. [14].

The input of our procedure for the construction of polygon P (see Table 2), consists of (*i*) the set of vertices V and (*ii*) the distinguished extreme points v^1, v^2, v^3 and v^4. Abusing slightly the notation, in the description below, we use: (*i*) P, for the array of the points that form the girding polygon, and (*ii*) k for the last vertex included so far into the array P. Initially, $P := (v^1)$ and $k := v^1$.

Table 2. Procedure *polygon*.

PROCEDURE *polygon*(V, v^1, v^2, v^3, v^4)	
1 $P := (v^1)$	*//Initializing variables*
2 $k := v^1$	
3 **WHILE** $k \neq v^2$ **DO**	*//Step 1*
4 form a subset of vertices $V^* := \{\, i \mid x_i < x_k \wedge y_i \geq y_{v^2} \,;\, i \in V \,\}$	*//$V^* \subset V$*
5 form a subset of edges $E^* := \{(k, j); j \in V^*\}$	*//$E^* \subset E$*
6 form a set of angles $\Theta^* := \{\theta(k, j); (k, j) \in E^*\}$	
7 get the minimum angle $\theta(k, l)$ from Θ^*	
8 append the vertex l to P and update k equal to l.	
9	
10 **WHILE** $k \neq v^3$ **DO**	*//Step 2*
11 form a subset of vertices $V^* := \{\, i \mid x_i \leq x_{v^3} \wedge y_i < y_k \,;\, i \in V \,\}$	
12 form a subset of edges $E^* := \{(k, j); j \in V^*\}$	
13 form a set of angles $\Theta^* := \{\theta(k, j); (k, j) \in E^*\}$	
14 get the minimum angle $\theta(k, l)$ from Θ^*	
15 append the vertex l to P and update k equal to l.	
16	
17 **WHILE** $k \neq v^4$ **DO**	*//Step 3*
18 form a subset of vertices $V^* := \{\, i \mid x_i > x_k \wedge y_i \leq y_{v^4} \,;\, i \in V \,\}$	
19 form a subset of edges $E^* := \{(k, j); j \in V^*\}$	
20 form a set of angles $\Theta^* := \{\theta(k, j); (k, j) \in E^*\}$	
21 get the minimum angle $\theta(k, l)$ from Θ^*	
22 append the vertex l to P and update k equal to l.	
23	
24 **WHILE** $k \neq v^1$ **DO**	*//Step 4*
25 form a subset of vertices $V^* := \{\, i \mid x_i \geq x_{v^1} \wedge y_i > y_k \,;\, i \in V \,\}$	
26 form a subset of edges $E^* := \{(k, j); j \in V^*\}$	
27 form a set of angles $\Theta^* := \{\theta(k, j); (k, j) \in E^*\}$	
28 get the minimum angle $\theta(k, l)$ from Θ^*	
29 append the vertex l to P and update k equal to l.	

Lemma 2. *The time complexity of Procedure polygon is $O(n^2)$.*

Proof of Lemma 2. There are four independent *while* statements with similar structure, each of which can be repeated at most n times. In the first line of each of these *while* statements, in lines 4, 11, 18, and 25, the set of points V^* is formed that yields $\{f(n) = 2n\}$ operations. In lines 5, 12, 19, and 26, the set of $n-1$ edges E^* is formed in time $\{f(n) = n-1\}$. In lines 6, 13, 20, and 27, the set of angles Θ^* consisting of at most $n-1$ elements is formed in time $\{f(n) = n-1\}$. In lines 7, 14, 21, and 28 to find the minimum angle in set Θ^* at most $n-1$ comparisons are needed and the lemma follows. $\quad\square$

In Figure 3, we illustrate an example with $V = \{1, 2, \cdots, 6\}$ with coordinates $X = \{x_1, x_2, \cdots, x_6\}$ and $Y = \{y_1, y_2, \cdots, y_6\}$. The extreme points are: $v^1 = 4$, $v^2 = 2$, $v^3 = 5$ and $v^4 = 5$ and $P = (4, 2, 5, 4)$. Initially, $P = (4)$. Then, vertex 2 is added to polygon in Step 1, vertex 5 is added in Step 2; Step 3 is not carried out because $v^3 = v^4$; vertex 4 is added at Step 4.

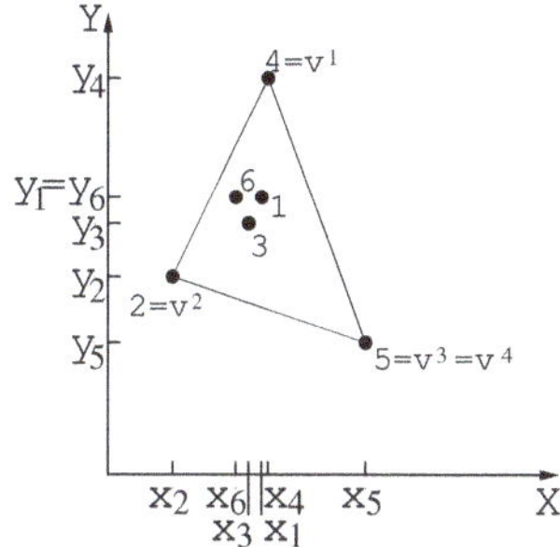

Figure 3. Example that shows the extreme vertices and girding polygon.

Using polygon $P(V)$ constructed by the Procedure Polygon, we obtain our initial, yet infeasible (partial) tour $T_0 = (t_1, t_2, \cdots, t_m, t_1)$ that is merely formed by all the points $t_1, t_2, \cdots, t_m$ of that polygon, where $t_1 = v^1$ and m is the number of the points.

In the example of Figure 3, P is the initial infeasible tour $T_0 = (4, 2, 5, 4)$. $V \setminus T_0 = \{1, 3, 6\}$ is the set of points that will be inserted into the final tour.

2.2. Phase 2

The initial tour of Phase 1 is iteratively extended with new points from the internal area of polygon $P(V)$ using the cheapest insertion strategy at Phase 2 [15].

Let $l \notin T_{h-1}$ be a candidate point to be included in tour T_{h-1}, resulting in an extended tour T_h of iteration $h > 0$, and let $t_i \in T_{h-1}$. Due to the triangle inequality, $w(t_i, l) + w(l, t_{i+1}) \geq w(t_i, t_{i+1})$; i.e., *the insertion of point l between points t_i and t_{i+1}, will increase the current total cost $C(T_{h-1})$ by* $w(t_i, l) + w(l, t_{i+1}) - w(t_i, t_{i+1}) \geq 0$ (see Figure 4). Once point l is included between points t_i and t_{i+1}, for the convenience of the presentation, we let $t_m := t_{m-1}, t_{m-1} := t_{m-2}, \cdots, t_{i+3} := t_{i+2}, t_{i+2} := t_{i+1}$ and $t_{i+1} := l$ (due to the way in which we represent our tours, this re-indexing yields no extra cost in our algorithm).

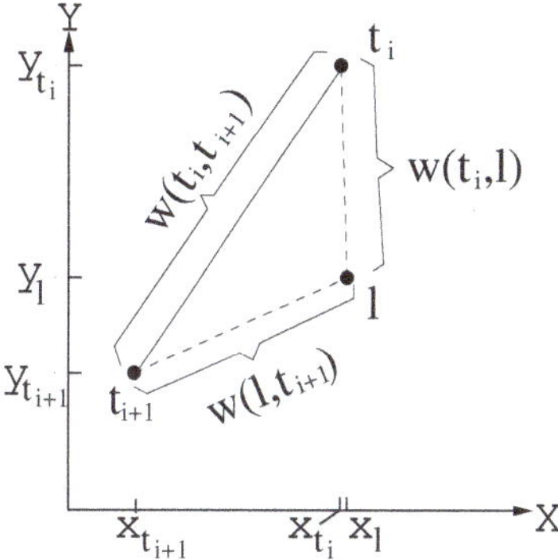

Figure 4. The triangle inequality.

In Table 3, we give a formal description of our procedure that inserts point l between points t_i and t_{i+1} in tour T.

Table 3. Procedure *insert_point_in_tour*.

PROCEDURE *insert_point_in_tour*(T, l, i)
1    $p := \|T\|$
2   **IF** $i < p$ **THEN**
3      $j := p + 1$
4   **WHILE** $j > i + 1$ **DO**
5        $t_j := t_{j-1}$
6        $j := j - 1$
7    $t_{i+1} := l$
8   **RETURN** T

Procedure construc_tour

At each iteration h, the current tour T_{h-1} is extended by point $l^h \in V \setminus T_{h-1}$ yielding the minimum cost c_l^h (defined below), which represents the increase in the the current total cost $C(T_{h-1})$ if that point is included into the current tour T_{h-1}. The cost for point $l \in V \setminus T_{h-1}$ is defined as follows:

$$c_l^h = \min_{t_i \in T_{h-1}} \{w(t_i, l) + w(l, t_{i+1}) - w(t_i, t_{i+1})\}. \tag{9}$$

For further references, we denote by $i(l)$ the index of point t_i for which the above minimum for point l is reached, i.e., $w(t_{i(l)}, l) + w(l, t_{i(l)+1}) - w(t_{i(l)}, t_{i(l)+1}) = \min_{t_i \in T_{h-1}} \{w(t_i, l) + w(l, t_{i+1}) - w(t_i, t_{i+1})\}$.

Thus, l^h is a point that attains the minimum

$$\min\{c_l^h | l \in V \setminus T_{h-1}\}, \tag{10}$$

whereas the ties can be broken arbitrarily.

To speed up the procedure, we initially calculate the minimum cost for each point $l \in V \setminus T_{h-1}$. After the insertion of point l^h, the minimum cost c_l^h is updated as follows:

$$c_l^h := \min\{c_l^{h-1}, w(t_i, l) + w(l, t_{i+1}) - w(t_i, t_{i+1}), \; w(t_{i+1}, l) + w(l, t_{i+2}) - w(t_{i+1}, t_{i+2})\}. \tag{11}$$

We can describe now Procedure *construct_tour* as shown in Table 4.

Table 4. Procedure *construct_tour*.

PROCEDURE *construct_tour*(V, T_0)
1    $h := 1$
2   **FOR** each point $l \in V \setminus T_{h-1}$ **DO**
3      $c_l^h := \min_{t_i \in T_{h-1}} \{w(t_i, l) + w(l, t_{i+1}) - w(t_i, t_{i+1})\}$
4   **WHILE** exists a vertex $l \in V \setminus T_{h-1}$ **DO**
5   get l^h
6   *insert_point_in_tour* $(T_{h-1}, l^h, i(l^h))$
7   **FOR** each point $l \in V \setminus T_h$ **DO**
8        $c_l^{h+1} := \min\{c_l^h, \; w(t_i, l) + w(l, t_{i+1}) - w(t_i, t_{i+1}), \; w(t_{i+1}, l) + w(l, t_{i+2}) - w(t_{i+1}, t_{i+2})\}$
9    $h := h + 1$

Lemma 3. *The time complexity of the Procedure construct_tour is $O(n^2)$.*

Proof of Lemma 3. In lines 2–3, there is a *for* statement with $n - (m + h - 1)$ repetitions. To calculate c_l^h in line 3, the same number of repetitions is needed and the total cost of the *for* statement is $[n - (m + h - 1)][n - (m + h - 1)] = [n^2 - 2(m + h - 1)n + (m + h - 1)^2]$. The *while* statement in lines 4–9 is repeated at most $n - (m + h - 1)$ times. In line 5, to calculate c_{lh}^h (Equation (10)) $n - (m + h - 1)$ comparisons are required. In lines 7–8, there is a *for* statement nested in the above *while* statement with $n - (m + h)$ repetitions. Hence, the total cost is $[n^2 - 2(m + h - 1)n + (m + h - 1)^2] + [n - (m + h - 1)]\{[n - (m + h - 1)] + [n - (m + h)]\} = [n^2 - 2(m + h - 1)n + (m^2 - 2m - 2h + h^2 + 1)] + [n - (m + h - 1)][2n - (2m + 2h - 1)] = [n^2 - (2m + 2h - 2)n + (m^2 - 2m - 2h + h^2 + 1)] + [2n^2 - (4m + 4h - 3)n + (2m^2 + 4mh - 3m - 3h + 2h^2 + 1)] = 3n^2 - (6m + 6h - 5)n + (3m^2 + 4mh - 5m - 5h + 3h^2 + 2) = O(n^2)$. $\square$

In the example of Figure 5, $T_0 = (4, 2, 5)$. The costs $c_l^1, l \in V \setminus T_0$, are calculated as follows:

$$c_1^1 = \min\{w(4,1) + w(1,2) - w(4,2),\ w(2,1) + w(1,5) - w(2,5),\ w(5,1) + w(1,4) - w(5,4)\}$$
$$= w(5,1) + w(1,4) - w(5,4),$$
$$c_3^1 = \min\{w(4,3) + w(3,2) - w(4,2),\ w(2,3) + w(3,5) - w(2,5),\ w(5,3) + w(3,4) - w(5,4)\}$$
$$= w(4,3) + w(3,2) - w(4,2),$$
$$c_6^1 = \min\{w(4,6) + w(6,2) - w(4,2),\ w(2,6) + w(6,5) - w(2,5),\ w(5,6) + w(6,4) - w(5,4)\}$$
$$= w(4,6) + w(6,2) - w(4,2).$$

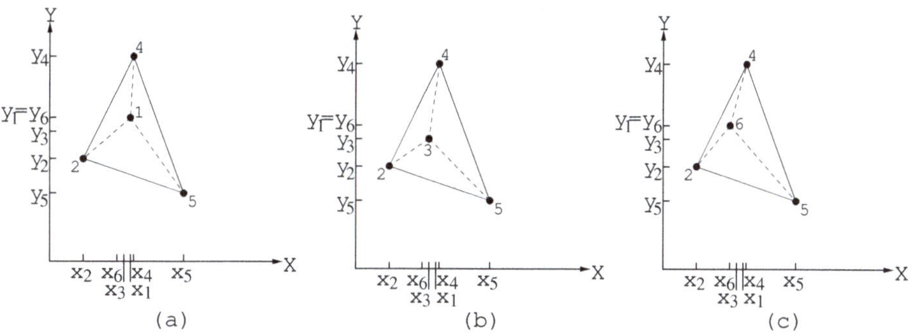

Figure 5. Points 1, 3, and 6 that can be inserted between point 4 and 2, 2 and 5, or 5 and 4 from partial tour T_0 are depicted in Figures (**a**), (**b**), and (**c**), respectively.

Hence, $\min\{c_1^1, c_3^1, c_6^1\} = c_6^1 = w(4,6) + w(6,2) - w(4,2);\ l^1 = 6$ and $i(6) = 4$. Therefore, point 6 will be included in tour T_1 between points 4 and 2 (Figure 6).

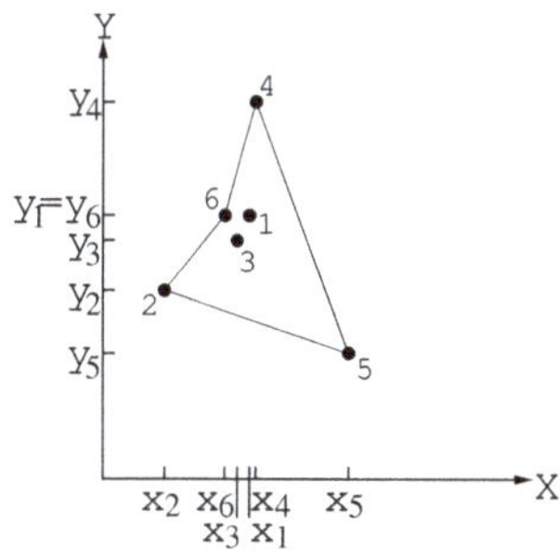

Figure 6. Point 6 was inserted in the tour T_0 between points 4 and 2.

Now, $T_1 = (4, 6, 2, 5, 4)$ and the minimum costs c_l^2 for each point $l \in V \setminus T_1$ are:

$$c_1^2 = \{c_1^1, w(4,1) + w(1,6) - w(4,6), w(6,1) + w(1,2) - w(6,2)\}$$
$$= w(4,1) + w(1,6) - w(4,6).$$

$$c_3^2 = \{c_3^1, w(4,3) + w(3,6) - w(4,6), w(6,3) + w(3,2) - w(6,2)\}$$
$$= w(6,3) + w(3,2) - w(6,2).$$

Hence, $\min\{c_1^2, c_3^2\} = c_3^2 = w(6,3) + w(3,2) - w(6,2); l^2 = 3$ and $i(3) = 6$. Therefore, point 3 will be included in tour T_2 between points 6 and 2 (Figure 7).

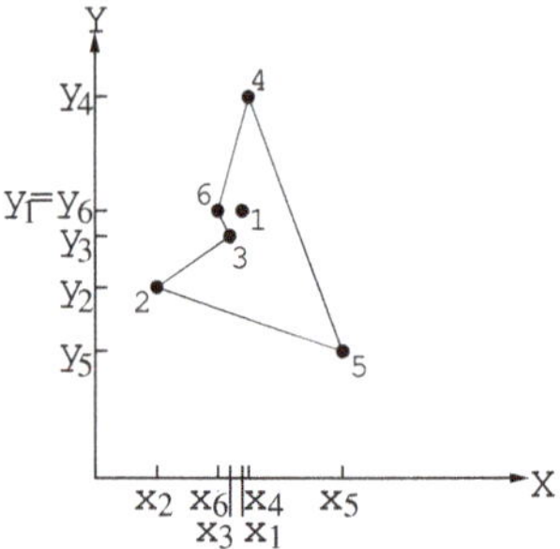

Figure 7. Point 3 was inserted in the tour T_1 between points 6 and 2.

Now, $T_2 = (4,6,3,2,5,4)$ and the minimum costs $c_l^3, l \in V \setminus T_2$ are
$$c_1^3 = \{c_1^2, w(6,1) + w(1,3) - w(6,3), w(3,1) + w(1,2) - w(3,2) = c_1^2.$$

Hence, $\min\{c_1^3\} = c_1^2 = w(4,1) + w(1,6) - w(4,6); l^3 = 1$ and $= i(1) = 4$. Therefore, point 1 will be included in tour T_3 between points 4 and 6 (Figure 8).

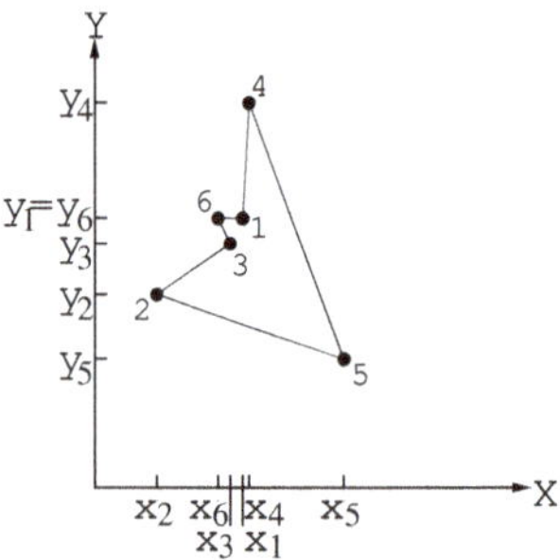

Figure 8. Point 1 be inserted in the tour T_2 between points 4 and 6.

The resultant tour $T = T_3 = (4,1,6,3,2,5,4)$ includes all points from set V and Procedure *construct_tour* halts.

2.3. Phase 3

At Phase 3, we iteratively improve the feasible tour T delivered by Phase 2. We use two heuristic algorithms. The first one is called *2-Opt*, which is a local search algorithm proposed by Croes [13]. The second one is based on our *construct_tour* procedure, named *improve_tour*. The current solution (initially, it is the tour delivered by Phase 2) is repeatedly improved first by *2-Opt*-heuristics and then by Procedure *improve_tour*, until there is an improvement. Phase 3 halts if either the output of one of the heuristics has the same objective value as the input (by the construction, the output cannot be worse than the input) or the following condition is satisfied:

$$C(T_{in}) - C(T_{out}) \leq dif_{min}, \tag{12}$$

where dif_{min} is a constant (for instance, we let $dif_{min} = 0.0001$). Thus, initially, *2-Opt*-heuristics runs with input T. Repeatedly, Condition (12) is verified for the the output of every call of each of the heuristics. If it is satisfied, Phase 3 halts; otherwise, for the output of the last called heuristics, the other one is invoked and the whole procedure is repeated; see Figure 9.

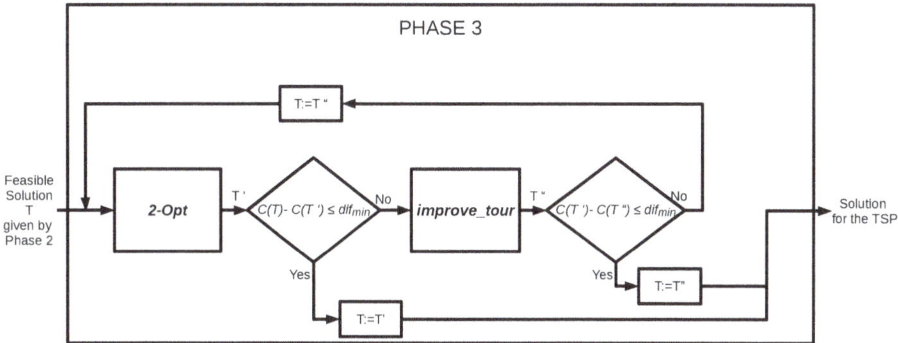

Figure 9. Block diagram of Phase 3.

2.3.1. Procedure 2-Opt

Procedure *2-Opt* is a local search algorithm improving feasible solution $T = (t_1, t_2, \cdots, t_n, t_1)$ ($n = |V|$). It is well-known that the time complexity of this procedure is $O(n^2)$. For the completeness of our presentation, we give a formal description of this procedure in Table 5.

Table 5. Procedure *2-Opt*.

PROCEDURE *2-Opt*(V,T)
1   $i := 1$
2   $n := |V|$
3   **WHILE** $i < n - 2$ **DO**
4     $j := i + 1;$
5   **WHILE** $j < n - 1$ **DO**
6   **IF** $w(t_i, t_j) + w(t_{i+1}, t_{j+1}) < w(t_i, t_{i+1}) + w(t_j, t_{j+1})$ **THEN**
7         $x := i + 1$
8         $y := j$
9   **WHILE** $x < y$ **DO**
10           $t_{aux} := t_x$
11           $t_x := t_y$
12           $t_y := t_{aux}$
13           $x := x + 1$
14           $y := y - 1$
15       $j := j + 1$
16     $i := i + 1$
17   **RETURN** T

The result of a local replacement carried out by the procedure is represented schematically in the Figure 10).

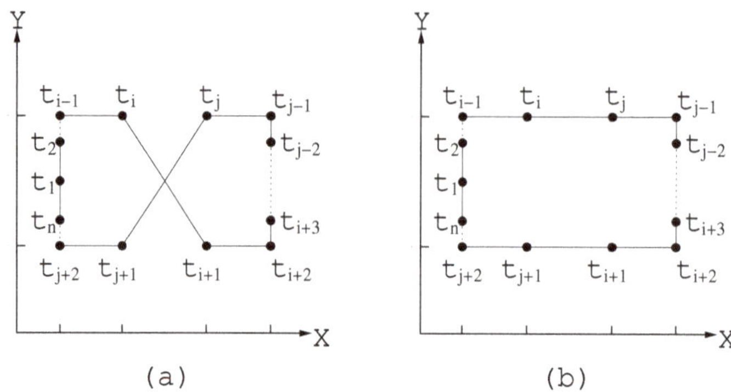

Figure 10. (a) a fragment of a solution before applying the algorithm *2-Opt*; (b) the corresponding fragment after applying algorithm *2-Opt*.

2.3.2. Procedure improve_tour

We also use our algorithm *construct_tour* to improve a feasible solution $T = (t_1, t_2, \cdots, t_n, t_1)$, $n = |V|$. Iteratively, point t_{i+1}, $1 \leq i < n$, is removed from the tour T and is reinserted by a call of procedure *construct_tour*$(V, T \setminus \{t_{i+1}\})$. If a removed point gets reinserted in the same position, then $i := i + 1$ and the procedure continues until $i \leq n$ (see Table 6).

Table 6. Procedure *improve_tour*.

PROCEDURE *improve_tour(V,T)*	
1 $i := 1$	
2 **WHILE** $i < n$ **DO**	
3 $t_j := t_{i+1}$	
4 remove t_{i+1} from the tour T	//now T is infeasible
5 *construct_tour*$(V, T \setminus \{t_{i+1}\})$	//T is feasible again
6 **IF** $t_{i+1} = t_j$ **THEN**	
7 $i := i + 1$	
8 **RETURN** T	

Figure 11 illustrates the iterative improvement in the cost of the solutions obtained at Phase 3 for a sample problem instance *usa115475*. The initial solution T_0 of Phase 2 is iteratively improved as shown in the diagram.

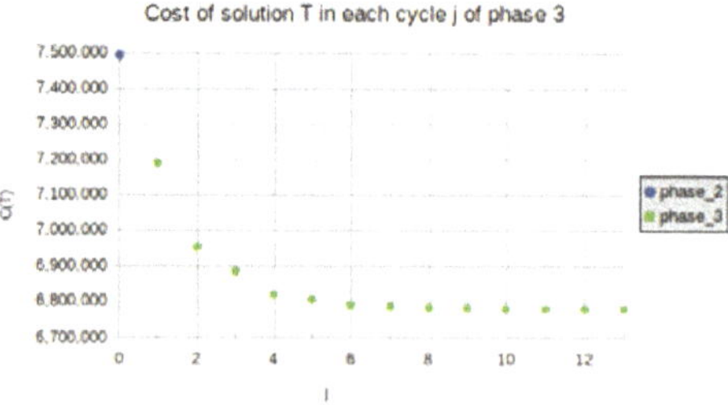

Figure 11. The improvement rate at Phase 3 for instance *usa115475*.

Lemma 4. *The time complexity of the Procedure improve_tour is $O(n^2)$.*

Proof of Lemma 4. In lines 2–7, there is a *while* statement with $n - 1$ repetitions. The call of Procedure *construct_tour* in line 5 yields the cost $O(n)$ since with $m = n - 1, h = 1$; see the proof of Lemma 3 (m is the number of points in the current partial tour). The lemma follows. $\square$

3. Implementation and Results

CII-algorithm was coded in C++ and compiled in g++ on a server with processor 2x Intel Xeon E5-2650 0 @ 2.8 GHz (Cuernavaca, Mor., Mexico), 32 GB in RAM and Ubuntu 18.04 (bionic) operating system (we have used only one CPU in our experiments). We did not keep the cost matrix in computer memory, but we have rather calculated the costs using the coordinates of the points. This does not increase the computation time too much and saves considerably the required computer memory.

We have tested the performance of *CII*-algorithm for 85 benchmark instances from TSPLIB [16] library and for 135 benchmark instances from TSP Test Data [17] library. The detailed results are presented in the Appendix. In our tables, parameter "Error" specifies the approximation factor of algorithm H compared to cost of the best known solution (C(BKS)):

$$Error_H = \left| \frac{C(BKS) - C(T_H)}{C(BKS)} \right| 100\%. \tag{13}$$

In Table 7 below, we give the data on the average performance of our heuristics. The average error percentage of our heuristics is calculated using Formula (13). It shows, for each group of instances, the average error of the solutions delivered by Phase 2 and, at Phase 3, the number of cycles at Phase 3 and the average decrease in the cost of the solution decreased at Phase 3 compared to that Phase 3.

Table 7. Statistics about the solutions delivered by *CII*.

Description	TSPLIB	NATIONAL	ART GALLERY	VLSI	All
Number of instances	83	27	6	102	218
Average error percentage of the solutions at Phase 2	11.8%	17.7%	6.7%	18.4%	15.4%
Average number of cycles performed at Phase 3	7	11	11	10	9
Average decrease in error at Phase 3	6.5%	9.6%	3.1%	9.8%	8.3%
Final average error percentage	5.3%	8.2%	3.6%	8.6%	7.2%
Average memory usage	0.8 MiB	1.6 MiB	10.9 MiB	2.3 MiB	1.88 MiB

In the diagrams below (on the left hand-side), we illustrate the dependence of the approximation given by our algorithm on the size of the tested instances, and the dependence of the execution time of our algorithm on the size of the instances (right hand-side diagrams). We classify the tested instance into three groups: the small ones (from 1 to 199 points in Figure 12), the middle-sized ones (from 200 to 9999 points in Figure 13), and large instances (from 10,000 to 250,000 in Figure 14). We do not include the data for the largest two problem instances *lra498378* and *lrb744710* because of the visualization being technically complicated. The error for these instances is 12.5% and 15.9%, respectively, and the CPU time was limited to two weeks for both instances. As we can see, at Phase 3, there is an improvement in the quality of the solutions delivered by Phase 2.

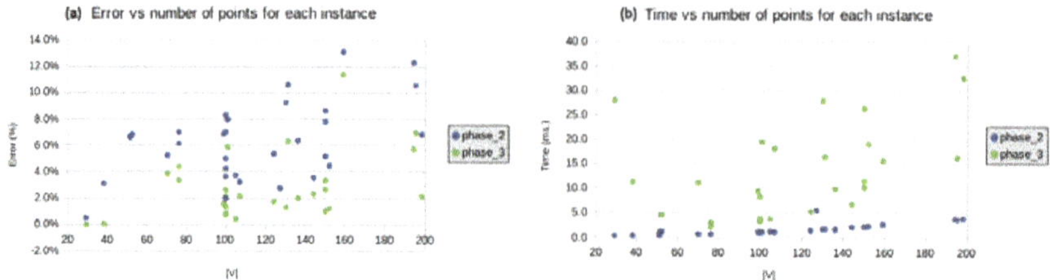

Figure 12. (**a**) error vs. number of points, and (**b**) processing time vs. number of points, where $1 \leq |V| < 200$.

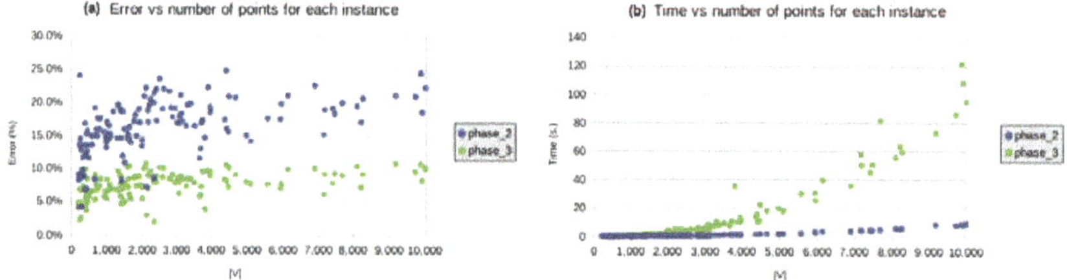

Figure 13. (**a**) error vs. number of points, and (**b**) processing time vs. number of points, where $200 \leq |V| < 10,000$.

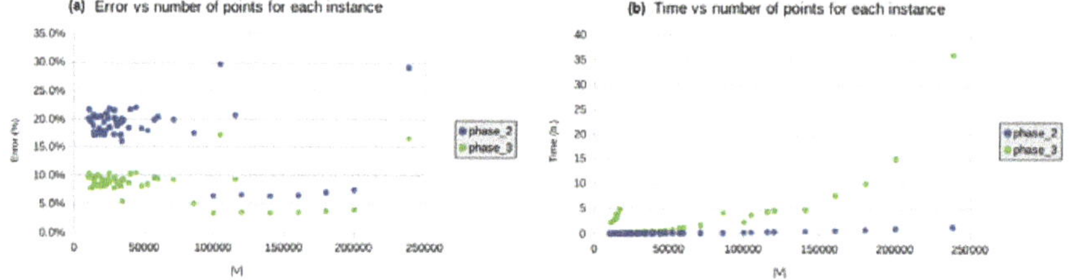

Figure 14. (**a**) error vs. number of points, and (**b**) processing time vs. number of points, where $10,000 \leq |V| < 250,000$.

Table 8 shows the summary of the comparison statistics of the solutions delivered by our algorithm *CII* with the solutions obtained by the heuristics that we have mentioned in the introduction (namely, DFACO [10], ACO-3Opt [10], ESACO [7], PACO-3Opt [8], DPIO [12], ACO-RPMM [9], Partial ACO [6], and PRNN [11]). We may observe in Table 9 that algorithm *CII* has attained an improved approximation for 17 instances. At the same time, in terms of the execution time, our heuristic dominates the other heuristics.

Table 8. Statistics between *CII* and other heuristics.

Description	TSPLIB	NATIONAL	ART GALLERY	VLSI	All
Number of instances	83	27	6	102	218
Number of the known results from other heuristics	142	0	10	12	164
Number of time *CII* gave a better error than other heuristics	2	0	4	12	18
Number of times *CII* has improved the earlier known best execution time	140	0	0		140

In the Table 9, we specify the problem instances for which our algorithm provided a better relative error than some of the earlier cited algorithms.

Table 9. Comparative relative errors for some problem instances.

Description	$Error_{CII}$	$Error_H$
TSPLIB/*rat783*	7.4%	19.1% and 19.5% (DFACO [10] and ACO-3Opt [10])
ART/*Mona-lisa100K*	3.4%	5.5% (Partial ACO [6])
ART/*Vangogh120K*	3.5%	5.8% (Partial ACO [6])
ART/*Venus140K*	3.4%	5.8% (Partial ACO [6])
ART/*Earring200K*	3.9%	7.2% (Partial ACO [6])
VLSI/*dca1376*	7.6%	19.6% (PRNN [11])
VLSI/*djb2036*	10.0%	23.4% (PRNN [11])
VLSI/*xqc2175*	9.1%	21.4% (PRNN [11])
VLSI/*xqe3891*	9.7%	21.7% (PRNN [11])
VLSI/*bgb4355*	8.4%	22.8% (PRNN [11])
VLSI/*xsc6880*	10.1%	21.9% (PRNN [11])
VLSI/*bnd7168*	9.2%	21.7% (PRNN [11])
VLSI/*ida8197*	7.2%	23.2% (PRNN [11])
VLSI/*dga9698*	9.6%	21.1% (PRNN [11])
VLSI/*xmc10150*	9.6%	20.3% (PRNN [11])
VLSI/*xvb13584*	9.5%	23.6% (PRNN [11])
VLSI/*frh19289*	9.3%	22.5% (PRNN [11])

In terms of the CPU time comparison, see Table 10.

Table 10. Comparative CPU time for the problem instances for which the other heuristics were faster.

Description	$Time_{CII}$	$Time_H$
TSPLIB/*pla33810*	25.7 m	21.0 m (DPIO [12])
TSPLIB/*pla85900*	4.1 h	1.4 h (DPIO [12])
Art Gallery/*mona-lisa100K*	2.3 h	1.4 h and 1.1 h (ACO-RPMM [9] and Partial ACO [6])
Art Gallery/*vangogh120K*	4.6 h	1.9 h and 1.5 h (ACO-RPMM [9] and Partial ACO [6])
Art Gallery/*venus140K*	4.8 h	2.6 h and 2.1 h (ACO-RPMM [9] and Partial ACO [6])
Art Gallery/*pareja160K*	7.7 h	3.5 h (ACO-RPMM [9])
Art Gallery/*coubert180K*	10.1 h	4.5 h (ACO-RPMM [9])
Art Gallery/*earring200K*	15.1 h	6.0 h and 5.1 h (ACO-RPMM [9] and Partial ACO [6])

In the diagram below (Figure 15), we illustrate the dependence of the memory used by our algorithm of all tested instances.

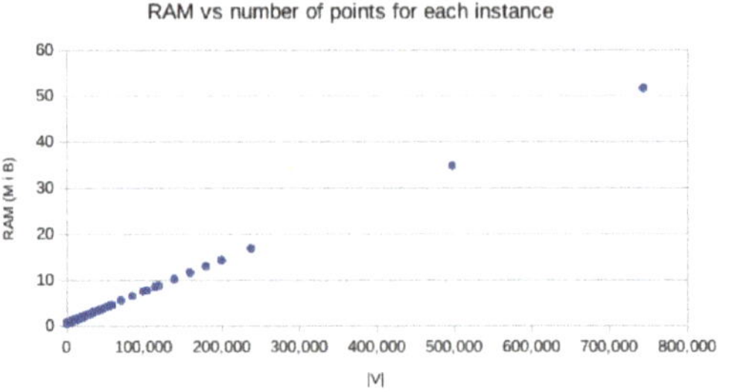

Figure 15. RAM vs. number of points for all the tested instances.

4. Conclusions and Future Work

We have presented a simple, easily implementable and fast heuristic algorithm for the Euclidean traveling salesman problem that solves both small and large scale instances with an acceptable approximation and consumes a little computer memory. Since the algorithm uses simple geometric calculations, it is easily implementable. The algorithm is fast, the first two phases run in time $O(n^2)$, whereas the number of the improvement repetitions in the third phase, in practice, is not large. The first two phases might be used independently from the third phase, for instance, for the generation of an initial tour in more complex loop improvement heuristics. The quality of the solution delivered already by Phase 2 is acceptable and is expected to greatly outperform that of a random solution used normally to initiate meta-heuristic algorithms. We have implemented NN (Nearest Neighborhood) heuristics and run the code for the benchmark instances (the initial vertex for NN heuristic was selected randomly). Phase 2 gave essentially better results. In average, for the tested 135 instances (6 large, 32 Medium and 97 small ones), the difference between the approximation factor obtained by the procedure of Phase 2 and that of Nearest Neighbor heuristic was 9.65% (the average error of Phase 2 was 16.89% and that of NN was 26.55%, whereas the standard deviations were similar, 0.05% and 0.04%, respectively). As for the overall algorithm, it uses a negligible computer memory. Although for most of the tested benchmark instances it did not improve the best known results, the execution time of our heuristic, on average, was better than the earlier reported best known times. For future work, we intend to create a more powerful, yet more complex, *CII*-algorithm by augmenting each of the three phases of our algorithm with alternative ways for the creation of the initial tour and alternative insertion and improvement procedures.

Author Contributions: Conceptualization, N.V. and J.M.S.; Methodology, V.P.-V.; Validation, N.V.; Formal Analysis, N.V. and J.M.S.; Investigation, V.P.-V.; Resources, UAEMor administrated by J.A.H.; Writing—original draft preparation, V.P.-V.; Writing—review and editing, N.V.; Visualization, V.P.-V. and N.V.; Supervision, N.V.; Project administration, N.V.; All authors have read and agreed to the published version of the manuscript.

Funding: This research received no external funding.

Conflicts of Interest: The authors declare no conflict of interest.

Appendix A

In the table below (Table A1), we give some details on the earlier mentioned heuristics with which we compare our results (the entries in the column "Runs" specify the number of the reported runs of the corresponding heuristic).

Table A1. Heuristics used to compare the *CII*-algorithm.

Heuristic Id	Heuristic Name	Number of Reported Instances	Runs
ACO-RPMM [9]	ACO - Restricted Pheromone Matrix Method	6 Large	10
Partial ACO [6]	Partial ACO	4 Large and 5 Small	100
DFACO [10]	Dynamic Flying ACO	30 Small	100
ACO-3Opt [10]	ACO-3Opt	30 Small	100
DPIO [12]	Discrete Pigeon-inspired optimization with Metropolis acceptance	1 Large, 6 Medium and 28 Small	25
PACO-3Opt [8]	Parallel Cooperative Hybrid Algorithm ACO	21 Small	20
ESACO [7]	Effective Strategies + ACO	5 Medium and 17 Small	20
PRNN [11]	Parallel Repetitive Nearest Neighbor	3 Medium and 9 Small	$n = |V|$
NN	Nearest Neighbor Algorithm	4 Large, 25 Medium and 61 Small	1

The next table (Table A2) discloses the headings of our tables.

Table A2. Description of the headings of Tables A3–A6.

Header	Header Description
$|V|$	the number of vertices in the instance
Opt?	"yes" if Best Known Solution (BKS) is optimal, "no" otherwise
$C(BKS)$	the cost of BKS
$C(T)$	Cost of the solution constructed by *CII* heuristic
RAM	RAM used by *CII* heuristics
#	the number of cycles at Phase 3 of *CII* heuristic
Error	as defined in Formula (13)
$C_{avg}(T_H)$	the average cost of the solution obtained by heuristic H
Heuristic Id	nomenclature used in Table A1
Time	the processing time of a heuristic
ms, s, m, h, d	time units for milliseconds, seconds, minutes, hours and days respectively.

In the tables below, each line corresponds to a particular benchmark instance. For each of these instances, we indicate the performance of Phase 2 and Phase 3, separately, and that of the other heuristics reporting the results for that instance. In addition, 85 benchmark instances were taken from TSPLIB [16] and 135 instances are from TSP Test Data [17] libraries. Tables A3, A4, and A6 include the earlier known results.

In some lines of our tables (e.g., line 1, Table A5), a slight difference in the approximation errors of our algorithm and those of the algorithms from the "Results for National TSP Benchmarks" table can be seen due to the way the distances in the obtained solutions are represented in our algorithm (we do not round the distances represented as decimal numbers, whereas the distances in the best known solutions are rounded).

Table A3. Results for **TSPLIB** benchmarks.

Instance			CII Heuristic (Phase 2)				CII Heuristic (Phase 3)						Other Heuristics		
	V	Opt?	$C(BKS)$	$C(T)$	$Error_{CII}$	Time	$C(T)$	$Error_{CII}$	Time	RAM	#	$C_{min}(T_H)$	$Error_H$	Time	Heuristic Id
eil51	51	yes	426	454	6.6%	0.4 ms	454	6.6%	1.0 ms	0.5 MiB	1	426 426 426	0.0% 0.0% 0.0%	1.0 s 1.0 s 1.1 s	DFACO ACO-3Opt ESACO
berlin52	52	yes	7542	8058	6.8%	1.1 ms	8058	6.8%	4.5 ms	0.6 MiB	3	7542 7542	0.0% 0.0%	1.0 s 1.0 s	DFACO ACO-3Opt
st70	70	yes	675	710	5.2%	0.6 ms	701	3.8%	11.1 ms	0.6 MiB	3	826	22.3%	0.4 ms	NN
eil76	76	yes	538	576	7.0%	0.7 ms	556	3.4%	2.2 ms	0.6 MiB	3	538 538 538	0.0% 0.0% 0.0%	3.0 s 3.0 s 1.4 s	DFACO ACO-3Opt ESACO
pr76	76	yes	108,159	114,808	6.1%	0.7 ms	112,911	4.4%	3.0 ms	0.6 MiB	4	148,348	37.2%	0.5 ms	NN
rat99	99	yes	1211	1294	6.9%	1.0 ms	1230	1.5%	9.4 ms	0.6 MiB	3	1442	19.1%	0.8 ms	NN
kroA100	100	yes	21,282	23,050	8.3%	1.1 ms	21,443	0.8%	3.5 ms	0.6 MiB	3	21,282 21,282 21,282	0.0% 0.0% 0.0%	2.0 s 2.0 s 2.6 s	DFACO ACO-3Opt ESACO
kroB100	100	yes	22,141	23,247	5.0%	1.1 ms	22,716	2.6%	3.3 ms	0.6 MiB	3	22,141 22,141	0.0% 0.0%	2.0 s 2.0 s	DFACO ACO-3Opt
kroC100	100	yes	20,749	21,632	4.3%	1.1 ms	20,922	0.8%	3.8 ms	0.6 MiB	3	20,749 20,749	0.0% 0.0%	2.0 s 2.0 s	DFACO ACO-3Opt
kroD100	100	yes	21,294	21,712	2.0%	1.1 ms	21,582	1.4%	3.4 ms	0.6 MiB	3	21,294 21,294	0.0% 0.0%	3.0 s 3.0 s	DFACO ACO-3Opt
kroE100	100	yes	22,068	22,870	3.6%	1.0 ms	22,528	2.1%	8.3 ms	0.6 MiB	3	22,068 22,068	0.0% 0.0%	2.0 s 2.0 s	DFACO ACO-3Opt
rd100	100	yes	7910	8465	7.0%	1.2 ms	8245	4.2%	3.8 ms	0.6 MiB	3	7910 7910	0.0% 0.0%	2.0 s 2.0 s	DFACO ACO-3Opt
eil101	101	yes	629	679	7.9%	1.1 ms	666	5.9%	19.5 ms	0.6 MiB	3	629 629	0.0% 0.0%	12.0 s 10.0 s	DFACO ACO-3Opt
lin105	105	yes	14,379	14,913	3.7%	1.2 ms	14,440	0.4%	3.8 ms	0.6 MiB	3	14,379 14,379 14,379	0.0% 0.0% 0.0%	2.0 s 2.0 s 2.0 s	DFACO ACO-3Opt ESACO
pr107	107	yes	44,303	45,730	3.2%	1.1 ms	45,262	2.2%	18.1 ms	0.6 MiB	5	54,121	22.2%	0.9 ms	NN
pr124	124	yes	59,030	62,193	5.4%	1.4 ms	60,055	1.7%	5.3 ms	0.6 MiB	3	73,008	23.7%	1.3 ms	NN
bier127	127	yes	118,282	121,544	2.8%	5.4 ms	121,544	2.8%	5.6 ms	0.6 MiB	3	118,282 118,282	0.0% 0.0%	47.0 s 56.0 s	DFACO ACO-3Opt
ch130	130	yes	6110	6676	9.3%	1.7 ms	6190	1.3%	27.9 ms	0.6 MiB	9	6110 6110	0.0% 0.0%	13.0 s 16.0 s	DFACO ACO-3Opt
pr136	136	yes	96,772	102,934	6.4%	1.7 ms	98,711	2.0%	9.9 ms	0.6 MiB	5	125,458	29.6%	1.2 ms	NN
pr144	144	yes	58,537	60,625	3.6%	2.1 ms	59,902	2.3%	6.8 ms	0.6 MiB	3	64,886	10.8%	1.4 ms	NN
ch150	150	yes	6528	7038	7.8%	2.1 ms	6746	3.3%	11.5 ms	0.6 MiB	3	6,528 6528	0.0% 0.0%	24.0 s 17.0 s	DFACO ACO-3Opt

Table A3. *Cont.*

Instance				CII Heuristic (Phase 2)			CII Heuristic (Phase 3)					Other Heuristics			
	V	Opt?	$C(BKS)$	$C(T)$	$Error_{CII}$	Time	$C(T)$	$Error_{CII}$	Time	RAM	#	$C_{min}(T_H)$	$Error_H$	Time	Heuristic Id
kroA150	150	yes	26,524	28,814	8.6%	2.2 ms	27,230	2.7%	10.2 ms	0.6 MiB	5	26,524 26,524	0.0% 0.0%	57.0 s 1.4 m	DFACO ACO-3Opt
kroB150	150	yes	26,130	27,476	5.2%	2.2 ms	26,399	1.0%	26.4 ms	0.6 MiB	5	26,130 26,130	0.0% 0.0%	7.0 s 9.0 s	DFACO ACO-3Opt
pr152	152	yes	73,682	76,952	4.4%	2.3 ms	74,605	1.3%	19.0 ms	0.6 MiB	5	86,906	17.9%	1.4 ms.	
u159	159	yes	42,080	47,591	13.1%	2.6 ms	46,875	11.4%	15.7 ms	0.6 MiB	3	53,918	28.1%	1.6 ms	NN
rat195	195	yes	2323	2569	10.6%	3.7 ms	2485	7.0%	16.2 ms	0.6 MiB	4	2826	21.7%	2.0 ms	NN
d198	198	yes	15,780	16,862	6.9%	3.9 ms	16,119	2.1%	32.6 ms	0.6 MiB	4	15,780	0.0%	6.5 s	ESACO
kroA200	200	yes	29,368	31,792	8.3%	3.9 ms	30,767	4.8%	17.5 ms	0.6 MiB	5	29,368 29,379 29,368	0.0% 0.04% 0.0%	2.8 m 3.5 m 4.7 s	DFACO ACO-3Opt ESACO
kroB200	200	yes	29,437	32,123	9.1%	3.7 ms	30,631	4.1%	11.8 ms	0.6 MiB	3	29,442 29,443	0.02% 0.02%	3.1 m 2.3 m	DFACO ACO-3Opt
ts225	225	yes	126,643	157,163	24.1%	4.5 ms	132,803	4.9%	30.6 ms	0.6 MiB	7	151,685	19.8%	2.5 ms	NN
tsp225	225	yes	3916	4442	13.4%	4.9 ms	4183	6.8%	22.9 ms	0.6 MiB	5	4733	20.9%	2.7 ms	NN
pr226	226	yes	80,369	83,637	4.1%	4.8 ms	82,151	2.2%	18.2 ms	0.6 MiB	3	94,258	17.3%	2.5 ms	
gil262	262	yes	2378	2681	12.8%	6.5 ms	2539	6.8%	45.4 ms	0.6 MiB	6	3102	30.5%	3.4 ms	NN
pr264	264	yes	49,135	53,416	8.7%	6.4 ms	50,402	2.6%	41.4 ms	0.6 MiB	5	58,615	19.3%	3.6 ms	NN
a280	280	yes	2579	2686	4.1%	33.6	ms 2686	4.1%	52.9 ms	0.6 MiB	5	2579	0.0%	4.5 s	ESACO
pr299	299	yes	48,191	52,912	9.8%	8.1 ms	50,225	4.2%	43.6 ms	0.6 MiB	5	63,254	31.3%	4.3 ms	NN
lin318	318	yes	42,029	46,904	11.6%	9.4 ms	45,063	7.2%	38.8 ms	0.6 MiB	4	42,228 42,244 42,054	0.5% 0.5% 0.06%	6.4 m 5.8 m 10.2 s	DFACO ACO-3Opt ESACO
linhp318	318	yes	41,345	46,904	13.4%	9.4 ms	45,063	9.0%	37.3 ms	0.6 MiB	4	50,299	21.7%	5.1 ms	NN
rd400	400	yes	15,281	17,146	12.2%	14.7 ms	16,158	5.7%	92.8 ms	0.6 MiB	6	15,384 15,614	0.7% 2.2%	2.2 m 24.9 m	PACO-3Opt DFACO
fl417	417	yes	11,861	12,680	6.9%	14.6 ms	12,295	3.7%	119 ms	0.6 MiB	8	11,880 11,987	0.2% 1.1%	1.6 m 34.1 m	PACO-3Opt DFACO
pr439	439	yes	107,217	120,679	12.6%	17.8 ms	112,531	5.0%	66.7 ms	0.6 MiB	3	107,516 108,702	0.3% 1.4%	2.4 m 35.5 m	PACO-3Opt DFACO
pcb442	442	yes	50,778	58,746	15.7%	17.7 ms	53,275	4.9%	126 ms	0.7 MiB	7	51,047 52,202 50,804	0.5% 2.8% 0.05%	2.2 m 34.8 m 11.5 s	PACO-3Opt DFACO ESACO
d493	493	yes	35,002	39,050	11.6%	21.8 ms	37,045	5.8%	129 ms	0.6 MiB	5	35,266 35,841	0.8% 2.4%	2.3 m 52.9 m	PACO-3Opt DFACO
u574	574	yes	36,905	42,435	15.0%	29.7 ms	39,355	6.6%	247 ms	0.6 MiB	9	37,367 38,031	1.3% 3.0%	1.9 m 1.5 h	PACO-3Opt DFACO

Table A3. *Cont.*

Instance				CII Heuristic (Phase 2)			CII Heuristic (Phase 3)					Other Heuristics		
	V	Opt?	$C(BKS)$	$C(T)$	$Error_{CII}$	Time	$C(T)$	$Error_{CII}$	Time	RAM	#	$C_{min}(T_H)$	$Error_H$	Time	Heuristic Id
rat575	575	yes	6773	7692	13.6%	29.4 ms	7215	6.5%	231 ms	0.7 MiB	8	7012	3.5%	1.4 h	PACO-3Opt
p654	654	yes	34,643	37,542	8.4%	37.6 ms	36,441	5.2%	179 ms	0.6 MiB	5	34,741 35,075	0.3% 1.2%	1.7 m 2.5 h	DFACO PACO-3Opt
d657	657	yes	48,912	56,268	15.0%	36.7 ms	51,553	5.4%	265 ms	0.6 MiB	7	49,463 50,277	1.1% 2.8%	2.3 m 2.4 h	DFACO PACO-3Opt
u724	724	yes	41,910	48,198	15.0%	60.9 ms	44,748	6.8%	264 ms	0.7 MiB	6	42,438 43,122	1.3% 2.9%	2.3 m 3.2 h	DFACO PACO-3Opt
rat783	783	yes	8806	10,218	16.0%	54.1 ms	9454	7.4%	332 ms	0.7 MiB	6	10,492 10,525 9127 8810	19.1% 19.5% 3.6% 0.04%	2.5 m 15.4 m 4.0 h 22.6 s	DFACO ACO-3Opt PACO-3Opt ESACO
dsj1000	1000	yes	18,659,688	21,836,514	17.0%	83.6 ms	20,225,584	8.4%	460 ms	0.7 MiB	5	18,732,088	0.4%	16.6 s	DPIO
dsj1000ceil	1000	yes	18,660,188	21,836,514	17.0%	83.5 ms	20,225,584	8.4%	452 ms	0.6 MiB	5	23,813,050	27.6%	39 ms	NN
pr1002	1002	yes	259,045	295,879	14.2%	87.7 ms	276,122	6.6%	744 ms	0.7 MiB	5	260,426 259,509 260,366	0.5% 0.2% 0.5%	14.3 s 35.8 s 14.1 s	DPIO ESACO DPIO
u1060	1060	yes	224,094	261,093	16.5%	99.5 ms	239,705	7.0%	1.0 s	0.7 MiB	11	224,932	0.4%	15.3 s	DPIO
vm1084	1084	yes	239,297	275,989	15.3%	104 ms	257,399	7.6%	901 ms	0.6 MiB	9	240,079	0.3%	17.4 s	DPIO
pcb1173	1173	yes	56,892	67,497	18.6%	124 ms	60,792	6.9%	775 ms	0.7 MiB	7	57,243	0.6%	17.8 s	DPIO
d1291	1291	yes	50,801	58,230	14.6%	136 ms	54,285	6.9%	927 ms	0.7 MiB	7	51,459	1.3%	19.4 s	DPIO
rl1304	1304	yes	252,948	302,661	19.7%	148 ms	277,193	9.6%	1.2 s	0.7 MiB	9	253,740	0.3%	21.5 s	DPIO
rl1323	1323	yes	270,199	322,964	19.5%	157 ms	288,501	6.8%	1.3 s	0.7 MiB	9	273,368 273,970 271,245 271,301	1.2% 1.4% 0.4% 0.4%	38.1 m 37.8 m 22.2 s 22.0 s	DFACO ACO-3Opt ACO-3Opt DPIO
nrw1379	1379	yes	56,638	64,925	14.6%	168 ms	59,905	5.8%	1.2 s	0.7 MiB	8	56,932	0.5%	23.2 s	DPIO
fl1400	1400	yes	20,127	21,800	8.3%	162 ms	21,071	4.7%	1.8 s	0.7 MiB	10	20,301 20,292 20,342 20,211	0.9% 0.8% 1.1% 0.4%	40.9 m 41.2 m 24.6 s 24.5 s	DFACO ACO-3Opt ACO-3Opt DPIO
u1432	1432	yes	152,970	171,179	11.9%	181 ms	160,260	4.8%	1.1 s	0.7 MiB	7	153,564	0.4%	23.9 s	DPIO
fl1577	1577	yes	22,249	25,513	14.7%	210 ms	24,518	10.2%	1.4 s	0.7 MiB	7	22,289 22,293	0.2% 0.2%	25.3 s 46.4 s	DPIO ESACO
d1655	1655	yes	62,128	70,779	13.9%	225 ms	65,520	5.5%	1.5 s	0.7 MiB	7	63,708 63,722 62,769	2.5% 2.6% 1.0%	25.4 m 29.2 m 27.5 s	DFACO ACO-3Opt ACO-3Opt

Table A3. *Cont.*

Instance			CII Heuristic (Phase 2)				CII Heuristic (Phase 3)						Other Heuristics		
	V	Opt?	$C(BKS)$	$C(T)$	$Error_{CII}$	Time	$C(T)$	$Error_{CII}$	Time	RAM	#	$C_{min}(T_H)$	$Error_H$	Time	Heuristic Id
vm1748	1748	yes	336,556	394,389	17.2%	267 ms	365,608	8.6%	2.0 s	0.7 MiB	7	338,118	0.5%	34.3 s	DPIO
u1817	1817	yes	57,201	65,783	15.0%	395 ms	61,453	7.4%	1.8 s	0.7 MiB	7	57,522	0.6%	30.3 s	DPIO
rl1889	1889	yes	316,536	376,715	19.0%	319 ms	344,514	8.8%	2.1 s	0.8 MiB	7	318,714	0.7%	36.6 s	DPIO
d2103	2103	yes	80,450	86,286	7.3%	373 ms	82,856	3.0%	2.5 s	0.7 MiB	7	80,567	0.1%	23.8 s	DPIO
u2152	2152	yes	64,253	75,216	17.1%	516 ms	68,766	7.0%	2.7 s	0.7 MiB	7	64,791	0.8%	25.9 s	DPIO
u2319	2319	yes	234,256	254,420	8.6%	501 ms	238,785	1.9%	3.1 s	0.7 MiB	7	236,158	0.8%	34.2 s	DPIO
pr2392	2392	yes	378,032	443,372	17.3%	495 ms	408,237	8.0%	3.0 s	0.7 MiB	6	380,346	0.6%	29.7 s	DPIO
pcb3038	3038	yes	137,694	160,909	16.9%	807 ms	146,378	6.3%	6.2 s	0.8 MiB	9	138,684	0.7%	43.5 s	DPIO
fl3795	3795	yes	28,772	33,002	14.7%	1.2 s	29,882	3.9%	35.6 s	0.9 MiB	34	29,209 28,883	1.5% 0.4%	1.1 m 2.0 m	DPIO ESACO
fnl4461	4461	yes	182,566	211,064	15.6%	1.9 s	195,786	7.2%	11.1 s	0.9 MiB	7	184,560 183,446	1.1% 0.5%	44.2 s 3.2 m	DPIO ESACO
rl5915	5915	yes	565,530	664,788	17.6%	3.1 s	605,687	7.1%	31.2 s	1.0 MiB	11	571,214 568,935	1.0% 0.6%	1.1 m 3.6 m	DPIO ESACO
rl5934	5934	yes	556,045	666,295	19.8%	3.2 s	599,066	7.7%	25.8 s	1.0 MiB	9	561,878	1.0%	48.7 s	DPIO
pla7397	7397	yes	23,260,728	27,709,175	19.1%	4.4 s	25,075,678	7.8%	45.3 s	1.1 MiB	11	23,605,219 23,389,341	1.5% 0.6%	1.8 m 3.6 m	DPIO ESACO
rl11849	11,849	yes	923,288	1,103,854	19.6%	12.4 s	994,606	7.7%	2.3 m	1.4 MiB	11	933,093 930,338	1.1% 0.8%	5.0 m 9.6 m	DPIO ESACO
usa13509	13,509	yes	19,982,859	24,125,443	20.7%	16.2 s	21,907,190	9.6%	2.8 m	1.5 MiB	10	20,217,458 20,195,089	1.2% 1.1%	4.5 m 15.2 m	DPIO ESACO
brd14051	14,051	yes	469,385	552,658	17.7%	15.9 s	506,668	7.9%	3.1 m	1.5 MiB	11	474,788 474,087	1.1% 1.0%	5.1 m 11.4 m	DPIO ESACO
d15112	15,112	yes	1,573,084	1,847,377	17.4%	19.2 s	1,705,664	8.4%	3.6 m	1.6 MiB	11	1,588,563 1,589,288	1.0% 1.0%	8.7 m 12.9 m	DPIO ESACO
d18512	18,512	yes	645,238	756,668	17.3%	28.1 s	696,542	8.0%	5.8 m	1.9 MiB	12	652,613 653,154	1.1% 1.2%	8.3 m 11.4 m	DPIO ESACO
pla33810	33,810	yes	66,048,945	76,625,752	16.0%	1.6 m	69,626,380	5.4%	25.7 m	2.9 MiB	17	67,185,647	1.7%	21.0 m	DPIO
pla85900	85,900	yes	142,382,641	167,355,049	17.5%	10.5 m	149,546,776	5.0%	4.1 h	6.5 MiB	27	144,334,707	1.4%	1.4 h	DPIO

Table A4. Results for **Art TSP** benchmarks.

| Instance | | | | CII Heuristic (Phase 2) | | | CII Heuristic (Phase 3) | | | | | | Other Heuristics | | |
	V	Opt?	$C(BKS)$	$C(T)$	$Error_{CII}$	Time	$C(T)$	$Error_{CII}$	Time	RAM	#	$C_{min}(T_H)$	$Error_H$	Time	Heuristic Id
mona-lisa 100K	100,000	no	5,757,191	6,123,262	6.4%	14.4 m	5,951,462	3.4%	2.3 h	7.5 MiB	9	5,855,063	1.7%	1.4 h	ACO-RPMM
												6,070,958	5.5%	1.1 h	Partial ACO
vangogh 120K	120,000	no	6,543,610	6,971,470	6.5%	20.8 m	6,773,421	3.5%	4.6 h	8.8 MiB	12	6,661,395	1.8%	1.9 h	ACO-RPMM
												6,924,448	5.8%	1.5 h	Partial ACO
venus 140K	140,000	no	6,810,665	7,245,012	6.4%	28.0 m	7,043,702	3.4%	4.8 h	10.2 MiB	9	6,933,257	1.8%	2.6 h	ACO-RPMM
												7,206,365	5.8%	2.1 h	Partial ACO
pareja 160K	160,000	no	7,619,953	8,113,501	6.5%	37.3 m	7,888,641	3.5%	7.7 h	11.6 MiB	11	7,760,922	1.9%	3.5 h	ACO-RPMM
courbet 180K	180,000	no	7,888,733	8,439,701	7.0%	48.2 m	8,179,440	3.7%	10.1 h	13.0 MiB	11	8,038,619	1.9%	4.5 h	ACO-RPMM
earring 200K	200,000	no	8,171,677	8,781,766	7.5%	58.7 m	8,493,724	3.9%	15.1 h	14.3 MiB	12	8,335,111	2.0%	6.0 h	ACO-RPMM
												8,760,038	7.2%	5.1 h	Partial ACO

Table A5. Results for **National TSP** benchmarks.

Instance			CII Heuristic (Phase 2)				CII Heuristic (Phase 3)						Other Heuristics			
	V	Opt?	$C(BKS)$	$C(T)$	$Error_{CII}$	Time	$C(T)$	$Error_{CII}$	Time	RAM	#	$C_{min}(T_H)$	$Error_H$	Time	Heuristic Id
wi29	29	yes	27,603	27,739	0.5%	0.3 ms	27,601	0.0%	28.0 ms	0.6 MiB	3	35,474	28.5%	0.2 ms	NN
dj38	38	yes	6656	6863	3.1%	0.3 ms	6659	0.1%	11.2 ms	0.6 MiB	5	8165	22.7%	0.3 ms	NN
qa194	194	yes	9352	10,505	12.3%	3.8 ms	9886	5.7%	37.1 ms	0.6 MiB	7	12,481	33.5%	2.6 ms	NN
zi929	929	yes	95,345	110,187	15.6%	73.5 ms	100,842	5.8%	630 ms	0.7 MiB	8	119,685	25.5%	36.7 ms	NN
lu980	980	yes	11,340	12,834	13.2%	86.4 ms	12,077	6.5%	404 ms	0.6 MiB	5	14,284	26.0%	29.4 ms	NN
rw1621	1621	yes	26,051	30,315	16.4%	233 ms	28,771	10.4%	1.6 s	0.7 MiB	8	33,493	28.6%	71.5 ms	NN
mu1979	1979	yes	86,891	99,356	14.3%	350 ms	91,684	5.5%	3.8 s	0.8 MiB	10	113,362	30.5%	112 ms	NN
nu3496	3496	yes	96,132	111,981	16.5%	1.1 s	103,717	7.9%	9.2 s	0.8 MiB	10	121,713	26.6%	327 ms	NN
ca4663	4663	yes	1290319	1,557,923	20.7%	1.9 s	1,407,891	9.1%	18.2 s	0.9 MiB	10	1,637,468	26.9%	564 ms	NN
tz6117	6117	no	394,718	477,869	21.1%	3.5 s	433,784	9.9%	40.0 s	1 MiB	14	494,624	25.3%	843 ms	NN
eg7146	7146	no	172,386	198,566	15.2%	4.5 s	182,979	6.1%	57.9 s	1.1 MiB	14	219,365	27.3%	1.1 s	NN
ym7663	7663	yes	238,314	285,881	20.0%	5.0 s	259,780	9.0%	1.4 m	1.1 MiB	18	308,219	29.3%	1.1 s	NN
pm8079	8079	no	114,855	137,182	19.4%	5.7 s	126,746	10.4%	55.6 s	1.2 MiB	10	148,936	29.7%	1.2 s	NN
ei8246	8246	yes	206,171	248,695	20.6%	6.0 s	225,178	9.2%	1.0 m	1.1 MiB	11	254,553	23.5%	1.2 s	NN
ar9152	9152	no	837,479	1,014,041	21.1%	8.4 s	927,348	10.7%	1.2 m	1.2 MiB	10	1,063,376	27.0%	1.5 s	NN
ja9847	9847	yes	491,924	611,959	24.4%	8.7 s	544,411	10.7%	2.0 m	1.2 MiB	16	630,169	28.1%	1.9 s	NN
gr9882	9882	yes	300,899	356,753	18.6%	8.5 s	325,599	8.2%	1.8 m	1.3 MiB	14	395,267	31.4%	2.3 s	NN
kz9976	9976	no	1,061,881	1,298,405	22.3%	8.9 s	1,168,843	10.1%	1.6 m	1.3 MiB	12	1,344,845	26.6%	1.8 s	NN
fi10639	10639	yes	520,527	633,623	21.7%	9.8 s	574,001	10.3%	2.1 m	1.3 MiB	14	659,800	26.8%	2.0 s	NN

Table A5. *Cont.*

| Instance | | | CII Heuristic (Phase 2) | | | CII Heuristic (Phase 3) | | | | | | Other Heuristics | | |
	V	Opt?	C(BKS)	C(T)	$Error_{CII}$	Time	C(T)	$Error_{CII}$	Time	RAM	#	$C_{min}(T_H)$	$Error_H$	Time	Heuristic Id
mo14185	14185	no	427,377	516,028	20.7%	17.3 s	465,202	8.9%	3.8 m	1.6 MiB	14	529,396	23.9%	4.6 s	NN
ho14473	14473	no	177,092	207,322	17.1%	18.5 s	193,672	9.4%	3.0 m	1.6 MiB	10	216,776	22.4%	4.0 s	NN
it16862	16862	yes	557315	670,706	20.3%	24.8 s	613,132	10.0%	4.8 m	1.7 MiB	12	706,420	26.8%	6.2 s	NN
vm22775	22775	yes	569,288	688,981	21.0%	44.3 s	617,703	8.5%	11.0 m	2.1 MiB	16	720,288	26.5%	9.9 s	NN
sw24978	24978	yes	855,597	1,042,499	21.8%	53.5 s	944,536	10.4%	10.2 m	2.3 MiB	12	1,073,993	25.5%	12.2 s	NN
bm33708	33708	no	959,289	1,151,420	20.0%	1.6 m	1,046,776	9.1%	22.1 m	2.9 MiB	14	1,209,682	26.1%	21.5 s	NN
ch71009	71009	no	4,566,506	5,475,575	19.9%	7.4 m	4,986,973	9.2%	1.7 h	5.5 MiB	14	5,629,331	23.3%	1.6 m	NN
usa115475	115475	no	6,204,999	7,492,272	20.7%	19.0 m	6,779,417	9.3%	4.3 h	8.5 MiB	13	7,691,402	24.0%	4.1 m	NN

Table A6. Results for the **VLSI TSP** benchmark.

| Instance | | | CII Heuristic (Phase 2) | | | CII Heuristic (Phase 3) | | | | | | Other Heuristics | | |
	V	Opt?	C(BKS)	C(T)	$Error_{CII}$	Time	C(T)	$Error_{CII}$	Time	RAM	#	$C_{min}(T_H)$	$Error_H$	Time	Heuristic Id
xqf131	131	yes	564	624	10.6%	1.8 ms	600	6.3%	16.4 ms	0.6 MiB	3	712	26.3%	1.0 ms	NN
xqg237	237	yes	1019	1166	14.4%	5.0 ms	1064	4.4%	31.9 ms	0.6 MiB	7	1325	30.0%	3.0 ms	NN
pma343	343	yes	1368	1490	8.9%	10.0 ms	1425	4.2%	58.9 ms	0.6 MiB	5	1846	35.5%	7.4 ms	NN
pka379	379	yes	1332	1422	6.8%	12.1 ms	1391	4.4%	66.4 ms	0.6 MiB	4	1606	20.6%	7.5 ms	NN
bcl380	380	yes	1621	1894	16.9%	12.4 ms	1781	9.9%	97.0 ms	0.6 MiB	6	2055	26.8%	6.6 ms	NN
pbl395	395	yes	1281	1432	11.8%	13.6 ms	1349	5.3%	95.9 ms	0.6 MiB	7	1581	23.5%	7.9 ms	NN
pbk411	411	yes	1343	1505	12.1%	14.4 ms	1431	6.6%	111 ms	0.6 MiB	7	1789	33.2%	7.7 ms	NN
pbn423	423	yes	1365	1573	15.2%	15.1 ms	1460	7.0%	77.1 ms	0.6 MiB	5	1811	32.6%	9.2 ms	NN
pbm436	436	yes	1443	1638	13.5%	16.4 ms	1565	8.4%	93.5 ms	0.6 MiB	5	1783	23.6%	9.0 ms	NN

Table A6. *Cont.*

Instance				CII Heuristic (Phase 2)			CII Heuristic (Phase 3)					Other Heuristics			
	V	Opt?	$C(BKS)$	$C(T)$	$Error_{CII}$	Time	$C(T)$	$Error_{CII}$	Time	RAM	#	$C_{min}(T_H)$	$Error_H$	Time	Heuristic Id
xql662	662	yes	2513	2995	19.2%	36.4 ms	2742	9.1%	269 ms	0.6 MiB	8	3147	25.2%	19 ms	NN
rbx711	711	yes	3115	3612	16.0%	42.8 ms	3348	7.5%	312 ms	0.6 MiB	8	3748	20.3%	22 ms	NN
rbu737	737	yes	3314	3899	17.6%	45.0 ms	3557	7.3%	230 ms	0.6 MiB	5	4090	23.4%	24 ms	NN
dkg813	813	yes	3199	3763	17.6%	53.7 ms	3470	8.5%	369 ms	0.6 MiB	5	4126	29.0%	26 ms	NN
lim963	963	yes	2789	3199	14.7%	78.8 ms	2974	6.6%	929 ms	0.6 MiB	10	3583	28.5%	37 ms	NN
pbd984	984	yes	2797	3189	14.0%	80.8 ms	2950	5.5%	641 ms	0.6 MiB	9	3521	25.9%	36 ms	NN
xit1083	1083	yes	3558	4082	14.7%	98.8 ms	3800	6.8%	763 ms	0.7 MiB	8	4781	34.4%	42 ms	NN
dka1376	1376	yes	4666	5546	18.8%	167 ms	5082	8.9%	1.0 s	0.7 MiB	7	5924	27.0%	65 ms	NN
dca1389	1389	yes	5085	6045	18.9%	156 ms	5471	7.6%	1.0 s	0.7 MiB	7	6080	19.6%	'NR'	PRNN
dja1436	1436	yes	5257	6236	18.6%	168 ms	5628	7.1%	1.3 s	0.7 MiB	8	6656	26.6%	72 ms	NN
icw1483	1483	yes	4416	5124	16.0%	180 ms	4761	7.8%	1.1 s	0.7 MiB	5	5572	26.2%	75 ms	NN
fra1488	1488	yes	4264	4728	10.9%	179 ms	4479	5.1%	1.6 s	0.6 MiB	8	5578	30.8%	76 ms	NN
rbv1583	1583	yes	5387	6207	15.2%	205 ms	5777	7.2%	2.2 s	0.7 MiB	11	6876	27.6%	80 ms	NN
rby1599	1599	yes	5533	6345	14.7%	215 ms	5999	8.4%	1.9 s	0.7 MiB	10	6809	23.1%	83 ms	NN
fnb1615	1615	yes	4956	5675	14.5%	213 ms	5259	6.1%	1.6 s	0.7 MiB	8	6377	28.7%	83 ms	NN
djc1785	1785	yes	6115	7225	18.2%	261 ms	6656	8.9%	2.1 s	0.7 MiB	9	7719	26.2%	103 ms	NN
dcc1911	1911	yes	6396	7484	17.0%	296 ms	6872	7.4%	2.0 s	0.7 MiB	7	8045	25.8%	116 ms	NN
dkd1973	1973	yes	6421	7280	13.4%	302 ms	6892	7.3%	2.1 s	0.7 MiB	7	8502	32.4%	119 ms	NN
djb2036	2036	yes	6197	7495	20.9%	337 ms	6819	10.0%	2.2 s	0.7 MiB	7	7645	23.4%	'NR'	PRNN
dcb2086	2086	yes	6600	8066	22.2%	354 ms	7307	10.7%	2.9 s	0.7 MiB	9	8335	26.3%	124 ms	NN
bva2144	2144	yes	6304	7494	18.9%	362 ms	6870	9.0%	2.6 s	0.7 MiB	7	8264	31.1%	129 ms	NN
xqc2175	2175	yes	6830	8167	19.6%	386 ms	7453	9.1%	5.2 s	0.7 MiB	13	8291	21.4%	'NR'	PRNN
bck2217	2217	yes	6764	8153	20.5%	398 ms	7408	9.5%	3.3 s	0.7 MiB	9	8515	25.9%	141 ms	NN

Table A6. *Cont.*

Instance			CII Heuristic (Phase 2)			CII Heuristic (Phase 3)						Other Heuristics		
	V	Opt?	$C(BKS)$	$C(T)$	$Error_{CII}$	Time	$C(T)$	$Error_{CII}$	Time	RAM	#	$C_{min}(T_H)$	$Error_H$	Time	Heuristic Id
xpr2308	2308	yes	7219	8663	20.0%	434 ms	7837	8.6%	3.3 s	0.7 MiB	8	9130	26.5%	155 ms	NN
ley2323	2323	yes	8352	10,146	21.5%	439 ms	9014	7.9%	4.9 s	0.7 MiB	11	10,330	23.7%	148 ms	NN
dea2382	2382	yes	8017	9782	22.0%	455 ms	8726	8.8%	4.4 s	0.7 MiB	9	9962	24.3%	157 ms	NN
rbw2481	2481	yes	7724	9548	23.6%	495 ms	8511	10.2%	4.1 s	0.7 MiB	9	9867	27.7%	169 ms	NN
pds2566	2566	yes	7643	9100	19.1%	523 ms	8310	8.7%	4.2 s	0.8 MiB	8	9867	29.1%	190 ms	NN
mlt2597	2597	yes	8071	9850	22.0%	547 ms	8889	10.1%	5.0 s	0.8 MiB	10	10,295	27.6%	183 ms	NN
bch2762	2762	yes	8234	10,020	21.7%	614 ms	8934	8.5%	5.0 s	0.7 MiB	9	10,394	26.2%	205 ms	NN
irw2802	2802	yes	8423	10,044	19.2%	625 ms	9131	8.4%	5.9 s	0.7 MiB	9	11,087	31.6%	210 ms	NN
lsm2854	2854	yes	8014	9445	17.9%	658 ms	8753	9.2%	5.6 s	0.7 MiB	9	10,105	26.1%	218 ms	NN
dbj2924	2924	yes	10,128	12,069	19.2%	676 ms	10,922	7.8%	4.6 s	0.7 MiB	7	12,935	27.7%	229 ms	NN
xva2993	2993	yes	8492	9936	17.0%	719 ms	9226	8.6%	5.9 s	0.8 MiB	9	10,821	27.4%	237 ms	NN
pia3056	3056	yes	8258	9749	18.1%	757 ms	8918	8.0%	8.2 s	0.8 MiB	11	10,585	28.2%	245 ms	NN
dke3097	3097	yes	10,539	12,767	21.1%	766 ms	11,481	8.9%	5.1 s	0.8 MiB	7	3249	25.7%	247 ms	NN
lsn3119	3119	yes	9114	10,784	18.3%	803 ms	9895	8.6%	8.0 s	0.8 MiB	11	11,467	25.8%	260 ms	NN
lta3140	3140	yes	9517	11,160	17.3%	805 ms	10,330	8.5%	7.5 s	0.8 MiB	10	12,455	30.9%	260 ms	NN
fdp3256	3256	yes	10,008	11,661	16.5%	908 ms	10,749	7.4%	7.1 s	0.8 MiB	8	12,677	26.7%	276 ms	NN
beg3293	3293	yes	9772	11,693	19.7%	877 ms	10,598	8.5%	10.2 s	0.7 MiB	13	12,636	29.3%	283 ms	NN
dhb3386	3386	yes	11,137	13,349	19.9%	932 ms	12,082	8.5%	8.0 s	0.7 MiB	9	13,894	24.8%	302 ms	NN
fjs3649	3649	yes	9272	10,345	11.6%	1.1 s	9812	5.8%	7.3 s	0.7 MiB	7	12,786	37.9%	326 ms	NN
fjr3672	3672	yes	9601	10,854	13.1%	1.1 s	10,181	6.0%	8.7 s	0.7 MiB	8	12,840	33.7%	331 ms	NN
dlb3694	3694	yes	10,959	12,818	17.0%	1.2 s	11,763	7.3%	10.4 s	0.7 MiB	10	13,986	27.6%	344 ms	NN
ltb3729	3729	yes	11,821	13,874	17.4%	1.1 s	12,948	9.5%	10.3 s	0.7 MiB	9	15,259	29.1%	361 ms	NN
xqe3891	3891	yes	11,995	14,672	22.3%	1.3 s	13,153	9.7%	10.0 s	0.8 MiB	9	14,592	21.7%	'NR'	PRNN

Table A6. *Cont.*

| Instance | | | CII Heuristic (Phase 2) | | | CII Heuristic (Phase 3) | | | | | | Other Heuristics | | |
	V	Opt?	$C(BKS)$	$C(T)$	$Error_{CII}$	Time	$C(T)$	$Error_{CII}$	Time	RAM	#	$C_{min}(T_H)$	$Error_H$	Time	Heuristic Id
xua3937	3937	yes	11,239	13,412	19.3%	1.2 s	12,285	9.3%	13.3 s	0.8 MiB	11	14,520	29.2%	373 ms	NN
dkc3938	3938	yes	12,503	14,817	18.5%	1.3 s	13,619	8.9%	10.5 s	0.7 MiB	9	15,932	27.4%	396 ms	NN
dkf3954	3954	yes	12,538	14,939	19.1%	1.3 s	13,728	9.5%	11.6 s	0.8 MiB	10	15,679	25.1%	412 ms	NN
bgb4355	4355	yes	12,723	14,948	17.5%	1.5 s	13,789	8.4%	14.0 s	0.9 MiB	10	15,623	22.8%	'NR'	PRNN
bgd4396	4396	yes	13,009	16,239	24.8%	1.6 s	14,385	10.6%	15.7 s	0.8 MiB	11	16,726	28.6%	472 ms	NN
frv4410	4410	yes	10,711	12,440	16.1%	1.5 s	11,587	8.2%	10.3 s	0.8 MiB	7	13,756	28.4%	518 ms	NN
bgf4475	4475	yes	13,221	15,989	20.9%	1.6 s	14,562	10.1%	22.6 s	0.8 MiB	15	16,439	24.3%	487 ms	NN
xqd4966	4966	yes	15,316	17,630	15.1%	2.0 s	16,545	8.0%	19.8 s	0.8 MiB	10	19,807	29.3%	571 ms	NN
fqm5087	5087	yes	13,029	14,877	14.2%	2.1 s	14,041	7.8%	18.1 s	0.8 MiB	9	17,554	34.7%	586 ms	NN
fea5557	5557	yes	15,445	18,171	17.6%	2.4 s	16,629	7.7%	30.4 s	0.9 MiB	13	19,738	27.8%	688 ms	NN
xsc6880	6880	yes	21,535	26,404	22.6%	3.9 s	23,704	10.1%	36.0 s	1.1 MiB	10	26,243	21.9%	'NR'	PRNN
bnd7168	7168	yes	21,834	25,963	18.9%	4.1 s	23,848	9.2%	50.3 s	1.1 MiB	13	26,574	21.7%	'NR'	PRNN
lap7454	7454	yes	19,535	23,107	18.3%	4.5 s	21,345	9.3%	50.7 s	1 MiB	12	24,184	23.8%	1.1 s	NN
ida8197	8197	yes	22,338	26,152	17.1%	5.4 s	23,954	7.2%	1.1 m	1.2 MiB	13	27,513	23.2%	'NR'	PRNN
dga9698	9698	yes	27,724	33,533	21.0%	7.9 s	30,374	9.6%	1.4 m	1.3 MiB	12	33,564	21.1%	'NR'	PRNN
xmc10150	10,150	yes	28,387	34,071	20.0%	8.8 s	31,124	9.6%	1.1 m	1.3 MiB	8	34,147	20.3%	'NR'	PRNN
xvb13584	13,584	yes	37,083	44,129	19.0%	15.8 s	40,591	9.5%	2.6 m	1.5 MiB	11	45,835	23.6%	'NR'	PRNN
xrb14233	14,233	no	45,462	54,786	20.5%	17.1 s	49,593	9.1%	3.2 m	1.4 MiB	12	57,034	25.5%	3.6 s	NN
xia16928	16,928	no	52,850	62,195	17.7%	24.0 s	57,220	8.3%	3.4 m	1.6 MiB	9	66,398	25.6%	5.3 s	NN
pjh17845	17,845	no	48,092	56,892	18.3%	27.5 s	51,934	8.0%	5.3 m	1.7 MiB	13	60,797	26.4%	5.4 s	NN
frh19289	19,289	no	55,798	67,243	20.5%	32.3 s	61,007	9.3%	5.3 m	1.9 MiB	11	68,360	22.5%	'NR'	PRNN
fnc19402	19,402	no	59,287	69,912	17.9%	32.0 s	64,170	8.2%	5.3 m	1.8 MiB	11	74,447	25.6%	6.5 s	NN
ido21215	21,215	no	63,517	75,879	19.5%	38.4 s	69,205	9.0%	8.0 m	1.9 MiB	14	79,469	25.1%	7.6 s	NN

Table A6. *Cont.*

Instance			CII Heuristic (Phase 2)				CII Heuristic (Phase 3)						Other Heuristics			
	V	Opt?	$C(BKS)$	$C(T)$	$Error_{CII}$	Time	$C(T)$	$Error_{CII}$	Time	RAM	#	$C_{min}(T_H)$	$Error_H$	Time	Heuristic Id
fma21553	21,553	no	66,527	77,951	17.2%	41.0 s	71,929	8.1%	6.6 m	2.0 MiB	11	83,449	25.4%	8.3 s	NN
lsb22777	22,777	no	60,977	71,997	18.1%	44.6 s	66,298	8.7%	7.3 m	2.0 MiB	11	76,551	25.5%	8.8 s	NN
xrh24104	24,104	no	69,294	83,300	20.2%	49.1 s	75,766	9.3%	6.8 m	2.1 MiB	9	87,747	25.2%	10.2 s	NN
bbz25234	25,234	no	69,335	82,214	18.6%	55.6 s	75,492	8.9%	10.5 m	2.2 MiB	13	87,345	26.0%	11.1 s	NN
irx28268	28,268	no	72,607	85,130	17.2%	1.2 m	78,250	7.8%	15.2 m	2.4 MiB	15	90,936	25.2%	13.3 s	NN
fyg28534	28,534	no	78,562	95,525	21.6%	1.2 m	85,843	9.3%	13.4 m	2.4 MiB	13	97,260	23.8%	14.0 s	NN
icx28698	28,698	no	78,087	93,828	20.2%	1.2 m	85,562	9.6%	11.8 m	2.4 MiB	11	96,987	24.2%	13.6 s	NN
boa28924	28,924	no	79,622	95,729	20.2%	1.2 m	86,834	9.1%	13.9 m	2.5 MiB	13	99,881	25.4%	14.4 s	NN
ird29514	29,514	no	80,353	96,206	19.7%	1.4 m	87,565	9.0%	14.6 m	2.5 MiB	13	100,617	25.2%	15.4 s	NN
pbh30440	30,440	no	88,313	104,985	18.9%	1.3 m	95,949	8.6%	13.5 m	2.6 MiB	11	110,335	24.9%	16.6 s	NN
xib32892	32,892	no	96,757	113,361	17.2%	1.6 m	104,523	8.0%	15.4 m	2.7 MiB	11	120,736	24.8%	19.2 s	NN
fry33203	33,203	no	97,240	116,014	19.3%	1.6 m	105,745	8.7%	20.8 m	2.8 MiB	15	120,664	24.1%	19.4 s	NN
bby34656	34,656	no	99,159	118,792	19.8%	1.7 m	108,423	9.3%	17.0 m	2.9 MiB	11	124,834	25.9%	22.3 s	NN
pba38478	38,478	no	108,318	128,315	18.5%	2.1 m	117,712	8.7%	24.4 m	3.1 MiB	13	134,770	24.4%	25.4 s	NN
ics39603	39,603	no	106,819	130,049	21.7%	2.2 m	117,804	10.3%	26.2 m	3.2 MiB	13	133,660	25.1%	26.9 s	NN
rbz43748	43,748	no	125,183	152,817	22.1%	2.6 m	138,235	10.4%	29.4 m	3.5 MiB	11	157,173	25.6%	33.2 s	NN
fht47608	47,608	no	125,104	148,051	18.3%	3.2 m	135,216	8.1%	39.4 m	3.7 MiB	13	155,972	24.7%	39.2 s	NN
fna52057	52,057	no	147,789	174,317	18.0%	3.8 m	160,231	8.4%	46.9 m	4.1 MiB	13	187,336	26.8%	51.6 s	NN
bna56769	56,769	no	158,078	189,521	19.9%	4.6 m	173,074	9.5%	1.0 h	4.4 MiB	14	200,198	26.6%	56.8 s	NN
dan59296	59,296	no	165,371	199,175	20.4%	5.0 m	180,850	9.4%	1.2 h	4.5 MiB	15	206,775	25.0%	1.0 m	NN
sra104815	104,815	no	251,761	326,561	29.7%	15.6 m	295,092	17.2%	3.7 h	7.7 MiB	14	329,120	30.7%	3.2 m	NN
ara238025	238,025	no	578,761	747,619	29.2%	1.4 h	674,559	16.6%	1.5 d.	16.8 MiB	22	759,882	31.3%	16.5 m	NN
lra498378	498,378	no	2,168,039	2,710,116	25.0%	5.8 h	2,438,410	12.5%	15.0 d.	34.7 MiB	49	2,688,804	24.0%	1.2 h	NN
lrb744710	744,710	no	1,611,232	2,076,966	28.9%	13.7 h	1,867,273	15.9%	15.0 d.	51.6 MiB	18	2,104,585	30.6%	2.7 h	NN

References

1. Papadimitriou, C.H. The Euclidean travelling salesman problem is NP-complete. *Theor. Comput. Sci.* **1977**, *4*, 237–244. [CrossRef]
2. Garey, M.R.; Graham, R.L.; Jhonson, D.S. Some NP-Complete geometric problems. In Proceedings of the Eight Annual ACM Symposium on Theory of Computing, Hershey, PA, USA, 3–5 May 1976; ACM: New York, NY, USA, 1976; pp. 10–22.
3. Lawler, E.L.; Lenstra, J.K.; Rinnooy Kan, A.H.; Shmoys, D.B. (Eds.) *The Traveling Salesman Problem: A Guided Tour of Combinatorial Optimization*; Wiley: Chichester, UK, 1985.
4. Jünger, M.; Reinelt, G.; Rinaldi, G. The traveling salesman problem. In *Handbooks in Operations Research and Management Science*; Elsevier Science B.V.: Amsterdam, The Netherlands, 1995; Volume 7, pp. 225–330.
5. Lin, S.; Kernighan, B.W. An effective heuristic algorithm for the traveling-salesman problem. *Oper. Res.* **1973**, *21*, 498–516. [CrossRef]
6. Chitty, D.M. Applying ACO to large scale TSP instances. In *UK Workshop on Computational Intelligence*; Springer: Cham, Switzerland, 2017; pp. 104–118.
7. Ismkhan, H. Effective heuristics for ant colony optimization to handle large-scale problems. *Swarm Evol. Comput.* **2017**, *32*, 140–149. [CrossRef]
8. Gülcü, Ş.; Mahi, M.; Baykan, Ö.K.; Kodaz, H. A parallel cooperative hybrid method based on ant colony optimization and 3-Opt algorithm for solving traveling salesman problem. *Soft Comput.* **2018**, *22*, 1669–1685.
9. Peake, J.; Amos, M.; Yiapanis, P.; Lloyd, H. Scaling Techniques for Parallel Ant Colony Optimization on Large Problem Instances. In Proceedings of the Gecco'19—The Genetic and Evolutionary Computation Conference 2019, Prague, Czech Republic, 13–17 July 2019.
10. Dahan, F.; El Hindi, K.; Mathkour, H.; AlSalman, H. Dynamic Flying Ant Colony Optimization (DFACO) for Solving the Traveling Salesman Problem. *Sensors* **2019**, *19*, 1837. [CrossRef] [PubMed]
11. Al-Adwan, A.; Mahafzah, B.A.; Sharieh, A. Solving traveling salesman problem using parallel repetitive nearest neighbor algorithm on OTIS-Hypercube and OTIS-Mesh optoelectronic architectures. *J. Supercomput.* **2008**, *74*, 1–36. [CrossRef]
12. Zhong, Y.; Wang, L.; Lin, M.; Zhang, H. Discrete pigeon-inspired optimization algorithm with Metropolis acceptance criterion for large-scale traveling salesman problem. *Swarm Evol. Comput.* **2019**, *48*, 134–144. [CrossRef]
13. Croes, G.A. A method for solving traveling-salesman problems. *Oper. Res.* **1958**, *6*, 791–812. [CrossRef]
14. Vakhania, N.; Hernandez, J.A.; Alonso-Pecina, F.; Zavala, C. A Simple Heuristic for Basic Vehicle Routing Problem. *J. Comput. Sci.* **2016**, *3*, 39. [CrossRef]
15. Sahni, S.; Horowitz, E. *Fundamentals of Computer Algorithms*; Computer Science Press, Inc.: Rockville, MD, USA, 1978; pp. 174–179.
16. Universität Heidelberg, Institut für Informatik; Reinelt, G. Symmetric Traveling Salesman Problem (TSP). Available online: https://www.iwr.uni-heidelberg.de/groups/comopt/software/TSPLIB95/ (accessed on 8 June 2019).
17. Natural Sciences and Engineering Research Council of Canada (NSERC) and Department of Combinatorics and Optimization at the University of Waterloo. TSP Test Data. Available online: http://www.math.uwaterloo.ca/tsp/data/index.html (accessed on 8 June 2019).

Article

A Generalized MILP Formulation for the Period-Aggregated Resource Leveling Problem with Variable Job Duration

Ilia Tarasov [1,2,]*, Alain Haït [1] and Olga Battaïa [3]

[1] ISAE-SUPAERO, University of Toulouse, 10 avenue Edouard Belin-BP 54032, 31055 Toulouse CEDEX 4, France; alain.hait@isae-supaero.fr

[2] V. A. Trapeznikov Institute of Control Sciences of Russian Academy of Sciences, 65 Profsoyuznaya street, Moscow 117997, Russia

[3] Kedge Business School (Talence), 680 cours de la Liberation, 33405 Talence CEDEX, France; olga.battaia@kedgebs.com

* Correspondence: Ilia.TARASOV@isae-supaero.fr

Received: 14 November 2019; Accepted: 16 December 2019; Published: 23 December 2019

Abstract: We study a resource leveling problem with variable job duration. The considered problem includes both scheduling and resource management decisions. The planning horizon is fixed and separated into a set of time periods of equal length. There are several types of resources and their amount varies from one period to another. There is a set of jobs. For each job, a fixed volume of work has to be completed without any preemption while using different resources. If necessary, extra resources can be used at additional costs during each time period. The optimization goal is to minimize the total overload costs required for the execution of all jobs by the given deadline. The decision variables specify the starting time of each job, the duration of the job and the resource amount assigned to the job during each period (it may vary over periods). We propose a new generalized mathematical formulation for this optimization problem. The formulation is compared with existing approaches from the literature. Theoretical study and computational experiments show that our approach provides more flexible resource allocation resulting in better final solutions.

Keywords: resource leveling problem; project scheduling

---

1. Introduction

In the field of operations research, project management remains a topic of intensive research from various angles such as scheduling and resource allocation. For both cases, there exist such well-known formulations as a resource-constrained project scheduling problem (RCPSP) and a resource leveling problem (RLP). The former aims to minimize the completion time of the set of jobs with given precedence relations and the set of limited required resources. The latter deals with resource allocation with the objective to minimize the cost of resource usage. Both problems were proved to be NP-hard (see [1] for RCPSP and [2] for the resource leveling problem). To respond to practical requirements, these basic models undergo numerous modifications. For example, the model that optimizes the distribution of man hours for workforce or throughput on a production line was presented by Neubert and Savino [3].

In this paper, we focus on the resource leveling problem with several practical enhancements. However, we also refer to RCPSP modeling techniques and approaches that are used for the RLP and applicable to our case. The RLP was actively studied from both the theoretical and practical sides. In the paper of Rieck and Zimmermann [4] three basic objective function types from the existing literature were presented with properties and existing solution approaches. The first objective function type was described as a total amount of variations in resource utilization within the project duration. The second case arises when available resource utilization is exceeded; it is total (squared) overload cost. The authors also presented a total adjustment cost function, which is formed by costs arising from increasing and decreasing resource utilization. In all cases, the planning horizon was fixed with the reference that for mid-term planning usually there is a determined project deadline. There are some papers involving multi-criteria optimization techniques with attention to both project duration and resource utilization. However, RLP with a fixed deadline and the objective functions that were presented above still attracts attention: The majority of the following literature references are focused on this formulation.

From the practical side, these RLP types are found in a wide range of industries, especially in construction. For these particular cases, researchers apply different heuristics to provide good solutions. An overview of existing RLP heuristic techniques was prepared by Christodoulou et al. [5]. We can point out recent results with references to industrial problems. For example, a meta-heuristic genetic algorithm was applied to RLP based on a construction project in the paper of Selvam and Tadepalli [6]. Simulated annealing was also tested in the construction area by Piryonesi et al. [7]. Li et al. [8] implemented a genetic algorithm for the RLP with uncertain activity durations and possible overlapping. Construction area and project management usually include discrete resources such as equipment, machines, materials or manpower. Cherkaoui et al. [9] studied a tactical level of construction and large-scale engineering planning. To achieve robust tactical planning and resource allocation, they created a proactive approach providing lower variations in project costs in case of resource capacity uncertain data. The problem was denoted as rough cut capacity planning (RCCP), but it is also similar to the classical RLP with several practical changes. The same notation was also used by Baydoun et al. [10].

There is another point that inspires us to study and develop RLP solution methods. Energy is the main continuous resource, and the energy management problem becomes more and more crucial for any industry. According to this, the scheduling of operations with attention to careful energy consumption remains actual. Artigues et al. [11] considered the industrial case of the energy scheduling problem for a pipe-manufacturing plant. The goal was to minimize the electricity bill, which was raised by penalties for power overrun. The two-step solution approach was proposed with a constraint programming assignment and sequencing part and an MILP scheduling and energy part. Many industrial management problems can be represented as RLP, especially if the parameters are defined and well formalized. Sometimes the energy is not the main resource. For example, for computer embedded systems it can be CPU power, and for data transfer networks it is channel capacity. An example of an optimization problem from the space industry could be found in the paper of Capelle et al. [12]. In this practical study, the goal was to maximize a data transfer from satellites to a network of optical ground stations. While we see that this formulation is closer to an assignment problem, further research may fit in RLP formulation with satellite data buffer capacity, energy, and data transfer limits.

For the resource leveling problem, the case with variable job duration was considered by Hans, who proposed a branch and cut algorithm [13]; furthermore, Kis developed an improved branching scheme [14]. In these models, the precedence relations are defined on periods and job duration depends on the resources allocated to each job. The resource consumption is calculated in an aggregated way for each period. The concept of variable job duration has also been modeled in the following studies (see [4,15]) but with the objective of makespan minimization or balanced profile usage. Bianco et al. [16] also considered the resource leveling problem and proposed a lower bound based on Lagrangian relaxation, and a branch and bound algorithm for the suggested model. The resource leveling model with the overload cost and overlapping of jobs with precedence relations was presented by Baydoun et al. [10] with a focus on different overlapping rules such as overlapping after implementation of some essential predecessor part.

We follow these ideas and study the generalization of models presented by Baydoun et al. [10] and Bianco et al. [16]. The goal is to minimize overload cost when an extra resource amount is required beyond the available limit. In contrast to these studies, we consider a more flexible resource distribution. The main difference could be presented in the following way. In these papers, there is one decision variable (denoted as assigned workload [10] or the fraction [16] of activity in a period) describing the progress made by a given job in a given period, and it defines the requirement for each resource type. In our case, we enrich the model to make the allocation of resources per period independently with additional decision variables.

For example, suppose there is one job j which requires overall one unit of resource r_1 and two units of resource r_2. In the models considered by [10,16] the solution defines only job fractions (these models are denoted as 'aggregate fraction' in the following). Suppose in the solution the job is implemented in two time periods with equal fractions in both periods (50%/50%). Then according to the models in the first period, we involve 0.5 units of resource r_1 and 1 unit of resource r_2, and the same in the second period. The involved resource amount equals the multiplication of this decision variable and a resource amount required to carry out the job (fixed input parameter). So there exists a constant ratio between the involved renewable resource amount (or efforts made by different resources) for a given job in all periods. We point out that particular fraction values are not important. With any fraction of the job j in any period, there will be the same ratio 0.5 between the involved resources (caused by the total required amount ratio of resource r_1 to resource r_2). According to our model, it is possible to make a more flexible allocation without any fixed relations between the resources involved to complete the job in some period. For example, it is possible to involve 0.75 units of r_1 and 0.5 units of r_2 in the first period and the remaining 0.25 units of r_1 and 1.5 units of r_2 in the second period. In this solution, resources are allocated with an independent ratio: 1.5 in the first period and 0.16 in the second period.

The main contribution of this paper is in the new generalized formulation for this type of resource leveling problem. We analyze the difference between this new generalized model and aggregated fraction models from the literature, with a comparison of performance and solution quality for the same instance set. Several particular formulations of this new approach were studied. We also study the case with discrete overloading of resources, which is compliant with human resources allocation. However, we do not consider job overlapping of predecessors and successors.

The rest of the paper is organized as follows. In Section 2 we present the new generalized formulation idea with several modeling implementations. We consider three different formulations of scheduling constraints and decision variables. After some experiments, we choose the binary step start/end formulation of variables for scheduling constraints. In Section 3 we describe a theoretical difference and relations between our generalized formulation and aggregated fraction and analyze the results of computational experiments. Section 4 presents the case of the discrete overload objective function and discrete resources. After, we make some concluding remarks.

2. Mathematical Model

2.1. Problem Parameters

The planning horizon is represented by the set of periods $T = \{1, \ldots, m\}$ with given length d, the last time period available defines the strict deadline for the execution of all the jobs. There is a set of jobs J and a set of resources R. For each job $j \in J$, some given work volume W_{jr} must be executed with resource $r \in R$, while in each period $t \in T$ the available capacity L_{rt} is known for each resource $r \in R$ is fixed. The allocation of an extra unit of resource capacity is subject to extra cost e_r. Work volume is the result of the multiplication of the assigned resource amount by the duration of its usage. Each job $j \in J$ has a given time window with lower $p_{min,jr}$ and upper $p_{max,jr}$ period for each resource $r \in R$. The pairs of predecessors and successors are defined in set P. The list of all parameters in our mathematical model is presented in Table 1.

Table 1. Model nomenclature: Parameters notation.

Parameters	
T	planning horizon, $T = \{1, \ldots, m\}$
d	period length
R	resources set
L_{rt}	availability of resource $r \in R$ in period $t \in T$
e_r	extra resource cost
J	jobs set
W_{jr}	job $j \in J$ work volume with resource $r \in R$
$p_{min,jr}$	job $j \in J$ minimal requirement per period in resource $r \in R$
$p_{max,jr}$	job $j \in J$ maximal requirement per period in resource $r \in R$
P	set of arcs in the given precedence graph

2.2. Generalized Model Description

For any kind of RLP, it is possible to separate the set of variables into two subsets, related to job scheduling and resource allocation. The subset of scheduling variables for discrete-time period models is well studied in project scheduling (starting from pioneer introduction of a 1-0 variable approach by Pritsker et al. [17]). Usually, it is a set of binaries that configure staring and ending periods. If we consider precedence relations and allow the successor to start we also use continuous duration variables responsible for the duration inside the period. We define binary variables responsible for the job starts and ends as S_{jt} and E_{jt}, respectively. They may be used in several ways.

- As a step pointer, i.e., if job j starts at period t, then $\forall t_1 < t \; S_{jt_1} = 0$ and $\forall t_2 \geq t \; S_{jt_2} = 1$. In this case the same logic is implemented for E_{jt}, if a job ends at period t, than for $\forall t_1 \leq t \; E_{jt} = 0$ and $\forall t_2 > t \; E_{jt} = 1$. To the best of our knowledge, in similar resource leveling models step pointer to job start and end periods is more common (for example, it was used by Baydoun et al. [10], Bianco et al. [16]).
- As an on–off (also referred as pulse) pointer, i.e., if job j starts at period t, then $\forall t_1 \neq t \; S_{jt_1} = 0$ and $S_{jt} = 1$ (the same logic for E_{jt}). This approach was considered by Tamas Kis and Marton Drotos, [18]. There is an alternative way, with an on–off function U_{jt}, where $U_{jt} = 1$ if job j is implemented at period t, and $U_{jt} = 0$ otherwise.

These approaches were also presented and compared in literature related to project scheduling, for example, see [19,20].

The second subset defines resource-allocation decision variables. We propose to generalize it and introduce separate fraction decision variables for each resource type instead of aggregated job fraction decision variables. In our case, these decision variables were defined as c_{jrt}. This approach makes an independent allocation of each resource type for every job. It leads to more flexible utilization of resources. However, it increases the size of the model and negatively affects performance, so it may require more computational resources to achieve a proofed optimal solution. We present a list of decision variable notations used in this paper in Table 2.

In the next subsections we describe three different approaches to represent scheduling constraints with binary variables. A general description of each type is presented in Figure 1.

Table 2. Model nomenclature: Decision variables notations.

Scheduling decision variables	
$U_{jt} \in \{0,1\}$	on/off function : Equals 1 of job $j \in J$ is implemented in period $t \in T$, 0 otherwise
$S_{jt} \in \{0,1\}$	pulse start function : Equals 1 if job $j \in J$ starts in period $t \in T$, 0 otherwise
$E_{jt} \in \{0,1\}$	pulse end function : Equals 1 if job $j \in J$ ends in period $t \in T$, 0 otherwise
$S_{jt}^* \in \{0,1\}$	step start function : Equals 1 if job $j \in J$ starts in $\forall t_1 \in T, t_1 \leq t$, 0 otherwise
$E_{jt}^* \in \{0,1\}$	step end function : Equals 1 if job $j \in J$ ends in $\forall t_1 \in T, t_1 < t$, 0 otherwise
$d_{jt} \in [0,d]$	duration of job $j \in J$ in period $t \in T$
Resource allocation decision variables	
$c_{jrt} \in [0, dp_{max,jr}]$	work volume of job $j \in J$ with resource $r \in R$ in period $t \in T$
$o_{rt} \in [0, \infty)$	extra cost of resource $r \in R$ in period $t \in T$

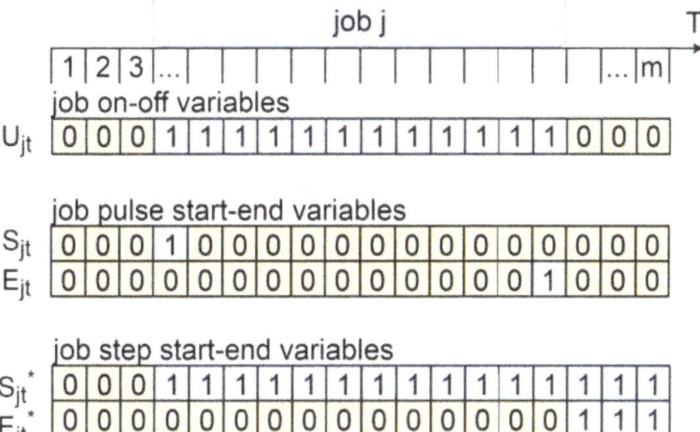

Figure 1. Three different ways to utilize binary scheduling variables.

2.2.1. Scheduling Constraints: Job On–Off Formulation

This set of constraints is defined as job on–off implementation formulation, as we use one binary variable implying that the job is implemented inside some period or not. This definition of binaries in project scheduling problems is denoted as pulse or on–off in literature. So we use a variable $U_{jt} \in \{0,1\}$, $U_{jt} = 1$ if job j is implemented in period t, $U_{jt} = 0$ otherwise. Constraints (1) imply that no preemptions are allowed.

$$(t_2 - t_1 - 1) \leq \sum_{l=t_1+1}^{t_2-1} U_{il} + m(U_{j,t_1} + U_{j,t_2} - 2), \ \forall t_1 \in T, \ t_2 \in T, \ t_1 < t_2, \ \forall j \in J. \tag{1}$$

If the job is implemented in some period $t_1 \in T$ and in some period $t_2 \in T, t_1 < t_2$, then it must be also implemented in any period $t_3 \in T, t_1 < t_3 < t_2$. In other words, if for some job j there exists $t_1 \in T, t_2 \in T, t_1 < t_2$, such that $U_{jt_1} = 1$ and $U_{jt_2} = 1$, then $\sum_{t_3=t_1+1}^{t_2-1} U_{jt_3} = t_2 - t_1 - 1$.

Constraints (2) and (3) set the minimal and maximal limit of periods when a job may be implemented:

$$\sum_{t \in T} U_{jt} \geq \lfloor \frac{d_{min,j}}{d} \rfloor, \ \forall j \in J; \tag{2}$$

$$\sum_{t \in T} U_{jt} \leq \lceil \frac{d_{max,j}}{d} \rceil, \ \forall j \in J; \tag{3}$$

where $d_{min,j}$ and $d_{max,j}$ are minimal and maximal allowed job duration variables, respectively. This variables will be described in Section 2.3.

Next two constraints make the correspondence between binary scheduling variable U_{jt} and job duration d_{jt}. Firstly, if job $j \in J$ is performed in three or more periods, then for all periods between start period and end period job duration is the same as period length, i.e., preemptions inside periods are not allowed. This constraints involve continuous duration and binaries variables $d_{jt} = d$ if $U_{j,t-1} = 1$, $U_{j,t} = 1, U_{j,t+1} = 1$:

$$(2 - U_{j,t+1} - U_{j,t-1})d + d_{jt} \geq dU_{jt}, \ \forall t \in T, \ \forall j \in J. \tag{4}$$

Secondly, if job $j \in J$ is not implemented in period $t \in T$, it must have zero duration inside this period. So $d_{jt} = 0$ is required if $U_{j,t} = 0$:

$$d_{jt} \leq dU_{jt}, \ \forall t \in T, \ \forall j \in J. \tag{5}$$

Precedence constraints are represented on two levels, on periods and inside each period. Firstly, we state that if there is a precedence $\{j_p, j_s\} \in P$ relation between two jobs, then it is impossible to implement the successor before the last period when predecessor is implemented (it is possible to start the successor at the last period of predecessor implementation). In mathematical form it is represented in the following condition: $\forall t_1 \in T$, if $U_{j_p,t_1} = 1$ then $\forall t_2 \in T, t_2 < t_1, U_{j_s,t_2} = 0$, and the corresponding constraints are:

$$U_{j_s,t_2} + U_{j_p,t_1} \leq 1, \ \forall \{j_p, j_s\} \in P, \ \forall t_1 \in T, \ \forall t_2 \in T, \ t_2 < t_1; \tag{6}$$

Secondly, we state that if these jobs are implemented in one period, total duration of both jobs is less than period duration. Otherwise it means that in this period j_p and j_s cross each other.

$$d_{j_1 t} + d_{j_2 t} \leq d, \ \forall t \in T, \ \forall (j_1, j_2) \in P. \tag{7}$$

We note that we use the Constraints (7) to take into account a case when in pair predecessor-successor both jobs are implemented in one period, which is the last period for predecessor and the first period for successor.

2.2.2. Scheduling Constraints: Job Pulse Start–End Formulation

We define this set of constraints as pulse start–end formulation because in this formulation we use binaries S_{jt} and E_{jt} which take the value 1 only in period of job start and end, respectively. We have variable job duration, so it is not sufficient to use only start pulse decision variable S_{jt}, we also need variables E_{jt}. Firstly, we imply Constraints (8) and (9) for S_{jt} and E_{jt} to start and end the job only once:

$$\sum_{t \in T} S_{jt} = 1, \ \forall j \in J; \tag{8}$$

$$\sum_{t \in T} E_{jt} = 1, \ \forall j \in J. \tag{9}$$

Secondly, we force the job duration variable d_{jt} to get zero value in periods t when job j is not implemented. The index of starting period for job j equals $\sum_{t \in T} t S_{jt}$, and ending period index is $\sum_{t \in T} t E_{jt}$. So we state that duration $d_{jt} = 0$ outside the interval $[\sum_{t \in T} t S_{jt}, \sum_{t \in T} t E_{jt}]$ with the following constraints. If job j is started in some period after the period t or it is finished in some period befor the period t, then $d_{jt} = 0$:

$$d_{jt} \leq d \left(1 - \sum_{k=t+1}^{m} S_{jk} - \sum_{l=1}^{t-1} E_{jl}\right), \ \forall j \in J, \ \forall t \in T; \tag{10}$$

Next constraints have the same sense as Constraints (4). In any period between start and end of the job j it must be implemented without preemptions inside the period, so $d_{jt} = d$ inside the interval $(\sum_{t \in T} tS_{jt}, \sum_{t \in T} tE_{jt})$. If job j was started before the period t and it was finished after the period t, $d_{jt} = d$:

$$d_{jt} \geq d \left(\sum_{k=1}^{t-1} S_{jk} + \sum_{l=t+1}^{m} E_{jl} - 1 \right), \ \forall j \in J, \ \forall t \in T. \tag{11}$$

The precedence constraints require two constraints for binaries and continuous job duration for each pair $\{j_p, j_s\} \in P$, representing precedence constraints on periods and inside each period. In this formulation we can just compare start period index of successor j_s and end period index of predecessor j_p:

$$\sum_{t_1 \in T} t_1 E_{j_p t_1} \leq \sum_{t_2 \in T} t_2 S_{j_s t_2}, \ \forall (j_p, j_s) \in P; \tag{12}$$

$$d_{j_p t} + d_{j_s t} \leq d, \ \forall t \in T, \ \forall (j_p, j_s) \in P. \tag{13}$$

2.2.3. Scheduling Constraints: Job Step Start–End Formulation

In this case we use step binaries S_{jt}^* and E_{jt}^* with a following rule:

- if job j starts at period t, then $\forall t_1 < t \ S_{jt_1}^* = 0$ and $\forall t_2 \geq t \ S_{jt_2}^* = 1$;
- if job j ends at period t, then $\forall t_1 \leq t \ E_{jt}^* = 0$ and $\forall t_2 > t \ E_{jt}^* = 1$.

This approach is defined as step formulation because for each job the plot with decision variables looks like a non decreasing step function. Firstly, we require proper values for all binary step variables. The job may be ended only if it was started in the same period or before, and the values of start and end step variables S_{jt}^* and E_{jt}^* must be non-decreasing for each job $j \in J$:

$$S_{jt}^* \geq E_{jt}^*, \ \forall j \in J, \ \forall t \in T; \tag{14}$$

$$S_{jt}^* \leq S_{j,t+1}^*, \ \forall j \in J, \ \forall t \in T; \tag{15}$$

$$E_{jt}^* \leq E_{j,t+1}^*, \ \forall j \in J, \ \forall t \in T. \tag{16}$$

Secondly, we set up the correspondence between binaries and decision variables $d_{jt} \in [0, d]$. As in previous cases, we imply that $d_{jt} = 0$ if the job was not started before period t or it was finished in some period before t:

$$d_{jt} \leq d \ (S_{jt}^* - E_{jt}^*), \ \forall j \in J, \ \forall t \in T. \tag{17}$$

If a job is implemented in three or more periods, inside all periods between the first one and the last one job duration is the same as period length:

$$d_{jt} \geq d \ (S_{jt}^* + S_{j,t-1}^* - 1 - E_{jt}^* - E_{j,t+1}^*), \ \forall j \in J, \ \forall t \in T. \tag{18}$$

It is also necessary to configure precedence constraints. Successor and predecessor both might be implemented in one period only if it is the last period of predecessor and the first period of successor:

$$S_{j_2 t}^* \leq E_{j_1, t+1}^*, \ \forall t \in T, \ \forall (j_1, j_2) \in P; \tag{19}$$

$$d_{j_1 t} + d_{j_2 t} \leq d, \ \forall t \in T, \ \forall (j_1, j_2) \in P. \tag{20}$$

2.2.4. Resource Allocations Constraints and Objective Function

In this set of constraints, we involve only d_{jt} from scheduling decision variables. We denote as $c_{jrt} \in [0, p_{max,jr} d]$ the volume of work related to job $j \in J$ in period $t \in T$ done by resource $r \in R$.

It has upper and lower limit defined by the minimal and maximal amount of assigned resources and job duration:

$$p_{min,jr}d_{jt} \leq c_{jrt} \leq p_{max,jr}d_{jt}, \ \forall j \in J, \ \forall r \in R, \ \forall t \in T. \tag{21}$$

All resource types must implement given total amount required to each job:

$$\sum_{t \in T} c_{jrt} = W_{jr}, \ \forall j \in J, \ \forall r \in R. \tag{22}$$

In the objective function, we use the amount of extra usage of each resource o_{rt}, defined by the following constraints:

$$o_{rt} \geq \sum_{j \in J} c_{jrt} - L_{rt}, \ \forall t \in T, \ \forall r \in R. \tag{23}$$

The objective function is the minimization of the extra resource allocation cost. We define the extra capacity of resource $r \in R$ needed in period $t \in T$ as o_{rt}. Therefore, the objective function is

$$Minimize \ \sum_{r \in R} \sum_{t \in T} e_r o_{rt}. \tag{24}$$

2.3. Reduction of Variable Domains

In our formulation job duration is a decision variable. However, it depends on the work volume W_{jr} and it is limited by the minimal and maximal resource requirement per period $p_{min,jr}$ and $p_{max,jr}$, respectively. These values are input parameters defined for each resource type and each job. These parameters are used to represent practical conditions. For example, with construction machines and equipment we can state that it is impossible to use less than one unit at any point in time. Resource usage usually has an upper limit inside each period. For example, the case when an assembly line does not allow to assign more than five workers to some operation simultaneously. The same condition is rational with continuous resources. Usually, there are some technical and management limits for any kind of power amount (electricity, heat, etc.) that might be applied to the job.

It is possible to use the classical approach with the earliest and latest starting times and completion time calculation. It is based on the precedence graph data which can provide critical paths and possibly additional data such as release times and deadlines of jobs. We can determine a lower bound and an upper bound for job duration to use these values in the critical path method. For each job j, we calculate the minimal allowed job duration $d_{min,j}$ based on the maximal amount of the resource usage per period:

$$d_{min,j} = \max_{r \in R} \frac{W_{jr}}{p_{max,jr}}; \tag{25}$$

where we note that we obtain minimal job duration if we allocate maximal allowed resource amount for this job in each period when it is implemented. The maximum function is applied since if multiple resources are required to execute a job we need to satisfy the minimal required duration constraint for each resource type $d_{min,j} \geq \frac{W_{jr}}{p_{max,jr}}$. The same logic is used for the maximal possible duration of job:

$$d_{max,j} = \min_{r \in R} \frac{W_{jr}}{p_{min,jr}}. \tag{26}$$

2.4. Best Scheduling Constraints Formulation: Comparison

We study three different versions of the mathematical model. Each version has the same Objective Function (24), and resource allocation constraints (21)–(23) and different type of scheduling constraints implementation:

1. job on–off variables, presented in Section 2.2.1;

2.   job pulse start–end variables, presented in Section 2.2.2;
3.   job step start–end variables, presented in Section 2.2.3.

In order to study the impact of these scheduling constraints on the performance of the model, we make numerical experiments on two datasets. Each dataset includes 100 instances. Parameters of these datasets are presented in Table 3. The instances were generated using a continuous uniform distribution of parameters. Precedence graphs were created with the given total number of directed edges under the condition of its acyclicity.

Table 3. Instance datasets parameters.

Data Set	$\lvert T \rvert$	d	$\lvert J \rvert$	$\lvert R \rvert$	$\lvert P \rvert$	L_{rt}	W_{jr}	$p_{min,jr}$	$p_{max,jr}$	e_r
inst_j10_r5	15	1	10	5	10	$[0.0, 70.0]$	$[30.0, 50.0]$	$[1.0, 5.0]$	$[6.0, 10.0]$	$[1.0, 4.0]$
inst_j15_r5	20	1	15	5	15	$[0.0, 70.0]$	$[30.0, 50.0]$	$[1.0, 5.0]$	$[6.0, 10.0]$	$[1.0, 4.0]$

We define three Mixed-Integer Linear Programming models. These models are implemented using the IBM ILOG CPLEX 12.8 mathematical programming solver with Java code on a workstation with 4 thread 2.70 GHz processor and 8 Gb RAM.

Figure 2 presents solution times obtained for these two datasets with three different model versions.

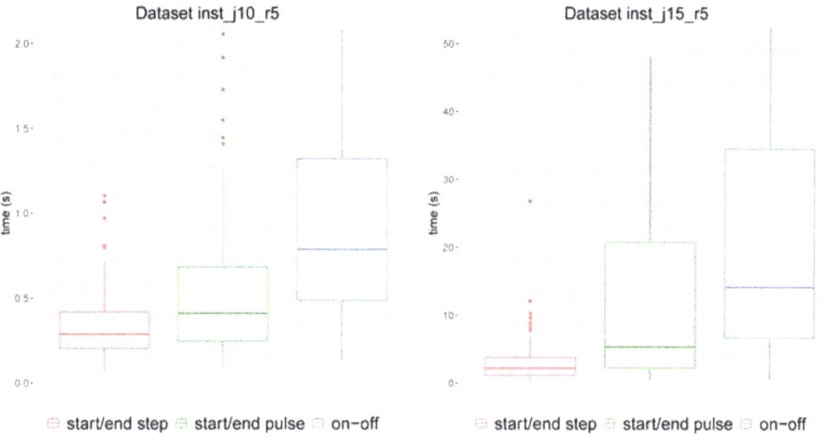

Figure 2. Different model formulation time boxplots for instances of datasets *inst_j10_r5* and *inst_j15_r5*.

We can conclude that the best performance in terms of solution time is obtained for generalized model with step formulation of scheduling constraints. This type of constraints has been also used in models of Baydoun et al. [10] and Bianco et al. [16]. In next sections, we use the generalized model with step start and end variables (presented in Section 2.2.3) to compare it with an aggregated fraction approach.

3. Aggregated Fraction Model and Generalized Model: Comparison

3.1. Structural Model Compliance

In order to obtain the model used by Baydoun et al. [10] and Bianco et al. [16], it is sufficient to replace all variables c_{jrt} by expression $W_{jr}f_{jt}$. Here $f_{jt} \in [0, 1]$ is an aggregated fraction continuous decision variable used to make decision about all resources involved in job implementation. There is another way to make our formulation equivalent to these models: We can strengthen the model with a constraint:

$$\frac{c_{jr_1t}}{W_{jr_1}} = \frac{c_{jr_2t}}{W_{jr_2}} \ \forall j \in J, \ \forall r_1, r_2 \in R, \ \forall t \in T. \tag{27}$$

In this case, all values of f_{jt} have to be identical for every pair of resource.

Further, we compare our generalized model with the aggregated fraction model. For this comparison, the aggregated fraction model is obtained by using $f_{jt} \in [0, 1]$ instead of c_{jrt} and with the transformation $c_{jrt} = W_{jr} f_{jt}$ in all corresponding constraints.

3.2. Time to Obtain the Proofed Optimal Solution

Firstly, we run both models and compare the time spent to construct the optimal solution. The results presented in Figure 3 confirm that the larger generalized model is slower.

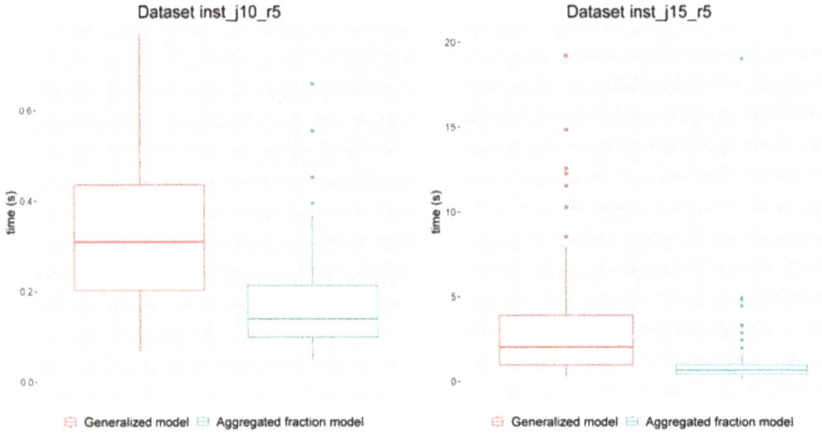

Figure 3. Time boxplots for datasets *inst_j10_r5* and *inst_j15_r5*.

3.3. Solution Quality

However, the generalized model provides more flexible solutions that allow to get much better solutions in terms of the objective function value. Denote an optimal solution objective function value for instance I of the generalized model and aggregated fraction model, respectively, as $V_g(I)$ and $V_{af}(I)$, and $X(I)$ as the ratio of these values,

$$X(I) = V_g(I)/V_{af}(I).$$

In Table 4 there is summary data of $X(I)$ for two datasets of instances. On average, in both cases the solution provided by generalized model has two times less objective function value.

Table 4. $X(I)$ values for instances of datasets *inst_j10_r5* and *inst_j15_r5*.

Data Set	Min	Q1	Median	Mean	Q3	Max
inst_j10_r5	0.17	0.41	0.49	0.48	0.55	0.71
inst_j15_r5	0.25	0.42	0.51	0.50	0.56	0.72

3.4. Reasons to Use the Generalized Model

Suppose there is a resource leveling problem without any special requirements for synchronized resource occupation for one job, i.e., both models are acceptable. Is it reasonable to use generalized model if it has worse performance? To answer this question, we make the following experiment: For each instance from dataset we run the generalized model with time limit which equals the execution time of aggregated fraction model. In other words, we compare the aggregated fraction model optimal solution with suboptimal solution of generalized model obtained in same time. Results are presented in Table 5.

Table 5. $X(I)$ value summary for instances of datasets *inst_j10_r5* and *inst_j15_r5*, with time limit equal to the optimal solution time for the aggregated fraction model.

Data Set	Min	Q1	Median	Mean	Q3	Max
inst_j10_r5	0.19	0.45	0.51	0.53	0.59	0.87
inst_j15_r5	0.25	0.46	0.55	0.55	0.61	0.9

We can conclude that for this dataset the ratio of aggregated fraction model optimal solution objective value to generalized model suboptimal solution objective function obtained remains about two times more as without time limits.

The value of $X(I)$ depends on the instance parameters. For example, if the availability of resources is higher in periods, then the objective function value decreases for both models. In this case, we obtain lower values of $X(I)$. We can demonstrate this on new datasets *inst_j10_r5_2* and *inst_j15_r5_2* with distribution $L_{rt} \in [0, 140]$ instead of $L_{rt} \in [0, 70]$, see Table 6.

Table 6. $X(I)$ values for instances of datasets *inst_j10_r5_2* and *inst_j15_r5_2*, in two cases: (a) without time limit and (b) with time limit equal to the optimal solution time for the aggregated fraction model.

Time limit	Data set	Min	Q1	Median	Mean	Q3	Max
(a) Not fixed	*inst_j10_r5_2*	0.1	0.3	0.36	0.36	0.42	0.63
	inst_j15_r5_2	0.04	0.26	0.32	0.32	0.39	0.52
(b) Aggregated fraction model	*inst_j10_r5_2*	0.1	0.33	0.4	0.41	0.47	0.97
optimal solution construction time	*inst_j15_r5_2*	0.09	0.27	0.37	0.36	0.43	0.59

We also make new experiments. In addition to datasets presented above, we generate larger instances and vary the number of resource types: We generate two datasets with 5 and 10 resource types and 30 jobs. In Table 7 we present the ranges used to generate each instance parameter. Each dataset contains 30 instances.

Table 7. Instance dataset parameters.

Data Set	$|T|$	d	$|J|$	$|R|$	$|P|$	L_{rt}	W_{jr}	$p_{min,jr}$	$p_{max,jr}$	e_r
inst_j30_r5	35	1	30	5	30	$[0.0, 70.0]$	$[30.0, 50.0]$	$[1.0, 5.0]$	$[6.0, 10.0]$	$[1.0, 4.0]$
inst_j30_r10	35	1	35	10	30	$[0.0, 70.0]$	$[30.0, 50.0]$	$[1.0, 5.0]$	$[6.0, 10.0]$	$[1.0, 4.0]$

We note that there are several RLP benchmarks already introduced in the literature. Bianco et al. [16] used two RLP benchmarks. The first benchmark was described by Kolish et al. [21] and contained datasets with 10 and 20 jobs. The second benchmark contained datasets with 10, 20, and 30 jobs. It was presented by Schwindt [22]. These datasets were prepared for the formulation with fixed job duration and intensity. For the variable duration case Bianco et al. enriched the data with the following assumption: Initial duration was considered as the maximal value, while the minimal duration was obtained by multiplying it by 0.75. Project deadline was calculated as the longest path from 0 to $n + 1$ job in the precedence graph. In some instances each job requires only one resource type. In both benchmarks the instances included 1, 3 or 5 resources.

In our research, we generate instances with the same size, but various parameters such as minimal and maximal duration of jobs and resource allocation limits. It is also important to demonstrate the difference in solution quality for the case when there are many resource types required to implement each job. For datasets *inst_j30_r5* and *inst_j30_r10* we set a 5 min time limit. In Table 8 we present the results for these datasets. We note that we compare suboptimal solutions, obtained in a given time limit.

Table 8. $X(I)$ value summary for instances of datasets *inst_j30_r5* and *inst_j30_r5*, with a 5 min time limit.

Data Set	Min	Q1	Median	Mean	Q3	Max
inst_j30_r5	0.33	0.45	0.48	0.49	0.53	0.62
inst_j30_r10	0.42	0.49	0.51	0.51	0.54	0.58

We point out that in the same time the generalized model provides better solutions in all cases. Worst-case ratio of generalized model solution objective to aggregated model solution value is around 0.6. It is reached by the flexible solution structure, even with worse performance evaluated by reached relative gap. We illustrate the gap values in Table 9 to make conclusions about the real optimal objective function value. The aggregated fraction model has a low relative gap in all cases. No significant progress can be achieved by the aggregated fraction model with higher time limits, so it cannot outperform generalized model in solution quality.

Table 9. Relative gap value summary for instances of datasets *inst_j30_r5* and *inst_j30_r5*, with 5 min time limit.

Data Set	Model	Min	Q1	Median	Mean	Q3	Max
inst_j30_r5	Aggregated fraction	0.004	0.011	0.016	0.018	0.023	0.053
	Generalized	0.05	0.11	0.16	0.16	0.19	0.3
inst_j30_r10	Aggregated fraction	0.007	0.01	0.012	0.014	0.019	0.032
	Generalized	0.07	0.12	0.14	0.16	0.18	0.38

Computational experiments confirm that we can achieve better solutions within the same solution time limit with the same solver if we apply the generalized formulation approach.

4. Discrete Resource Case

In the Objective Function (24), continuous overloading variables are used in order to calculate the cost of extra resources. It is compliant with such continuous resources as electricity or heat. However, in practice such resources as machines or human operators can be only available in discrete units. For this case, two possible models can be used.

- Firstly, decision variable o_{rt} can be defined as integer with the minimal unit of each resource q_r. New variables o_{rt}^* set the number of extra units used and they replace o_{rt} in Objective Function (24) and Constraint (23):

$$Minimize \sum_{r \in R} \sum_{t \in T} e_r q_r o_{rt}^*;$$ (28)

$$q_r o_{rt}^* \geq \sum_{j \in J} c_{jrt} - L_{rt}, \; \forall t \in T, \; \forall r \in R.$$ (29)

We define this model as *DO* (discrete objective). This case can be used not only for discrete resources, but it also suits the usual practice when additional resources could be demanded in some packages, for example, the batteries.
- Secondly, it is also possible to define other decision variables related to resource allocation as integers. This corresponds to the case when we have discrete resources and we allocate a discrete amount of workload to all periods. We define this model as *DO&R* (discrete objective and resources).

A computational experiment has been run to compare the behavior of continuous and discrete versions of the model. Seven models with different types and parameters of overload variables were considered: The original model with continuous overload variables, three versions of the discrete model with different resource unit size $q_r = q$, $\forall r \in R$, equal to 1, 3, and 5, defined as DO, and the same values of q for the discrete resource allocation case defined as $DO\&R$.

Figure 4 presents the computational results for dataset *inst_j10_r5* with a 90 s time limit for all models. Each column is a boxplot that aggregates the data about the solution time, defining the median, lower and upper values, and quartiles. In addition, Table 10 provides the mean values of the objective function, solution time, and gap which was not presented in Figure 4. Here we also compare optimal solution objective function value provided by discrete model and continuous generalized model (V_C), and calculate relative delta. For example, for instance I and model DO it is

$$\delta_{V_{DO}}(I) = \frac{V_{DO}(I) - V_C(I)}{V_C(I)}$$

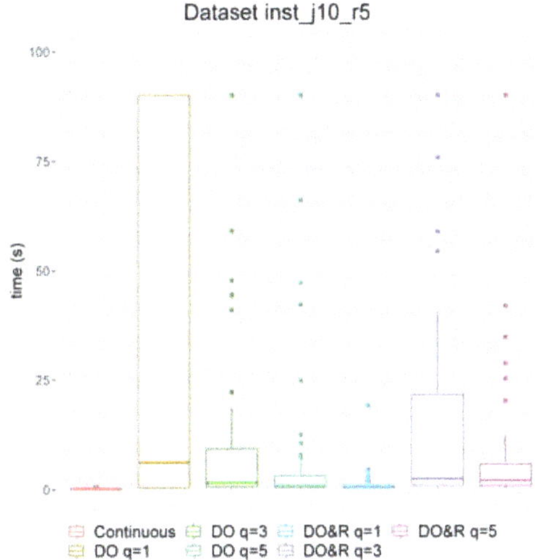

Figure 4. Time boxplot for different cases of discrete objective and allocation for dataset *inst_j10_r5*.

Table 10. Mean values for different models.

Model Overload Type	Objective	Time, s	Gap	δ
Continuous	471.5	0.18	0	-
DO q = 1	474.1	38.3	0.001	0.006
DO q=3	487.4	18.6	0.001	0.04
DO q = 5	502.6	8	0.0005	0.02
$DR\&O$ q = 1	482.5	0.9	0	0.06
$DR\&O$ q = 3	469.9	20.3	0.001	0.07
$DR\&O$ q = 5	471.5	12	0.001	0.09

As could be expected, the computation time for the discrete model is higher than for the basic continuous model. For some part of the instances, no optimal solution was reached in 90 s for the discrete model, while all the instances were solved for the basic continuous model faster than in a second. However, the information about the gap provided by the solver shows that the main issue is in the proof of optimality: All instances had a very small gap value when the time limit was reached. It is also interesting that the second model type $DO\&R$ with the discrete allocation of resources provides optimal solutions much faster than the first type *DO*. Consequently, it is possible to use these discrete models with some reasonable small gap tolerance.

5. Conclusions

In this paper, we propose a new mathematical formulation for a resource leveling problem with a variable duration of jobs. We consider the extra resource usage cost as the objective function which has to be minimized. Extra resources are required because of a lack of available resources during a fixed planning horizon with a deadline. The main idea behind this new formulation is to provide a more flexible allocation of different resources to jobs, which allows obtaining solutions with better objective function values. Moreover, we consider different models for scheduling decision variables and constraints.

This new formulation approach is compared to other RLP formulations with overload which were found in the literature. We defined them as aggregated fraction models to underline the main difference. The numerical experiments show that, even if the generalized formulation uses more variables and constraints, it provides much better solutions. In this paper the primary goal was to present and evaluate an improvement in solution quality. We also had a secondary goal to demonstrate that our generalized model can compete with the aggregated fraction model with the same solver computational resource and time limit. We compared the proofed optimal solution of the aggregated fraction formulation with a suboptimal solution of a generalized formulation obtained at the same time. We can state that with the same solver and time limit, the generalized formulation also provides better solution quality. However, we did not test any acceleration methods for this RLP model.

There are two directions for further research. Firstly, it would be valuable to provide a theoretical estimation of the difference in the value of the objective function for the generalized and fraction-aggregated models. This would allow determining some subsets of instances with a maximum difference between solution quality. Secondly, with good perspectives in solution quality, it is reasonable to focus on model improvements and adapt existing RLP solution algorithms for the new model. This will require more tests with instances and the application of some resource leveling problem benchmark datasets for model performance comparison. Some particular real case-based problems may also inspire further steps. For example, scheduling of workforce with additional constraints such as variable experience depending on previous actions [23].

In order to improve the solution procedure, the development of a Benders decomposition algorithm is planned for future research. Besides, approximation schemes can be quite efficient as we can estimate the objective function value difference for the discrete and the continuous resource models, which provides the estimated accuracy if the scheme is based on the continuous model.

Author Contributions: Conceptualisation, O.B. and A.H.; Methodology, I.T.; Software, I.T.; Writing—original draft, I.T.; Writing—review & editing, O.B. and A.H. All authors have read and agreed to the published version of the manuscript.

Funding: This research received no external funding.

Conflicts of Interest: The authors declare no conflict of interest.

References

1. Blazewicz, J.; Lenstra, J.; Kan, A. Scheduling subject to resource constraints: Classification and complexity. *Discret. Appl. Math.* **1983**, *5*, 11–24. [CrossRef]
2. Neumann, K.; Schwindt, C.; Zimmermann, J. Resource-Constrained Project Scheduling—Minimization of General Objective Functions. In *Project Scheduling with Time Windows and Scarce Resources: Temporal and Resource-Constrained Project Scheduling with Regular and Nonregular Objective Functions*; Springer: Berlin/Heidelberg, Germany, 2002; pp. 175–299. [CrossRef]
3. Neubert, G.; Savino, M.M. Flow shop operator scheduling through constraint satisfaction and constraint optimisation techniques. *Int. J. Prod. Qual. Manag.* **2009**, *4*, 549–568. [CrossRef]

4. Rieck, J.; Zimmermann, J. Exact Methods for Resource Leveling Problems. In *Handbook on Project Management and Scheduling*; Schwindt, C., Zimmermann, J., Eds.; Springer International Publishing: Cham, Switzerland, 2015; Volume 1, pp. 361–387. [CrossRef]

5. Christodoulou, S.E.; Michaelidou-Kamenou, A.; Ellinas, G. Heuristic Methods for Resource Leveling Problems. In *Handbook on Project Management and Scheduling*; Schwindt, C., Zimmermann, J., Eds.; Springer International Publishing: Cham, Switzerland, 2015; Volume 1, pp. 389–407. [CrossRef]

6. Tadepalli, T.C.M. Genetic algorithm based optimization for resource leveling problem with precedence constrained scheduling. *Int. J. Constr. Manag.* **2019**. [CrossRef]

7. Piryonesi, S.M.; Nasseri, M.; Ramezani, A. Resource leveling in construction projects with activity splitting and resource constraints: A simulated annealing optimization. *Can. J. Civ. Eng.* **2019**, *46*, 81–86. [CrossRef]

8. Li, H.; Wang, M.; Dong, X. Resource Leveling in Projects with Stochastic Minimum Time Lags. *J. Constr. Eng. Manag.* **2019**, *145*, 04019015. [CrossRef]

9. Cherkaoui, K.; Baptiste, P.; Pellerin, R.; Haït, A.; Perrier, N. Proactive tactical planning approach for large scale engineering and construction projects. *J. Mod. Proj. Manag.* **2017**, *5*, 96–105. [CrossRef]

10. Baydoun, G.; Haït, A.; Pellerin, R.; Cément, B.; Bouvignies, G. A rough-cut capacity planning model with overlapping. *OR Spectr.* **2016**, *38*, 335–364. [CrossRef]

11. Artigues, C.; Lopez, P.; Haït, A. The energy scheduling problem: Industrial case-study and constraint propagation techniques. *Int. J. Prod. Econ.* **2013**, *143*, 13–23. [CrossRef]

12. Capelle, M.; Huguet, M.J.; Jozefowiez, N.; Olive, X. Ground stations networks for Free-Space Optical communications: Maximizing the data transfer. *Electron. Notes Discret. Math.* **2018**, *64*, 255–264. [CrossRef]

13. Hans, E. Resource Loading by Branch-and-Price Techniques. Ph.D. Thesis, Twente University Press (TUP), Enschede, The Netherlands, 2001.

14. Kis, T. A branch-and-cut algorithm for scheduling of projects with variable-intensity activities. *Math. Program.* **2005**, *103*, 515–539. [CrossRef]

15. Bianco, L.; Caramia, M. Minimizing the completion time of a project under resource constraints and feeding precedence relations: A Lagrangian relaxation based lower bound. *4OR* **2011**, *9*, 371–389. [CrossRef]

16. Bianco, L.; Caramia, M.; Giordani, S. Resource levelling in project scheduling with generalized precedence relationships and variable execution intensities. *OR Spectr.* **2016**, *38*, 405–425. [CrossRef]

17. Pritsker, A.A.B.; Watters, L.J.; Wolfe, P.M. Multiproject Scheduling with Limited Resources: A Zero-One Programming Approach. *Manag. Sci.* **1969**, *16*, 93–108. [CrossRef]

18. Kis, T.; Drótos, M. Hard Planning and Scheduling Problems in the Digital Factory. In *Math for the Digital Factory*; Ghezzi, L., Hömberg, D., Landry, C., Eds.; Springer International Publishing: Cham, Switzerland, 2017; pp. 3–19. [CrossRef]

19. Naber, A. Resource-constrained project scheduling with flexible resource profiles in continuous time. *Comput. Oper. Res.* **2017**, *84*, 33–45. [CrossRef]

20. Artigues, C.; Koné, O.; Lopez, P.; Mongeau, M. Mixed-Integer Linear Programming Formulations. In *Handbook on Project Management and Scheduling*; Schwindt, C., Zimmermann, J., Eds.; Springer International Publishing: Cham, Switzerland, 2015; Volume 1, pp. 17–41. [CrossRef]

21. Kolisch, R.; Schwindt, C.; Sprecher, A. Benchmark Instances for Project Scheduling Problems. In *Project Scheduling: Recent Models, Algorithms and Applications*; Węglarz, J., Ed.; Springer: Boston, MA, USA, 1999; pp. 197–212. [CrossRef]

22. Schwindt, C. *Verfahren zur Lösung des ressourcenbeschränkten Projektdauerminimierungsproblems mit planungsabhängigen Zeitfenstern*; Berichte aus der Betriebswirtschaft, Shaker: Aachen, Germany, 1998.

23. Karam, A.; Attia, E.A.; Duquenne, P. A MILP model for an integrated project scheduling and multi-skilled workforce allocation with flexible working hours. *IFAC-PapersOnLine* **2017**, *50*, 13964–13969. [CrossRef]

MDPI
St. Alban-Anlage 66
4052 Basel
Switzerland
Tel. +41 61 683 77 34
Fax +41 61 302 89 18
www.mdpi.com

Algorithms Editorial Office
E-mail: algorithms@mdpi.com
www.mdpi.com/journal/algorithms